Multifractional Stochastic Fields

Wavelet Strategies in Multifractional Frameworks

Multifractional Stochastic Fields

Wavelet Strategies in Multifractional Frameworks

Antoine Ayache

Université de Lille, France

 **World Scientific**

NEW JERSEY · LONDON · SINGAPORE · BEIJING · SHANGHAI · HONG KONG · TAIPEI · CHENNAI · TOKYO

Published by

World Scientific Publishing Co. Pte. Ltd.

5 Toh Tuck Link, Singapore 596224

USA office: 27 Warren Street, Suite 401-402, Hackensack, NJ 07601

UK office: 57 Shelton Street, Covent Garden, London WC2H 9HE

British Library Cataloguing-in-Publication Data
A catalogue record for this book is available from the British Library.

MULTIFRACTIONAL STOCHASTIC FIELDS
Wavelet Strategies in Multifractional Frameworks

ISBN 978-981-4525-65-7

For any available supplementary material, please visit
https://www.worldscientific.com/worldscibooks/10.1142/8917#t=suppl

Printed in Singapore

To my mother Bahia, my father Sleiman and my life partner
Valérie

Preface

The fine study of path behavior of stochastic processes and fields is a very old, classical and wide research area in probability theory and some parts of harmonic analysis which is still very lively nowadays. As far as I know, its roots go back to the very fundamental article [Wiener (1923)] in which Norbert Wiener introduced through its Wiener measure the first mathematical formalization of the Brownian Motion $\{B(t), t \in \mathbb{R}\}$ and obtained a first result on the regularity of its paths: they satisfy, with probability 1, on any compact interval of the real line a uniform Hölder condition of an arbitrary order β strictly less than $1/2$. The paper [Wiener (1923)] also obtained a first result on the pointwise irregularity of paths of $\{B(t), t \in \mathbb{R}\}$: for all fixed point $u \in \mathbb{R}$, the probability that a path be non differentiable at u is equal to zero. Yet, the arguments given in [Wiener (1923)] for deriving the latter result do not show that it is valid on a event of probability 1 which does not depend on u. Therefore, one cannot conclude from it that paths of Browninan Motion are, with probability 1, nowhere differentiable functions; yet it seems that some people made this error in the 1920s.

Perhaps, one might assert that *in real analysis showing that a deterministic or random function has everywhere some irregularity (as for instance that it is nowhere differentiable) is in general more difficult than to show that it has everywhere some regularity (as for instance that it belongs to some Hölder space).* Brownian paths seem to be a good illustration of this assertion. Using, in the Gaussian frame, a strong version of the Kolmogorov's continuity Theorem (see Theorems A.6 and A.5), which was certainly unknown in the 1920s, one can easily get the result of [Wiener (1923)] on Hölder regularity of Brownian paths. While deriving, with probabiliy 1, their nowhere differentiability is still nowadays considerably more tricky. The first proof of the latter result was given in [Paley *et al.* (1933)], about ten years after the publication of the very famuous article [Wiener (1923)]. *How can one show that an erratic random function has, with probability 1, everywhere some irregularity?* The most important motivation of our book is to analyse this issue in the framework of some rather classical Multifractional stochastics Fields, namely the Gaussian Multifractional Brownian Field (MuBF) and the non-Gaussian Multifractional Field with Random Exponent (MuFRE). These two fields are natural extensions of the Gaussian Fractional Brownian Field (FBF) and consequently of Brownian Motion. More precisely, let N be an arbitrary positive fixed integer, the FBF with constant Hurst parameter $H \in (0,1)$, denoted by $\{B_H(t), t \in \mathbb{R}^N\}$, is the centered, real-valued, isotropic, H-self-similar Gaussian field with stationary increments whose covariance function is given, for all $(t', t'') \in \mathbb{R}^N \times \mathbb{R}^N$, by:

$$\mathbb{E}(B_H(t')B_H(t'')) = c_H \Big(|t'|^{2H} + |t''|^{2H} - |t' - t''|^{2H} \Big),$$

where $|\cdot|$ denotes the Euclidean norm over $\mathbb{R}^N$ and c_H a positive constant only depending on H. Observe that $\{B_{1/2}(t), t \in \mathbb{R}\}$ reduces to the usual Brownian Motion over the real line. Let $H(\cdot)$ be a deterministic continuous function over $\mathbb{R}^N$ with values in the open interval $(0,1)$, the MuBF with deterministic Hurst functional parameter $H(\cdot)$ is the centered Gaussian field denoted by $\{Y(t), t \in \mathbb{R}^N\}$ and defined, for each $t \in \mathbb{R}^N$, as $Y(t) := B_{H(t)}(t)$. Let $\{S(t), t \in \mathbb{R}^N\}$ be a random field with values in $(0,1)$ and with continuous paths, the MuFRE with random Hurst functional parameter $S(\cdot)$ is the non-Gaussian stochastic field denoted by $\{Z(t), t \in \mathbb{R}^N\}$ and defined, for each $t \in \mathbb{R}^N$, as $Z(t) := B_{S(t)}(t)$. In spite of the fact that the fields $\{Y(t), t \in \mathbb{R}^N\}$ and $\{Z(t), t \in \mathbb{R}^N\}$ are natural extensions of the FBF, they are in general with non-stationary increments. Local behavior of their paths may change considerably from point to point, and thus it is much more complex than that of paths of FBF which basically remains everywhere the same. Actually, in a neighborhood of any fixed point $u \in \mathbb{R}^N$, the behavior of a path of $\{Y(t), t \in \mathbb{R}^N\}$ (resp. $\{Z(t), t \in \mathbb{R}^N\}$) mainly depends on the deterministic (resp. random) value $H(u)$ (resp. $S(u)$).

The main goal of our book is to show that it is useful to employ a wavelet methodology in order to generalize to the Multifrational Fields $\{Z(t), t \in \mathbb{R}^N\}$ and $\{Y(t), t \in \mathbb{R}^N\}$ classical results on fine behavior of Brownian paths. A particular attention is given to the issue we have raised before.

The book is self-contained and most of the results in it are proved in a detailed way. It progressively allows the reader to get inside the domain of Multifractional Fields and maybe to do research works in this domain.

We believe that it is understandable by a student having a Master Degree in Mathematics and basic knowledge on stochastic processes and Fourier analysis. In the sequel, we briefly describe the content of each chapter of it.

The centered Gaussian field $\{X(u,v), (u,v) \in \mathbb{R}^N \times (0,1)\}$ defined, for every couple $(u,v) \in \mathbb{R}^N \times (0,1)$, as $X(u,v) := B_v(u)$ (that is, for each fixed v, $X(\cdot, v)$ is a FBF with Hurst parameter v), is called by us *the Generator of MuBF* and plays a key role in our book. After some recalls on FBF, Chapter 1 mainly presents some basic properties of the field $\{X(u,v), (u,v) \in \mathbb{R}^N \times (0,1)\}$ and of the associated MuBF $\{Y(t), t \in \mathbb{R}^N\} := \{X(t, H(t)), t \in \mathbb{R}^N\}$, as well as some important notions closely related to them namely, the notions of global, local and pointwise Hölder exponents and that of local asymptotic self-similarity.

The main goal of Chapter 2 is to introduce the notions of local times and local nondeterminism and to outline the main lines of a classical strategy due to [Berman (1972)], relying on them and Fourier analytic arguments, which allows to deal with issues similar to the one we have raised before. In fact, this strategy is based on the following very clever trick called the Berman's principle: *the more regular (smooth) is a path of a local time in the set variable, uniformly in the space variable, the more irregular (erratic) is the corresponding path of the stochastic field associated with this local time.* For instance, when the local time of a stochastic process is jointly continuous then the process has nowhere differentiable paths.

This classical strategy is general and powerful, yet it can hardly be applied to stochastic fields whose characteristic functions of finite-dimensional distributions are not given by explicit formulas which are exploitable. Therefore, applying it to the non-Gaussian MuFRE $\{Z(t), t \in \mathbb{R}^N\}$ seems to be difficult. Another difficulty one has to face when one uses the classical strategy is to show that the stochastic field one intends to study through it satisfies the local nondeterminism property. In the case of the Gaussian MuBF $\{Y(t), t \in \mathbb{R}^N\}$, this is a difficult problem which has been solved in [Ayache *et al.* (2011)] thanks to wavelet methods reminiscent to those described in Chapter 6. The latter methods are to a certain extent inspired by the ones, known since a long time (see [Kahane (1968 1st edition, 1985 2nd edition)] for instance) and presented in Chapter 3, which, among other things, allow to derive classical results on fine behavior of Brownian paths by using the random series representation of Brownian Motion in the Faber-Schauder system generated by dilations and translations of a function having a triangular shape. Basically, wavelet bases have the same nice structure.

Chapter 4 precisely explains what are orthonormal wavelet bases of the Hilbert space $L^2(\mathbb{R}^N)$ and how they can be contructed in a natural way starting from multiresolution analyses of this space. A particular attention is given to Lemarié-Meyer wavelet bases for which the wavelet functions offer the advantages to belong to the Schwartz class $S(\mathbb{R}^N)$ and to have compactly supported Fourier transforms vanishing in a neighborhood of 0 (the zero element of $\mathbb{R}^N$). Although this chapter is closely connected to the two following chapters, it can be viewed as a little introduction to the wavelet theory which has an interest in its own right.

Chapter 5 constructs a random wavelet series representation of the Generator of MuBF, that is the field $\{X(u, v), (u, v) \in \mathbb{R}^N \times (0, 1)\}$. Also, it shows that this series is, on each compact rectangle of $\mathbb{R}^N \times (0, 1)$, almost surely convergent uniformly in (u, v), and that its rate of convergence is optimal in the sense of [Kühn and Linde (2002)]. Notice that one can very easily deduce from such representation of $\{X(u, v), (u, v) \in \mathbb{R}^N \times (0, 1)\}$ wavelet series representations of the fields $\{Z(t), t \in \mathbb{R}^N\}$ and $\{Y(t), t \in \mathbb{R}^N\}$ themselves.

Chapter 6 makes use of the wavelet series representations of $\{X(u, v), (u, v) \in \mathbb{R}^N \times (0, 1)\}$, $\{Z(t), t \in \mathbb{R}^N\}$ and $\{Y(t), t \in \mathbb{R}^N\}$, introduced in the previous chapter, in order to derive several fine results on their path behavior. In particular, many classical results on Brownian paths are generalized to $\{Z(t), t \in \mathbb{R}^N\}$ and $\{Y(t), t \in \mathbb{R}^N\}$. Among other things, these generalizations concern the sharp Levy's global and local moduli of continuity (see [Lévy (1948 1st edition, 1965 2nd edition, 1937 1st edition, 1954 2nd edition)]) and the Dvoretzky's lower bound for oscillations (see [Dvoretzki (1963)]) which implies, with probability 1, nowhere differentiability of paths.

The Appendix A has been added mainly to allow the book to be self-contained and easier to understand. Among other things, it contains various auxiliary results on continuity and differentiability of Gaussian fields, on asymptotic behavior of sequences of independent $\mathcal{N}(0, 1)$ Gaussian random variables, and on Fourier analysis for L^p spaces on N-dimensional torus.

A. Ayache

Acknowledgments

I would like to express my deepest gratitude to my former PhD supervisor Professor Yves Meyer; since I was a third-year student at the University Paris-Dauphine in 1990, he has always kept explaining to me with great enthusiasm a lot of important notions and strategies in Mathematics. I am also very grateful to a very long list of persons that I can hardly provide here in an exhaustive manner. It includes Professors Patrice Abry, Jean-Marc Bardet, Pierre Bertrand, Sergio Bianchi, Aline Bonami, Youri Davydov, Paul Doukhan, Marco Dozzi, Anne Estrade, Erick Herbin, Yanick Heurteaux, Christian Houdré, Stéphane Jaffard, Davar Khoshnevisan, Michel Ledoux, Mikhail A. Lifshits, Werner Linde, Jacques Lévy Véhel, Yuliya Mishura, Andriy Olenko, Monique Pontier, Hervé Queffélec, François Roueff, Gennady Samorodnitsky, Stéphane Seuret, Narn-Rueih Shieh, Charles Suquet, Donatas Surgailis, Murad S. Taqqu, Nikolay Tzvetkov and Yimin Xiao. It also includes many of my colleagues at the University of Lille and of my former colleagues at the University Paul Sabatier in Toulouse, my three former PhD students Geoffrey Boutard, Julien Hamonier and Qidi Peng, Céline Esser who had a postdoctoral position under my supervision, Yassine Esmili who is currently my PhD student, and the editor Eng Huay Chionh from "World Scientific" who has followed with great patience during several years the very long process of writing this book.

Contents

Chapter 1

From Fractional Brownian Field to Multifractional Brownian Field

1.1 Introduction

Fractional Brownian Motion on $\mathbb{R}^N$, which we call Fractional Brownian Field (FBF), does not provide enough flexibility, since most of its properties are governed by a unique constant parameter, usually denoted by H, and called the Hurst parameter. A serious limitation of this basic model comes from the fact that its Hölder and self-similarity exponents are everywhere equal to H, which, among other things, obliges its path roughness to remain nearly the same, from place to place. In order to overcome the latter limitation, Multifractional Brownian Motion on $\mathbb{R}^N$, which we call Multifractional Brownian Field (MuBF), was introduced when $N = 1$ in [Peltier and Lévy Véhel (1995)] and for an arbitrary N in [Benassi *et al.* (1997)]. It is worth mentioning that these two pioneer works have been done almost at the same time and independently one from another. Roughly speaking, MuBF is obtained by replacing the constant Hurst parameter of FBF, by a function $H(\cdot)$, depending on the space variable t.

This first chapter plays a fundamental role in the book, this is why it is about 50 pages.

In section 1.2, we present three important concepts related to stochastic fields: *global self-similarity*, *isotropy* and *stationarity of increments*. Then we show how their combination in the Gaussian frame naturally leads to FBF: up to a multiplicative constant, FBF can be defined in distribution, as the unique self-similar, isotropic Gaussian field with stationary increments.

Section 1.3 is devoted to the presentation of global, local and pointwise Hölder exponents, which are three classical notions for describing path roughness globally and locally. Let us emphasize that even if local and pointwise Hölder exponents are both concerned by roughness in a neighborhood of a point, these two exponents are defined differently and should not be confused. The main message of section 1.3 is that stationarity of increments and/or global self-similarty and/or isotropy for a continuous nowhere differentiable stochastic field impose severe restrictions on its Hölder exponents. In other words, when such a stochastic field has Hölder exponents which change from place to place, then it is typically with non-stationary increments; also, in general, it is not globally self-similar. The concept of global self-similarity has to be weakened into *local asymptotic self-similarity* (see [Benassi *et al.* (1997); Falconer (2002, 2003)]), in order to fit with the non-stationary increments frame and also with the anisotropic one.

In section 1.4, first one defines the generator of MuBF; roughly speaking it is a Hölder continuous centered Gaussian field $\{X(u,v), (u,v) \in \mathbb{R}^N \times (0,1)\}$ such that, for each fixed $v \in (0,1)$, $X(\cdot,v) := \{X(u,v), u \in \mathbb{R}^N\}$ is a FBF with Hurst parameter v. Then $\{Y(t), t \in \mathbb{R}^N\}$, the MuBF with Hurst functional parameter $H(\cdot)$, is defined as $Y(t) := X(t, H(t))$, for every $t \in \mathbb{R}^N$. Under some Hölder conditions on $H(\cdot)$, a rather simple strategy, relying on estimates of incremental variance of $\{Y(t), t \in \mathbb{R}^N\}$, allows to show that: (*i*) the global Hölder exponent of MuBF on an arbitrary compact interval $I \subset \mathbb{R}^N$ equals almost surely $\min\{H(t) : t \in I\}$; (*ii*) the local and pointwise Hölder exponents of MuBF at an arbitrary point $\tau \in \mathbb{R}^N$ equal almost surely $H(\tau)$; (*iii*) MuBF is strongly locally asymptotically self-similar at τ with exponent $H(\tau)$. Yet, this strategy has a serious limitation: the result on pointwise Hölder exponent obtained through it is not *uniform* in the sense that one cannot conclude from the proof given in this first chapter that the result holds on *an event of probability* 1 *not depending on* τ. In fact, uniform results for pointwise Hölder exponents are valid, not only in the setting of MuBF but also in other Gaussian and non-Gaussian settings, as for instance in the case of Multifractional Field with Random Exponent (MuFRE) which is considerably more general than MuBF. Fine path behavior of MuFRE will be studied in detail in Chapter 6 by using a rather new strategy relying on wavelet methods.

1.2   Fractional Brownian Field

First we present the concepts of *global self-similarity, isotropy* and *stationarity of increments*, which are closely related to Fractional Brownian Field. It is worth mentioning that two well-known references on self-similar processes and fields are the two books [Samorodnitsky and Taqqu (1994); Embrechts and Maejima (2002)].

In a random framework, global self-similarity is a form of global statistical scale invariance, which, among other things, allows globally to view a stochastic field as a fractal objet.

Definition 1.1. A real-valued stochastic field $\{Z(t), t \in \mathbb{R}^N\}$ is said to be *globally self-similar*, if for some fixed positive real number H and for each positive real number [1] a, it satisfies:

$$\{a^{-H} Z(at), t \in \mathbb{R}^N\} \stackrel{d}{=} \{Z(t), t \in \mathbb{R}^N\}, \tag{1.1}$$

where $\stackrel{d}{=}$ means that the equality holds in the sense of the finite-dimensional distributions.

Remark 1.2. Except in the degenerate case [2], the scale invariance property (1.1) holds only for a unique H which we call *the global self-similarity exponent*.

Proof of Rermark 1.2. Assume that there exist $t \in \mathbb{R}^N$, and $0 < H_1 < H_2$, such that, for all $a > 0$, one has

$$a^{-H_1} Z(at) \stackrel{d_1}{=} Z(t) \stackrel{d_1}{=} a^{-H_2} Z(at), \tag{1.2}$$

where $\stackrel{d_1}{=}$ means that the real-valued random variables in (1.2), are equal in distribution. Thus, $\chi_{Z(t)}$, the characteristic function of $Z(t)$, satisfies, for all $\xi \in \mathbb{R}$, $\chi_{Z(t)}(\xi) = \chi_{Z(t)}(a^{H_2 - H_1} \xi)$; then, letting a goes to 0 and using the continuity property of $\chi_{Z(t)}$ one gets gets that $\chi_{Z(t)}(\xi) = \chi_{Z(t)}(0) = 1$ for all $\xi \in \mathbb{R}$, i.e. $Z(t) \stackrel{a.s.}{=} 0$. $\square$

Remark 1.3. Surely, global self-similarity is a nice property, but it is also somehow constraining; among other things, it implies that:

(i) Each typical path of the field $\{Z(t), t \in \mathbb{R}^N\}$ vanishes at the origin i.e.

$$Z(0) \stackrel{a.s.}{=} 0. \tag{1.3}$$

(ii) More importantly, the distribution of any of the random variables $Z(t)$, $t \neq 0$, is, up to rescale, only determined by $t/|t|$ the direction of the vector t:

$$Z(t) \stackrel{d_1}{=} |t|^H Z(t/|t|), \quad \text{for any } t \neq 0, \tag{1.4}$$

where $|\cdot|$ denotes the Euclidian norm [3].

---

[1] In fact a represents a given scale.

[2] That is the case where each random variable $Z(t)$ vanishes almost surely.

[3] Recall that the Euclidian norm $|t|$ is defined as $|t| := \left(\sum_{l=1}^{N} t_l^2 \right)^{1/2}$, where $t_1, \ldots, t_N$ denote the coordinates of the vector t in an arbitrary orthonormal basis of $\mathbb{R}^N$.

Proof of Rermark 1.3. The equality (1.4) is a straightforward consequence of (1.1). On the other hand, it follows from (1.1) and the elementary equality $a \cdot 0 = 0$, that $\chi_{Z(0)}$, the characteristic function of the random
variable $Z(0)$, satisfies, for all $\xi \in \mathbb{R}$ and $a > 0$, $\chi_{Z(0)}(\xi) = \chi_{Z(0)}(a^H \xi)$;
therefore letting a goes to 0 and using the continuity property of $\chi_{Z(0)}$, it
turns out that the values of the latter function are all equal to 1 i.e. (1.3)
holds. □

Let us now turn to the concept of isotropy. First, it is worth mentioning that the notation $Z(t)$, we have already used, is a little imprecise,
since it may mean that $Z(t)$ only depends on the vector t and not on the
choice of the orthonormal basis [4] of $\mathbb{R}^N$ which provides the coordinates of
t. Roughly speaking, isotropic real-valued stochastic fields on $\mathbb{R}^N$ are those
whose distributions do not depend on the choice of the latter basis. More
precisely:

Definition 1.4. One says that a real-valued stochastic field $\{Z(t), t \in \mathbb{R}^N\}$
is *isotropic* if the equality:

$$\{Z(Qt), t \in \mathbb{R}^N\} \overset{d}{=} \{Z(t), t \in \mathbb{R}^N\}, \tag{1.5}$$

holds, for all orthogonal matrix [5] Q of size N.

It follows immediately from Definition 1.4 that:

Remark 1.5.

(i) A real-valued stochastic process $\{Y(t), t \in \mathbb{R}\}$ is isotropic if and only if

$$\{Y(-t), t \in \mathbb{R}\} \overset{d}{=} \{Y(t), t \in \mathbb{R}\}.$$

(ii) Let $\{Z(t), t \in \mathbb{R}^N\}$ be a isotropic random field, then,

$$Z(e') \overset{d_1}{=} Z(e''), \quad \text{for all } e', e'' \in \mathcal{S}^{N-1},$$

---

[4] Usually, one chooses the canonical basis.
[5] Recall that orthogonal means that Q is a square matrix such that $QQ^* = I$, where Q^*
denotes the transpose of Q and I is the identity matrix. Also recall that an orthogonal
matrix of size 2 only depends on one parameter $\theta \in [0, 2\pi)$, and is always of the form:

$$\begin{pmatrix} \cos\theta & -\sin\theta \\ \sin\theta & \cos\theta \end{pmatrix} \quad \text{or} \quad \begin{pmatrix} \cos\theta & \sin\theta \\ \sin\theta & -\cos\theta \end{pmatrix};$$

thus, it can be viewed as a rotation of angle θ or as the composition of such a rotation
with an horizontal symmetry.

where $\mathcal{S}^{N-1} := \{e \in \mathbb{R}^N : |e| = 1\}$ denotes the unit sphere of $\mathbb{R}^N$; if, in addition, $\{Z(t), t \in \mathbb{R}^N\}$ is globally H-self-similar, then (1.4) reduces to

$$Z(t) \overset{d_1}{=} |t|^H Z(e_0), \quad \text{for any } t \neq 0, \tag{1.6}$$

where e_0 denotes an arbitrary fixed vector of $\mathcal{S}^{N-1}$.

Let us now turn to the concept of stationary increments. Roughly speaking, it means that distribution of increments is invariant under translation; more precisely:

Definition 1.6. One says that a real-valued stochastic field $\{Z(t), t \in \mathbb{R}^N\}$ is with *stationary increments* if the equality:

$$\{Z(t + \mathbf{t}) - Z(\mathbf{t}), t \in \mathbb{R}^N\} \overset{d}{=} \{Z(t) - Z(0), t \in \mathbb{R}^N\}, \tag{1.7}$$

holds, for all fixed $\mathbf{t} \in \mathbb{R}^N$.

Remark 1.7. When $\{Z(t), t \in \mathbb{R}^N\}$ is a second order [6] field with stationary increments, then one has,

$$\mathbb{E}(Z(t + \mathbf{t}) - Z(\mathbf{t}))^2 = \mathcal{V}_Z(t), \quad \text{for every } (t, \mathbf{t}) \in \mathbb{R}^N \times \mathbb{R}^N, \tag{1.8}$$

where $\mathcal{V}_Z$ denotes the *variogram* (function) of the field Z, defined as

$$\mathcal{V}_Z(t) := \mathbb{E}(Z(t) - Z(0))^2, \quad \text{for each } t \in \mathbb{R}^N. \tag{1.9}$$

The following proposition, whose proof is easy, provides in a setting of second order stochastic fields, an important motivation behind the concept of stationary increments.

Proposition 1.8. *Let* $\{Z(t), t \in \mathbb{R}^N\}$ *be a second order centered [7] stochastic field with stationary increments; also assume that* $Z(0) \overset{a.s.}{=} 0$. *Then the covariance structure of* $\{Z(t), t \in \mathbb{R}^N\}$ *is completely determined by the variogram* $\mathcal{V}_Z$, *namely, for all* $t', t'' \in \mathbb{R}^N$, *one has*

$$\text{Cov}(Z(t'), Z(t'')) = \mathbb{E}(Z(t')Z(t'')) \tag{1.10}$$
$$= 2^{-1}\big(\mathcal{V}_Z(t') + \mathcal{V}_Z(t'') - \mathcal{V}_Z(t' - t'')\big).$$

Thus, when in addition $\{Z(t), t \in \mathbb{R}^N\}$ *is Gaussian [8], its variogram fully characterizes its finite-dimensional distributions.*

[6]That is, for all $t \in \mathbb{R}^N$, one has $\mathbb{E}(Z(t))^2 < +\infty$.
[7]That is, for all $t \in \mathbb{R}^N$, one has $\mathbb{E}Z(t) = 0$.
[8]This means that any linear combination of random variables $Z(t)$ is a Gaussian random variable, namely, for each $m \in \mathbb{N}$, for all $t^1, \ldots, t^m \in \mathbb{R}^N$, and for every $\gamma_1, \ldots, \gamma_m \in \mathbb{R}$, the random variable $\sum_{l=1}^m \gamma_l Z(t^l)$ is Gaussian; with the convention that an almost surely vanishing random variable is $\mathcal{N}(0, 0)$ Gaussian.

Remark 1.9. Let $\{Z(t), t \in \mathbb{R}^N\}$ be a real-valued isotropic stochastic field with stationary increments; also assumes [9] that $Z(0) \overset{\text{a.s.}}{=} 0$. Then, for every $t \in \mathbb{R}^N$, the random variable $Z(t)$ has a symmetric distribution, namely $Z(t) \overset{d_1}{=} -Z(t)$; thus, when it exists, $\mathbb{E}(Z(t))$, the expectation of $Z(t)$, is necessarily equal to zero (i.e. $Z(t)$ is centered).

Proof of Remark 1.9. Using the equality $Z(0) \overset{\text{a.s.}}{=} 0$ and (1.7), one has, for all $t \in \mathbb{R}^N$,

$$Z(t) \overset{\text{a.s.}}{=} Z(t) - Z(0) \overset{d_1}{=} Z(0) - Z(-t) \overset{\text{a.s.}}{=} -Z(-t);$$

moreover $Z(-t) \overset{d_1}{=} Z(t)$, because of the isotropy of the field Z. $\square$

The following proposition provides a complete characterization of variograms of real-valued second order globally self-similar isotropic fields with stationary increments.

Proposition 1.10. *Let* $\{Z(t), t \in \mathbb{R}^N\}$ *be a non-degenerate real-valued second order globally H-self-similar isotropic field with stationary increments. Then, the corresponding variogram $\mathcal{V}_Z$ satisfies, for all $t \in \mathbb{R}^N$,*

$$\mathcal{V}_Z(t) = |t|^{2H} \operatorname{Var}(Z(e_0)), \qquad (1.11)$$

where e_0 denotes any arbitrary vector in the unit sphere $\mathcal{S}^{N-1}$. Moreover, the global self-similarity exponent H must necessarily belong to the interval $(0, 1]$.

Proof of Proposition 1.10. (1.9), the equality $Z(0) \overset{\text{a.s.}}{=} 0$, (1.6) and Remark 1.9 easily imply that (1.11) holds. The fact that $H \in (0, 1]$ follows from (1.8), (1.11) and the triangle inequality, which entail, if $H > 1$, that the function $t \mapsto \left(\mathbb{E}(Z(t))^2\right)^{1/2}$ has a vanishing differential at any point of $\mathbb{R}^N$; thus, for all $t \in \mathbb{R}^N$, one would have $\mathbb{E}(Z(t))^2 = \mathbb{E}(Z(0))^2 = 0$, which means that the field $\{Z(t), t \in \mathbb{R}^N\}$ is degenerate. $\square$

Remark 1.11. Propositions 1.8 and 1.10 show the uniqueness, up to positive multiplicative constants, of the distribution of non-degenerate real-valued Gaussian globally H-self-similar isotropic fields with stationary increments. More precisely, assume that $\{Z_1(t), t \in \mathbb{R}^N\}$ and $\{Z_2(t), t \in \mathbb{R}^N\}$ are two such fields, then setting $c := \sigma(Z_1(e_0))/\sigma(Z_2(e_0))$, where $\sigma(\cdot)$ denotes the standard deviation and e_0 an arbitrary vector in $\mathcal{S}^{N-1}$, it follows that $\{Z_1(t), t \in \mathbb{R}^N\} \overset{d}{=} \{cZ_2(t), t \in \mathbb{R}^N\}$.

[9] Notice that this assumption is automatically satisfied when $\{Z(t), t \in \mathbb{R}^N\}$ is self-similar.

So far, we have not shown the existence of a real-valued Gaussian globally H-self-similar isotropic field with stationary increments. This can easily be done when $H = 1$, indeed:

Proposition 1.12. *Let ε be a standard* [10] *Gaussian vector of size N and let c be a positive constant. Assume that $\{B_1(t), t \in \mathbb{R}^N\}$ is the centered Gaussian field defined as*

$$B_1(t) := c^{1/2} t \cdot \varepsilon, \quad \text{for all } t \in \mathbb{R}^N, \tag{1.12}$$

where " · " denotes the usual inner product [11] *on $\mathbb{R}^N$. Then $\{B_1(t), t \in \mathbb{R}^N\}$ is a globally 1-self-similar isotropic field with stationary increments; its variogram $\mathcal{V}_{B_1}$ is given by*

$$\mathcal{V}_{B_1}(t) = c|t|^2. \tag{1.13}$$

Proof of Proposition 1.12. It is quite easy to see that $\{B_1(t), t \in \mathbb{R}^N\}$ is globally 1-self-similar with stationary increments, since B_1 is a linear form with respect to t. Let us show that this field is isotropic. Denote by Q an arbitrary orthogonal matrix of size N, then (1.12) implies that, for every $t \in \mathbb{R}^N$, one has

$$B_1(Qt) := c^{1/2} t \cdot (Q^* \varepsilon);$$

thus, the equality

$$\{B_1(Qt), t \in \mathbb{R}^N\} \overset{d}{=} \{B_1(t), t \in \mathbb{R}^N\},$$

easily follows from the fact that $Q^* \varepsilon$ is a standard Gaussian vector. $\square$

When $H \in (0, 1)$, in order to show the existence of a real-valued Gaussian globally H-self-similar isotropic field with stationary increments, we will use the following two lemmas.

Lemma 1.13. *Let $\mathcal{K} := (\mathcal{K}_t)_{t \in \mathbb{R}^N}$ be a collection of deterministic functions belonging to the Hilbert space $L^2(\mathbb{R}^N)$* [12], *and let $\mathcal{R}_\mathcal{K}$ be the deterministic*

---

[10]That is a Gaussian vector whose coordinates are independent $\mathcal{N}(0, 1)$ Gaussian random variables.

[11]Recall that $t \cdot \varepsilon := \sum_{l=1}^{N} t_l \varepsilon_l$ where $t_1, \ldots, t_N$ are the coordinates of t and $\varepsilon_1, \ldots, \varepsilon_N$ those of ε.

[12]For each $p \in (0, +\infty)$, one denotes by $L^p(\mathbb{R}^N)$ the space of the complex-valued Borel functions f, such that the Lebesgue integral $\int_{\mathbb{R}^N} |f(s)|^p \, ds$ is finite; $L^2(\mathbb{R}^N)$ is a Hilbert space, it is equipped with the inner product $\langle \cdot, \cdot \rangle_{L^2(\mathbb{R}^N)}$, defined, for every $f, g \in L^2(\mathbb{R}^N)$, as

$$\langle f, g \rangle_{L^2(\mathbb{R}^N)} := \int_{\mathbb{R}^N} f(s) \overline{g(s)} \, ds.$$

8 Multifractional Stochastic Fields

function from $\mathbb{R}^N \times \mathbb{R}^N$ to $\mathbb{R}$, defined, for all $(t', t'') \in \mathbb{R}^N \times \mathbb{R}^N$, as

$$\mathcal{R}_\mathcal{K}(t', t'') := \mathrm{Re}\left(\langle \mathcal{K}_{t'}, \mathcal{K}_{t''}\rangle_{L^2(\mathbb{R}^N)}\right) = \mathrm{Re}\left(\int_{\mathbb{R}^N} \mathcal{K}_{t'}(s)\overline{\mathcal{K}_{t''}(s)}\,ds\right). \quad (1.14)$$

Then $\mathcal{R}_\mathcal{K}$ is symmetric [13] and positive-semidefinite [14]; therefore there exists a real-valued centered Gaussian field having $\mathcal{R}_\mathcal{K}$ as its covariance function.

Remark 1.14. Thanks to the Plancherel formula (see part (ii) of Proposition 4.5 in Chapter 4), $\mathcal{R}_\mathcal{K}$ can be expressed in the frequency domain, for all $(t', t'') \in \mathbb{R}^N \times \mathbb{R}^N$, as

$$\mathcal{R}_\mathcal{K}(t', t'') = \mathrm{Re}\left(\langle \widehat{\mathcal{K}}_{t'}, \widehat{\mathcal{K}}_{t''}\rangle_{L^2(\mathbb{R}^N)}\right) = \mathrm{Re}\left(\int_{\mathbb{R}^N} \widehat{\mathcal{K}}_{t'}(\xi)\overline{\widehat{\mathcal{K}}_{t''}(\xi)}\,d\xi\right), \quad (1.15)$$

where the functions $\widehat{\mathcal{K}}_{t'}$ and $\widehat{\mathcal{K}}_{t''}$ respectively denote the Fourier transforms [15] of $\mathcal{K}_{t'}$ and $\mathcal{K}_{t''}$.

Lemma 1.15. *For each fixed $H \in (0,1)$ and $t \in \mathbb{R}^N$, the function from $\mathbb{R}^N$ to $\mathbb{R}$, $s \mapsto |t-s|^{H-N/2} - |s|^{H-N/2}$, belongs to $L^2(\mathbb{R}^N)$.*

Proof of Lemma 1.13. In view of the definition of $\mathcal{R}_\mathcal{K}$ (see (1.14)), clearly this function is symmetric; moreover, its positivity easily results from the bilinearity and the positivity of the inner product $\langle \cdot, \cdot \rangle_{L^2(\mathbb{R}^N)}$. The existence of a centered Gaussian field having $\mathcal{R}_\mathcal{K}$ as its covariance function, is a very classical consequence of the fundamental Kolmogorov's consistency Theorem (see e.g. [Karatzas and Shreve (1987); Khoshnevisan (2002); Koralov and Sinai (2007); Revuz and Yor (1991)]). □

Proof of Lemma 1.15. It is clear that this function is Lebesgue square integrable, on each compact subset of $\mathbb{R}^N$. On the other hand, for some constant $c > 0$, one has

$$\left((1+x)^{H-N/2} - 1\right)^2 \leq cx^2, \quad \text{for all } x \in \mathbb{R} \text{ s.t. } |x| \leq 1/2. \quad (1.16)$$

---

[13]That is $\mathcal{R}_\mathcal{K}(t', t'') = \mathcal{R}_\mathcal{K}(t'', t')$, for all $(t', t'') \in \mathbb{R}^N \times \mathbb{R}^N$.

[14]That is for each $m \in \mathbb{N}$, for every $\gamma_1, \ldots, \gamma_m \in \mathbb{R}$, and for all $t^1, \ldots, t^m \in \mathbb{R}^N$, one has

$$\sum_{p=1}^m \sum_{q=1}^m \gamma_p \gamma_q \mathcal{R}_\mathcal{K}(t^p, t^q) \geq 0.$$

[15]When $f \in L^1(\mathbb{R}^N)$, its Fourier transform is defined, for all $\xi \in \mathbb{R}^N$, as

$$\widehat{f}(\xi) := \frac{1}{(2\pi)^{N/2}} \int_{\mathbb{R}^N} e^{-is\cdot\xi} f(s)\,ds;$$

thus $\mathcal{F}$, the Fourier transform map, is an isometry from $L^2(\mathbb{R}^N)$ to itself, the inverse map is denoted by $\mathcal{F}^{-1}$. Also, we mention that some useful properties of Fourier transform are recalled in Proposition 4.5 in Chapter 4.

Let us assume that $|s| > 2|t|$, using the inequalites

$$0 < |s| - |t| \leq |t - s| \leq |t| + |s|$$

and

$$(u + v)^2 \leq 2u^2 + 2v^2, \quad \text{for every } u, v \in \mathbb{R},$$

one gets that

$$\left(|t - s|^{H-N/2} - |s|^{H-N/2} \right)^2$$

$$\leq 2 \left(\left((|s| - |t|)^{H-N/2} - |s|^{H-N/2} \right)^2 + \left((|s| + |t|)^{H-N/2} - |s|^{H-N/2} \right)^2 \right)$$

$$= 2|s|^{2H-N} \left(\left(\left(1 - \frac{|t|}{|s|} \right)^{H-N/2} - 1 \right)^2 + \left(\left(1 + \frac{|t|}{|s|} \right)^{H-N/2} - 1 \right)^2 \right);$$

thus it follows from (1.16) that

$$\int_{|s|>2|t|} \left(|t-s|^{H-N/2} - |s|^{H-N/2} \right)^2 ds \leq 4c|t|^2 \int_{|s|>2|t|} |s|^{2H-N-2} ds < +\infty.$$

$$\square$$

One can derive from Lemmas 1.13 and 1.15 what follows.

Remark 1.16. For any fixed $H \in (0,1)$ and $c \in \mathbb{R} \setminus \{0\}$, there exists a real-valued centered Gaussian field, denoted by $\{B_H(t), t \in \mathbb{R}^N\}^{16}$, which satisfies, for all $(t', t'') \in \mathbb{R}^N \times \mathbb{R}^N$,

$$\text{Cov}(B_H(t'), B_H(t'')) = \mathbb{E}(B_H(t')B_H(t'')) \qquad (1.17)$$

$$= c^2 \int_{\mathbb{R}^N} \left(|t' - s|^{H-N/2} - |s|^{H-N/2} \right) \left(|t'' - s|^{H-N/2} - |s|^{H-N/2} \right) ds.$$

Observe that making in the last integral the change of variable $r = as$, where a is an arbitrary positive real number, it comes that, for every $(t', t'') \in \mathbb{R}^N \times \mathbb{R}^N$,

$$\text{Cov}(a^{-H}B_H(at'), a^{-H}B_H(at'')) = \text{Cov}(B_H(t'), B_H(t'')),$$

which means that $\{B_H(t), t \in \mathbb{R}^N\}$ is globally H-self-similar. Also observe that the change of variable $w = Qs$, where Q denotes an arbitrary orthogonal matrix, allows to show, for each $(t', t'') \in \mathbb{R}^N \times \mathbb{R}^N$, that

$$\text{Cov}(B_H(Qt'), B_H(Qt'')) = \text{Cov}(B_H(t'), B_H(t'')),$$

---

[16]We will very soon see that $\{B_H(t), t \in \mathbb{R}^N\}$ is in fact a Fractional Brownian Field.

which means that $\{B_H(t), t \in \mathbb{R}^N\}$ is isotropic. Finally observe that, when $t \in \mathbb{R}^N$ is arbitrary and fixed, setting $z = s + t$, in (1.17), one gets, for all $(t', t'') \in \mathbb{R}^N \times \mathbb{R}^N$, that

$$\text{Cov}\big(B_H(t' + t) - B_H(t), B_H(t'' + t) - B_H(t)\big) = \text{Cov}(B_H(t'), B_H(t'')),$$

which, in view of the equality $B_H(0) \overset{\text{a.s.}}{=} 0$, means that $\{B_H(t), t \in \mathbb{R}^N\}$ is with stationary increments.

The latter three results, combined with Proposition 1.10, (1.10) and Remark 1.11, allow us to state the following very important definition:

Definition 1.17 (Fractional Brownian Field (FBF)). For each fixed $H \in (0,1)$, there exists a real-valued globally H-self-similar isotropic (centered) Gaussian field with stationary increments; moreover, up to a multiplicative constant, this field is unique in distribution. It is called *Fractional Brownian Field (FBF)* of Hurst [17] parameter H, and denoted by $\{B_H(t), t \in \mathbb{R}^N\}$; the corresponding covariance function, is given, for all $(t', t'') \in \mathbb{R}^N \times \mathbb{R}^N$, by:

$$\mathbb{E}(B_H(t')B_H(t'')) = 2^{-1}\text{Var}(B_H(e_0))\Big(|t'|^{2H} + |t''|^{2H} - |t' - t''|^{2H}\Big), \quad (1.18)$$

where e_0 denotes an arbitrary vector of the unit sphere $\mathcal{S}^{N-1}$. We mention that in the particular case where $H = 1/2$, FBF is denoted by $\{B(t), t \in \mathbb{R}^N\}$ [18] and usually called *Lévy Brownian Motion* since it is the extension of Brownian Motion to $\mathbb{R}^N$, which was introduced by Paul Lévy (see for instance [Lévy (1948 1st edition, 1965 2nd edition)]). Also we mention that when $N = 1$, the Gaussian process $\{B_H(t), t \in \mathbb{R}\}$ is called *Fractional Brownian Motion (FBM)*.

It is worth noting that two very important references on Fractional Brownian Field and related topics are the two books [Samorodnitsky and Taqqu (1994); Embrechts and Maejima (2002)].

Remark 1.18. The equality (1.18) is equivalent to the equality: for all $(t', t'') \in \mathbb{R}^N \times \mathbb{R}^N$,

$$E\big|B_H(t') - B_H(t'')\big|^2 = \text{Var}(B_H(e_0))|t' - t''|^{2H}. \quad (1.19)$$

One can derive from (1.19) and the Kolmogorov's continuity Theorem, more precisely Theorem A.5, that FBF has a modification, also denoted

---

[17]This denomination makes reference to the hydrologist Hurst whose works, in the 50's, on a time series of Nile river flows over a very long period, have shown later the relevance of Fractional Brownian Motion and related self-similar processes, as models for such a data series.
[18]rather than $\{B_{1/2}(t), t \in \mathbb{R}^N\}$

by $\{B_H(t), t \in \mathbb{R}^N\}$, whose paths are, with probability 1, continuous functions on $\mathbb{R}^N$.

The most common expression for the covariance function of FBF is (1.18); however, its expression (1.17) as an integral in the time domain is also quite useful. Let us now give another important integral expression for this function.

Remark 1.19. The covariance function of FBF can be expressed, for each $(t', t'') \in \mathbb{R}^N \times \mathbb{R}^N$, as the following integral in the frequency domain:

$$\mathbb{E}(B_H(t')B_H(t'')) = c^2 \kappa_{H,N}^2 \int_{\mathbb{R}^N} \frac{(e^{-it' \cdot \xi} - 1)(e^{it'' \cdot \xi} - 1)}{|\xi|^{2H+N}} \, d\xi, \qquad (1.20)$$

where c is the same constant as in (1.17) and $\kappa_{H,N}$ is a constant only depending on H and N.

Proof of Remark 1.19. The equality (1.20) results from (1.15) and the fact [19] that, for every fixed $t \in \mathbb{R}^N$, $\mathcal{F}\big(|t - \cdot|^{H-N/2} - |\cdot|^{H-N/2}\big)$ the Fourier transform of the function $s \mapsto |t - s|^{H-N/2} - |s|^{H-N/2}$, satisfies:

$$\mathcal{F}\big(|t - \cdot|^{H-N/2} - |\cdot|^{H-N/2}\big)(\xi) = \kappa_{H,N} \frac{e^{-it \cdot \xi} - 1}{|\xi|^{H+N/2}}, \quad \text{for a.a. } \xi \in \mathbb{R}^N. \quad (1.21)$$

$\square$

Let us now provide an explicit construction of FBF through a Wiener integral. First, it is convenient to make some brief recalls on such stochastic integrals; to this end, we denote by $\mathcal{B}(\mathbb{R}^N)$ the σ-field of the Borel subsets of $\mathbb{R}^N$, and by $\mathcal{B}_0(\mathbb{R}^N)$ the δ-ring of the elements A of $\mathcal{B}(\mathbb{R}^N)$ with a finite Lebesgue measure $\lambda_N(A)$.

Definition 1.20 (Complex-valued random measures).
One says that a function ζ from $\mathcal{B}_0(\mathbb{R}^N)$ to $L^0(\Omega)$ [20] is a random mea-

---

[19] For obtaining (1.21), one mainly uses the formula (see [Schwartz (1966)]):

$$\mathcal{F}\big(|\cdot|^{H-N/2}\big) = \kappa_{H,N} |\cdot|^{-H-N/2},$$

where $\mathcal{F}(|\cdot|^{H-N/2})$ denotes the Fourier transform of the regular tempered distribution $|\cdot|^{H-N/2}$; notice that, if $H \geq N/2$ (which only occurs when $N = 1$ and $H \in [1/2, 1)$), then the right-hand side of the previous equality is no longer a regular tempered distribution and thus has to be understood in the sense of the principal value.

[20] $(\Omega, \mathcal{A}, \mathbb{P})$ is the underlying probability space, $L^0(\Omega)$ denotes the space of all the complex-valued random variables defined on Ω, and for each $p \in (0, +\infty)$, $L^p(\Omega)$ is the subspace of the random variables X satisfying $\mathbb{E}|X|^p < +\infty$; observe that $L^2(\Omega)$ can be equipped with a natural Hilbert space structure, where the inner product $\langle \cdot, \cdot \rangle_{L^2(\Omega)}$

sure [21] when it satisfies the following two properties:

(i) $\zeta(\emptyset) \stackrel{\text{a.s.}}{=} 0$;

(ii) ζ is *additive* i.e. for all *disjoint* sets A_1 and A_2 belonging to $\mathcal{B}_0(\mathbb{R}^N)$, one has

$$\zeta(A_1 \cup A_2) \stackrel{\text{a.s.}}{=} Z(A_1) + Z(A_2).$$

Definition 1.21 (Orthogonally scattered random measures).
Assume that ζ is with values in the Hilbert space $L^2(\Omega)$, one says that this random measure is *orthogonally scattered*, if for all *disjoint* sets A_1 and A_2 belonging to $\mathcal{B}_0(\mathbb{R}^N)$, the random variables $Z(A_1)$ and $Z(A_2)$ are orthogonal, namely:

$$\langle Z(A_1), Z(A_2)\rangle_{L^2(\Omega)} := \mathbb{E}\big(Z(A_1)\overline{Z(A_2)}\big) = 0.$$

Remark 1.22 (Control measure). Let ζ be an orthogonally scattered random measure, then thanks to the Carathéodory's extension Theorem (see e.g. [Koralov and Sinai (2007)]), the deterministic function μ from $\mathcal{B}_0(\mathbb{R}^N)$ to $\mathbb{R}_+$, defined, for all $A \in \mathcal{B}_0(\mathbb{R}^N)$ as $\mu(A) := \mathbb{E}|Z(A)|^2$, can be extended, in a unique way, to be a measure on $\mathcal{B}(\mathbb{R}^N)$, called *the control measure of ζ*.

Definition 1.23 (Stochastic integral over $L^2(\mathbb{R}^N, d\mu)$). Assume that ζ is an orthogonally scattered random measure with a control measure μ. One says that a deterministic complex-valued function h is *a step function*, if it is of the form [22]:

$$h = \sum_{j=1}^{k} a_j \mathbb{1}_{A_j},$$

where $A_1, \ldots, A_k$ are disjoint sets belonging to $\mathcal{B}_0(\mathbb{R}^N)$ and $a_1, \ldots, a_k$ are complex numbers. The stochastic integral of h with respect to ζ is the complex-valued random variable defined as:

$$\int_{\mathbb{R}^N} h(s)\, d\zeta(s) := \sum_{j=1}^{k} a_j \zeta(A_j).$$

is defined, for every $X, Y \in L^2(\Omega)$, as:

$$\langle X, Y\rangle_{L^2(\Omega)} := \mathbb{E}(X\overline{Y}).$$

[21] Note in passing that the concept of random measure can be defined in a much more general framework, where the measure space $\big(\mathbb{R}^N, \mathcal{B}(\mathbb{R}^N), \lambda_N\big)$ is replaced by any other measure space.

[22] Notice that the value of the positive integer k may change from one step function to another.

Observe that the following isometry property is satisfied:

$$\mathbb{E}\left|\int_{\mathbb{R}^N} h(s)\,d\zeta(s)\right|^2 = \int_{\mathbb{R}^N}|h(s)|^2\,d\mu(s). \tag{1.22}$$

In view of (1.22) and of the completeness of $L^2(\Omega)$, the stochastic integral with respect to ζ can be extended to any $f \in L^2(\mathbb{R}^N, d\mu)$, by using an arbitrary [23] sequence $(h_m)_{m\in\mathbb{N}}$ of step functions which approximate f in $L^2(\mathbb{R}^N, d\mu)$, and by setting

$$\int_{\mathbb{R}^N} f(s)\,d\zeta(s) := \lim_{m\to+\infty}\int_{\mathbb{R}^N} h_m(s)\,d\zeta(s),$$

where the convergence holds in $L^2(\Omega)$; the isometry property (1.22) remains valid, namely:

$$\mathbb{E}\left|\int_{\mathbb{R}^N} f(s)\,d\zeta(s)\right|^2 = \int_{\mathbb{R}^N}|f(s)|^2\,d\mu(s), \quad \text{for each } f \in L^2(\mathbb{R}^N, d\mu), \tag{1.23}$$

which is equivalent to

$$\mathbb{E}\left(\int_{\mathbb{R}^N} f(s)\,d\zeta(s) \times \overline{\int_{\mathbb{R}^N} g(s)\,d\zeta(s)}\right) \tag{1.24}$$

$$= \int_{\mathbb{R}^N} f(s)\overline{g(s)}\,d\mu(s), \quad \text{for all } f, g \in L^2(\mathbb{R}^N, d\mu).$$

Definition 1.24 (Brownian measure and Wiener integral). Let $\mathbb{W}$ be an orthogonally scattered random measure such that the control measure is the Lebesgue measure λ_N, one says that $\mathbb{W}$ is *a Brownian measure* [24], when $\{\mathbb{W}(A), A \in \mathcal{B}_0(\mathbb{R}^N)\}$ is a real-valued centered Gaussian process [25]. The stochastic integral corresponding to $\mathbb{W}$, is said to be *a Wiener integral*. One has that $\int_{\mathbb{R}^N} h(s)\,d\mathbb{W}(s)$ is a real-valued centered Gaussian variable, for any real-valued function $h \in L^2(\mathbb{R}^N)$. On the other hand, the isometry property:

$$\mathbb{E}\left|\int_{\mathbb{R}^N} f(s)\,d\mathbb{W}(s)\right|^2 = \int_{\mathbb{R}^N}|f(s)|^2\,ds, \quad \text{for each } f \in L^2(\mathbb{R}^N), \tag{1.25}$$

holds. Notice that (1.25) is equivalent to

$$\mathbb{E}\left(\int_{\mathbb{R}^N} f(s)\,d\mathbb{W}(s) \times \overline{\int_{\mathbb{R}^N} g(s)\,d\mathbb{W}(s)}\right) \tag{1.26}$$

$$= \int_{\mathbb{R}^N} f(s)\overline{g(s)}\,ds, \quad \text{for all } f, g \in L^2(\mathbb{R}^N).$$

---

[23] Thanks to (1.22), the definition of $\int_{\mathbb{R}^N} f(s)\,d\zeta(s)$ does not depend on the choice of the approximating sequence of step functions $(h_m)_{m\in\mathbb{N}}$.

[24] Sometimes a Brownian measure is called a white noise.

[25] That is for each $m \in \mathbb{N}$, for all $A_1, \ldots, A_m \in \mathcal{B}_0(\mathbb{R}^N)$ and for every $\gamma_1, \ldots, \gamma_m \in \mathbb{R}$, the random variable $\sum_{l=1}^m \gamma_l \mathbb{W}(A_l)$ is real-valued and Gaussian.

We are now in position to construct explicitly FBF; the following proposition provides a construction of it, through a Wiener integral in the time domain.

Proposition 1.25. *FBF* $\{B_H(t), t \in \mathbb{R}^N\}$ *can be obtained, for all* $t \in \mathbb{R}^N$, *as the Wiener integral:*

$$B_H(t) = c \int_{\mathbb{R}^N} \left(|t - s|^{H-N/2} - |s|^{H-N/2} \right) d\mathbb{W}(s), \qquad (1.27)$$

where c *is the same constant as in (1.17) and* $\mathbb{W}$ *denotes an arbitrary Brownian measure on* $\mathbb{R}^N$.

In order to rewrite the equality (1.27) in the Fourier domain, we need to define the orthogonally scattered random measure $\widehat{\mathbb{W}}$, which, roughly speaking, can be viewed as "the Fourier transform", of the Brownian measure $\mathbb{W}$.

Definition 1.26. We denote by $\widehat{\mathbb{W}}$ the orthogonally scattered random measure defined, for all $A \in \mathcal{B}_0(\mathbb{R}^N)$, as the Wiener integral:

$$\widehat{\mathbb{W}}(A) := \int_{\mathbb{R}^N} \widehat{\mathbb{1}}_A(s) \, d\mathbb{W}(s), \qquad (1.28)$$

where $\widehat{\mathbb{1}}_A$ is the Fourier transform of the indicator function $\mathbb{1}_A$. Observe that the isometry property (1.25) and the one of the Fourier transform imply that the stochastic integral with respect to $\widehat{\mathbb{W}}$ satisfies, for each $g \in L^2(\mathbb{R}^N)$,

$$\int_{\mathbb{R}^N} g(\xi) \, d\widehat{\mathbb{W}}(\xi) \overset{a.s.}{=} \int_{\mathbb{R}^N} \widehat{g}(s) \, d\mathbb{W}(s). \qquad (1.29)$$

As a consequence, it turns out that the stochastic integral $\int_{\mathbb{R}^N} (\,\cdot\,) d\widehat{\mathbb{W}}$ has the following isometry property:

$$\mathbb{E} \left| \int_{\mathbb{R}^N} g(\xi) \, d\widehat{\mathbb{W}}(\xi) \right|^2 = \int_{\mathbb{R}^N} |g(\xi)|^2 \, d\xi. \qquad (1.30)$$

The following proposition is a straightforward consequence of (1.29) and the equality $\overline{\widehat{f}} = \mathcal{F}^{-1}\left(\overline{f}\right)$, for each $f \in L^2(\mathbb{R}^N)$.

Proposition 1.27. *For every* $f \in L^2(\mathbb{R}^N)$, *one has:*

$$\int_{\mathbb{R}^N} \overline{\widehat{f}(\xi)} \, d\widehat{\mathbb{W}}(\xi) = \int_{\mathbb{R}^N} \overline{f(s)} \, d\mathbb{W}(s); \qquad (1.31)$$

which can be interpreted as a kind of Plancherel formula.

Putting together (1.27), (1.31) and (1.21), one gets the following proposition, which corresponds to the counterpart in the Fourier domain of Proposition 1.25.

Proposition 1.28. *Let* $\{B_H(t), t \in \mathbb{R}^N\}$ *be the FBF defined in (1.27), then one has, for all* $t \in \mathbb{R}^N$,

$$B_H(t) \overset{a.s.}{=} c\,\kappa_{H,N} \int_{\mathbb{R}^N} \frac{e^{it\cdot\xi} - 1}{|\xi|^{H+N/2}}\, d\widehat{W}(\xi), \qquad (1.32)$$

where $\kappa_{H,N}$ *is the same constant as in (1.20) and* $\widehat{W}$ *denotes the orthogonally scattered random measure introduced in (1.28).*

When $N = 1$, Fractional Brownian Motions (FBMs) can also be obtained through stochastic integrals in which the integrand,

$$s \mapsto |t-s|^{H-1/2} - |s|^{H-1/2} \quad \text{or} \quad \xi \mapsto \frac{e^{it\xi} - 1}{|\xi|^{H+1/2}},$$

has been somehow modified; among other things, the following two results holds.

Proposition 1.29. *For all fixed real number* β, *we denote by* $(\cdot)_+^\beta$, *the function such that, for each* $x \in \mathbb{R}$, *one has*

$$(x)_+^\beta := x^\beta \quad \text{if } x > 0 \text{ and } \quad (x)_+^\beta := 0 \quad \text{else.} \qquad (1.33)$$

For every fixed parameter $H \in (0,1)$, *let* $\{\widetilde{B}_H(t), t \in \mathbb{R}\}$ *be the real-valued centered Gaussian process defined, for each* $t \in \mathbb{R}$, *as the Wiener integral in the time domain:*

$$\widetilde{B}_H(t) := \frac{1}{\Gamma(H+1/2)} \int_{\mathbb{R}} \left((t-s)_+^{H-1/2} - (-s)_+^{H-1/2}\right) d\mathbb{W}(s), \qquad (1.34)$$

where Γ *is the Gamma function (see (A.7)). Then* $\{\widetilde{B}_H(t), t \in \mathbb{R}\}$ *is a FBM.*

Proof of Proposition 1.29. Similarly to the proof of Lemma 1.15, one can show that, for all fixed $t \in \mathbb{R}$, the function $s \mapsto (t-s)_+^{H-1/2} - (-s)_+^{H-1/2}$ belongs to $L^2(\mathbb{R})$, thus the Wiener integral in (1.34) is well-defined. Moreover, it follows from (1.34), (1.26) and some classical computations, that the covariance function of $\{\widetilde{B}_H(t), t \in \mathbb{R}\}$ is of the form (1.18). $\qquad\square$

Proposition 1.30. *For every fixed parameter* $H \in (0,1)$ *and* $t \in \mathbb{R}$, $\mathcal{F}\left((t-\cdot)_+^{H-1/2} - (-\cdot)_+^{H-1/2}\right)$ *the Fourier transform of the function* $s \mapsto (t-s)_+^{H-1/2} - (-s)_+^{H-1/2}$, *is given, for almost all* $\xi \in \mathbb{R}$, *by:*

$$\mathcal{F}\left((t-\cdot)_+^{H-1/2} - (-\cdot)_+^{H-1/2}\right)(-\xi) = \Gamma(H+1/2)\frac{e^{it\xi} - 1}{(i\xi)^{H+1/2}}, \qquad (1.35)$$

where Γ *is the Gamma function (see (A.7)), and for every* $\theta, \xi \in \mathbb{R} \setminus \{0\}$,

$$(i\xi)^\theta := |\xi|^\theta e^{i\theta \, \mathrm{sgn}(\xi)\pi/2}, \tag{1.36}$$

$\mathrm{sgn}(\cdot)$ *being the sign function (that is* $\mathrm{sgn}(\xi) := 1$ *if* $\xi > 0$*, and* $\mathrm{sgn}(\xi) := -1$ *if* $\xi < 0$*). As a consequence, the FBM* $\{\widetilde{B}_H(t), t \in \mathbb{R}\}$*, defined in the time domain through (1.34), can be expressed, in the Fourier domain, for all* $t \in \mathbb{R}$*, as the stochastic integral:*

$$\widetilde{B}_H(t) \stackrel{a.s.}{=} \int_{\mathbb{R}} \frac{e^{it\xi} - 1}{(i\xi)^{H+1/2}} \, \widehat{d\mathbb{W}}(\xi). \tag{1.37}$$

Proof of Proposition 1.30. The proof of (1.35) is similar to that of (1.21); it relies on the formula (see [Schwartz (1966)]):

$$\mathcal{F}\big((\cdot)_+^{H-1/2}\big) = \Gamma(H + 1/2)(i \cdot)^{-H-1/2},$$

where $\mathcal{F}\big((\cdot)_+^{H-1/2}\big)$ denotes the Fourier transform of the regular tempered distribution $(\cdot)_+^{H-1/2}$; notice that, if $H \geq 1/2$, then the right-hand side of the previous equality is no longer a regular tempered distribution and thus have to be understood in the sense of the principal value. The equality (1.37) can easily be obtained by combining (1.34) with (1.31) and (1.35). $\square$

Remark 1.31. Assume that $H = 1/2$, then FBM is a usual Brownian Motion (BM), moreover, in view of (1.33) and of the fact that $\Gamma(1) = 1$, the Wiener integral in (1.34) reduces to $\int_0^t d\mathbb{W}(s)$ when $t \geq 0$ and to $-\int_{-t}^0 d\mathbb{W}(s)$ otherwise; therefore, the BM $\{\widetilde{B}_{1/2}(t), t \in \mathbb{R}\}$ can be interpreted as a primitive of the white noise (the Brownian measure) $\mathbb{W}$. More generally, thanks to (1.34), for any $H \in (0,1)$, the FBM $\{\widetilde{B}_H(t), t \in \mathbb{R}\}$, can be interpreted, up to the "infrared" correction term $(-s)_+^{H-1/2}$, as a fractional primitive of order $H + 1/2$ of the white noise; this interpretation can be justified by the fact that I_+^α, the classical *left-sided fractional primitive operator* of an arbitrary order $\alpha > 0$, is defined as:

$$I_+^\alpha(f)(t) := \frac{1}{\Gamma(\alpha)} \int_{\mathbb{R}} (t - s)_+^{\alpha-1} f(s) \, ds, \quad t \in \mathbb{R}, \tag{1.38}$$

over some class of (generalized) functions f such that the integral (1.38) makes sense.

Remark 1.32. Even if FBM $\{B_H(t), t \in \mathbb{R}\}$ [26] with parameter $H \neq 1/2$, is a quite natural extension of BM, there are major differences between these

---

[26] Here $\{B_H(t), t \in \mathbb{R}\}$ denotes an arbitrary FBM which is not necessarily the one constructed through (1.27), where $N = 1$.

two processes; for instance, generally speaking, the increments of the first one are correlated while those of the second one are independent [27]. Also, it worth noticing that, standard computations relying on (1.18), allow to derive that the increments of FBM of the form [28] $B_H(m + 1) - B_H(m)$, $m \in \mathbb{Z}$, are negatively correlated when $H < 1/2$ and positively correlated when $H > 1/2$ [29], they even display long range dependence in the latter case [30].

1.3 Hölder exponents for continuous nowhere differentiable fields

We have already mentioned (see Remark 1.18) that FBF has a modification whose paths are, with probability 1, continuous functions on $\mathbb{R}^N$; on the other hand, it has been shown (see for instance [Xiao (1997)]) that they are, with probability 1, nowhere differentiable. Similarly to FBF, all the stochastic fields we will study in this book have modifications with almost surely continuous nowhere differentiable paths. Thus, we will define Hölder exponents, only in this setting; we refer to e.g. [Andersson (1997); Ayache and Jaffard (2010); Jaffard and Meyer (1996); Meyer (1997); Seuret and Lévy Véhel (2002)] for their definitions in the most general framework.

Definition 1.33. Let I be an arbitrary compact interval [31] of $\mathbb{R}^N$. *The*

---

[27] One says that *the increments* of a real-valued stochastic process $\{Y(t), t \in \mathbb{R}\}$ are *independent*, if, for all integer $m \geq 3$ and for every real numbers $t_1 \leq t_2 \leq \ldots \leq t_m$, the random variables $Y(t_2) - Y(t_1), Y(t_3) - Y(t_2), \ldots, Y(t_m) - Y(t_{m-1})$ are independent.

The notion of independent increments cannot be defined in a canonical way for stochastic fields, since, in contrast with $\mathbb{R}$, when $N \geq 2$, there is no canonical total order relation on $\mathbb{R}^N$; so, this notion has been defined for them in different (non-equivalent) ways. As far as we know, none of the classical definitions of it, allows to say that the extension of Brownian Motion to $\mathbb{R}^N$, provided by Lévy Brownian Motion (see Definition 1.17), is with independent increments.

[28] Observe that $\{B_H(m + 1) - B_H(m), m \in \mathbb{Z}\}$ is a centered stationary Gaussian time series called *the discrete-time Fractional Gaussian Noise*; its covariance function is given, for each $m \in \mathbb{Z}$ and $q \in \mathbb{Z}$, by

$$\text{Cov}\big(B_H(m + q + 1) - B_H(m + q), B_H(m + 1) - B_H(m)\big) = \rho_H(q),$$

where

$$\rho_H(q) := 2^{-1}\text{Var}(B_H(1))\Big(|q + 1|^{2H} - 2|q|^{2H} - |q - 1|^{2H}\Big).$$

[29] That is, for all $q \in \mathbb{Z}$, one has $\rho_H(q) < 0$ when $H < 1/2$ and $\rho_H(q) > 0$ when $H > 1/2$.

[30] More formally, this means that, when $H > 1/2$, one has $\sum_{q=0}^{+\infty} \rho_H(q) = +\infty$.

[31] This means that I is a compact rectangle of $\mathbb{R}^N$, namely a cartesian product

$$I = [M_1^1, M_1^2] \times \ldots \times [M_N^1, M_N^2],$$

global Hölder space $\mathcal{C}^\beta(I)$ of an arbitrary order $\beta \in [0, 1)$ is the Banach space of the continuous functions g, from the interval I to $\mathbb{R}$, satisfying *a uniform β-Hölder condition on this interval* i.e.

$$\dot{\mathbb{H}}_{I,\beta}(g) := \sup_{t^1, t^2 \in I} \frac{|g(t^1) - g(t^2)|}{|t^1 - t^2|^\beta} < +\infty; \tag{1.39}$$

it is equipped with the norm

$$\mathbb{H}_{I,\beta}(g) := \|g\|_{I,\infty} + \dot{\mathbb{H}}_{I,\beta}(g), \tag{1.40}$$

where

$$\|g\|_{I,\infty} := \sup_{t \in I} |g(t)| \tag{1.41}$$

denotes the uniform norm on I. Observe that one has

$$\mathcal{C}^{\beta_2}(I) \subset \mathcal{C}^{\beta_1}(I), \quad \text{for all } 0 \le \beta_1 < \beta_2 < 1. \tag{1.42}$$

In the sequel f denotes an arbitrary deterministic continuous nowhere differentiable function from $\mathbb{R}^N$ to $\mathbb{R}$. Later f will be identified with a typical path $Z(\cdot, \omega)$ of a stochastic field $\{Z(t), t \in \mathbb{R}^N\}$.

Definition 1.34. *The global Hölder exponent of f on the interval I is denoted by $\beta_f(I)$; it provides a measure of the critical global Hölder regularity of f on I,* namely, it is defined as:

$$\beta_f(I) := \sup \left\{ \beta \in [0, 1) : f \in \mathcal{C}^\beta(I) \right\} \tag{1.43}$$
$$= \sup \left\{ \beta \in [0, 1) \cap \mathbb{Q} : f \in \mathcal{C}^\beta(I) \right\},$$

where the last equality follows from (1.42).

The following remark is a straightforward consequence of Definitions 1.33 and 1.34.

Remark 1.35. The global Hölder exponent β_f satisfies the following monotonicity property: for any compact intervals I' and I'' of $\mathbb{R}^N$, one has

$$\beta_f(I') \ge \beta_f(I''), \quad \text{when } I' \subseteq I''. \tag{1.44}$$

The following lemma provides a useful expression for $\beta_f(I)$.

Lemma 1.36. *Let us set*

$$\delta_I := \operatorname{diam}(I) := \sup \left\{ |t^1 - t^2| : (t^1, t^2) \in I^2 \right\}, \tag{1.45}$$

where $M^1 = (M_1^1, \ldots, M_N^1)$ and $M^2 = (M_1^2, \ldots, M_N^2)$ are two points of $\mathbb{R}^N$ whose coordinates satisfy $M_l^1 < M_l^2$, for each $l = 1, \ldots, N$; sometimes I is denoted by $[M^1, M^2]$.

and, for every $n \in \mathbb{N} := \mathbb{Z}_+ \setminus \{0\}$,

$$\Theta_{n,f}(I) \tag{1.46}$$

$$:= \sup \left\{ \left| f(t^1) - f(t^2) \right| : (t^1, t^2) \in I \times I \text{ and } \frac{\delta_I}{2^n} \le |t^1 - t^2| \le \frac{\delta_I}{2^{n-1}} \right\}.$$

Then one has

$$\beta_f(I) = \liminf_{n \to +\infty} \left\{ -n^{-1} \log_2 \left(\Theta_{n,f}(I) + 2^{-n^2} \right) \right\}, \tag{1.47}$$

where $\log_2$ denotes the logarithm to the base 2.

The proof of Lemma 1.36 mainly relies on the following remark, which is a straightforward consequence of (1.46).

Remark 1.37. Assume that ζ is an arbitrary real number such that

$$\sup_{n \in \mathbb{N}} \left\{ 2^{n\zeta} \Theta_{n,f}(I) \right\} < +\infty;$$

then, one has

$$\sup_{t^1, t^2 \in I} \frac{|f(t^1) - f(t^2)|}{|t^1 - t^2|^\zeta} < +\infty.$$

Proof of Lemma 1.36. Let $\epsilon > 0$ be arbitrarily small. Using (1.43), (1.39) and Remark 1.37, one can show that

$$\sup_{n \in \mathbb{N}} 2^{n(\beta_f(I) - \epsilon)} \left(\Theta_{n,f}(I) + 2^{-n^2} \right) < +\infty$$

and

$$\sup_{n \in \mathbb{N}} 2^{n(\beta_f(I) + \epsilon)} \left(\Theta_{n,f}(I) + 2^{-n^2} \right) = +\infty.$$

Therefore, one has that

$$\beta_f(I) - \epsilon \le \liminf_{n \to +\infty} \left\{ -n^{-1} \log_2 \left(\Theta_{n,f}(I) + 2^{-n^2} \right) \right\} \le \beta_f(I) + \epsilon;$$

then letting ϵ goes to 0, one gets (1.47). $\qquad\square$

Remark 1.38. Definitions 1.33 and 1.34 also make sense when I is a closed ball of $\mathbb{R}^N$; of course, Remark 1.35 and Lemma 1.36 remain valid in this case.

Thanks to the monotonicity property given in Remark 1.35, one can define the notion of local Hölder exponent.

Definition 1.39. *The local Hölder exponent of f at a fixed point $\tau \in \mathbb{R}^N$ is denoted by $\widetilde{\alpha}_f(\tau)$; roughly speaking, it provides a measure of the critical global Hölder regularity of f, on an arbitrarily small compact interval centered at τ. More precisely, it is defined as:*

$$\widetilde{\alpha}_f(\tau) \tag{1.48}$$

$$:= \sup\big\{\beta_f(I) : \text{over all the compact intervals } I \text{ such that } \tau \in \overset{\circ}{I}\big\},$$

where $\overset{\circ}{I}$ is the interior of I. Observe that the supremum remains unchanged when one further assumes that I belongs to the countable set of compact intervals with vertices having rational coordinates. Also, observe that it remains unchanged when I is assumed to be a closed ball instead of a compact interval.

Proposition 1.40. *The local Hölder exponent $\widetilde{\alpha}_f$ is a lower semi-continuous function from $\mathbb{R}^N$ to $[0,1]$, that is one has*

$$\liminf_{h\to 0} \widetilde{\alpha}_f(\tau + h) \geq \widetilde{\alpha}_f(\tau), \quad \text{for each } \tau \in \mathbb{R}^N. \tag{1.49}$$

Proof of Proposition 1.40. Assume that $\epsilon > 0$ is arbitrary and fixed, for showing that (1.49) holds, it is enough to prove that there exists $\eta > 0$ such that, for all $h \in \mathbb{R}^N$ satisfying $|h| < \eta$, one has

$$\widetilde{\alpha}_f(\tau + h) \geq \widetilde{\alpha}_f(\tau) - \varepsilon. \tag{1.50}$$

In view of (1.48), there exists a compact interval I_ϵ such that:

$$\tau \in \overset{\circ}{I}_\epsilon \text{ and } \beta_f(I_\epsilon) \geq \widetilde{\alpha}_f(\tau) - \epsilon. \tag{1.51}$$

Next setting

$$\eta := N^{-1/2}\mathrm{dist}\big(\tau, \mathbb{R}^N \setminus \overset{\circ}{I}_\epsilon\big) := N^{-1/2} \inf\big\{|\tau - s| : s \in \mathbb{R}^N \setminus \overset{\circ}{I}_\epsilon\big\}$$

and denoting by $\langle\eta\rangle$ the vector of $\mathbb{R}^N$ whose coordinates are all equal to η, one gets that

$$\big[\tau - \langle\eta\rangle, \tau + \langle\eta\rangle\big] \subseteq I_\epsilon;$$

thus, (1.44) and (1.51) imply that

$$\beta_f\big(\big[\tau - \langle\eta\rangle, \tau + \langle\eta\rangle\big]\big) \geq \beta_f(I_\epsilon) \geq \widetilde{\alpha}_f(\tau) - \epsilon. \tag{1.52}$$

Finally, noticing that $\tau + h$ belongs to the interior of $\big[\tau - \langle\eta\rangle, \tau + \langle\eta\rangle\big]$ for any $h \in \mathbb{R}^N$ satisfying $|h| < \eta$, it follows from (1.48) and (1.52) that (1.50) holds.

$\square$

When $N = 1$, the following converse of Proposition 1.40 has been obtained in [Seuret and Lévy Véhel (2002)].

Proposition 1.41 ([Seuret and Lévy Véhel (2002)]).
Any lower semi-continuous function from $\mathbb{R}$ to $\mathbb{R}_+$ [32] is a local Hölder exponent for some continuous functions from $\mathbb{R}$ to itself.

Definition 1.42. Let $\tau \in \mathbb{R}^N$ be an arbitrary fixed point. *The pointwise Hölder space $\mathcal{C}^\alpha(\tau)$ of an arbitrary order $\alpha \in [0,1)$ is the space of the real-valued functions g defined on neighborhoods of τ, and such that, for some $\delta > 0$, a priori depending on g, one has*

$$\sup_{t \in \mathcal{B}(\tau,\delta)} \frac{|g(t) - g(\tau)|}{|t - \tau|^\alpha} < +\infty, \tag{1.53}$$

where

$$\mathcal{B}(\tau, \delta) := \{t \in \mathbb{R}^N : |t - \tau| \leq \delta\}$$

is the closed ball centered at τ of radius δ. Observe that one has

$$\mathcal{C}^{\alpha_2}(\tau) \subset \mathcal{C}^{\alpha_1}(\tau), \quad \text{for all } 0 \leq \alpha_1 < \alpha_2 < 1. \tag{1.54}$$

Let us now provide a characterization of $\mathcal{C}^\alpha(\tau)$ in terms of oscillations in vicinity of τ.

Definition 1.43. Let A be an arbitrary subset of $\mathbb{R}^N$ with nonempty interior, and let g be an arbitrary function from A to $\mathbb{R}$. *The oscillation of g on A is denoted by $\mathrm{osc}_g(A)$, and defined as:*

$$\mathrm{osc}_g(A) := \sup_{t^1, t^2 \in A} |g(t^1) - g(t^2)| = \sup_{t \in A} g(t) - \inf_{t \in A} g(t); \tag{1.55}$$

observe that the value of $\mathrm{osc}_g(A)$ can be any nonnegative real number as well as $+\infty$.

Proposition 1.44. *Let $\tau \in \mathbb{R}^N$ and $\alpha \in [0,1)$ be arbitrary and fixed. Assume that g is an arbitrary function defined on some neighborhood of τ, then the following two assertions are equivalent.*

(i) g belongs to the pointwise Hölder space $\mathcal{C}^\alpha(\tau)$;
(ii) there exists $\delta > 0$ such that

$$\sup\left\{\rho^{-\alpha} \mathrm{osc}_g\big(\mathcal{B}(\tau,\rho)\big) : \rho \in (0,\delta]\right\} < +\infty. \tag{1.56}$$

---

[32] Recall that one can define Hölder exponents for continuous and more general functions without requiring the latter to be nowhere differentiable (see e.g. [Andersson (1997); Ayache and Jaffard (2010); Jaffard and Meyer (1996); Meyer (1997); Seuret and Lévy Véhel (2002)]); then these exponents may take any nonnegative values.

Proof of Proposition 1.44. Let us first show that $(i) \Rightarrow (ii)$. Using (1.55), the triangle inequality and (1.53), one has, for all $\rho \in (0, \delta]$,

$$\rho^{-\alpha}\mathrm{osc}_g\big(\mathcal{B}(\tau, \rho)\big)$$

$$\leq \rho^{-\alpha} \sup_{t^1 \in \mathcal{B}(\tau,\rho)} |g(t^1) - g(\tau)| + \rho^{-\alpha} \sup_{t^2 \in \mathcal{B}(\tau,\rho)} |g(t^2) - g(\tau)|$$

$$\leq \sup_{t^1 \in \mathcal{B}(\tau,\rho)} \frac{|g(t^1) - g(\tau)|}{|t^1 - \tau|^\alpha} + \sup_{t^2 \in \mathcal{B}(\tau,\rho)} \frac{|g(t^2) - g(\tau)|}{|t^2 - \tau|^\alpha}$$

$$\leq 2 \sup_{t \in \mathcal{B}(\tau,\delta)} \frac{|g(t) - g(\tau)|}{|t - \tau|^\alpha}.$$

Then, the fact that the latter supremum is finite and does not depend on ρ implies that (1.56) holds. Let us now show that $(ii) \Rightarrow (i)$. Using (1.55), one has

$$\sup_{t \in \mathcal{B}(\tau,\delta)} \frac{|g(t) - g(\tau)|}{|t - \tau|^\alpha} \leq \sup_{t \in \mathcal{B}(\tau,\delta)} \big\{ |t - \tau|^{-\alpha}\mathrm{osc}_g\big(\mathcal{B}(\tau, |t - \tau|)\big) \big\}$$

$$= \sup \big\{ \rho^{-\alpha}\mathrm{osc}_g\big(\mathcal{B}(\tau, \rho)\big) : \rho \in (0, \delta] \big\} < +\infty.$$

$\square$

Definition 1.45. *The pointwise Hölder exponent* of f at a fixed point $\tau \in \mathbb{R}^N$ is denoted by $\alpha_f(\tau)$; it provides a measure of the critical pointwise Hölder regularity of f at τ. More precisely, it is defined as:

$$\alpha_f(\tau) := \sup \big\{ \alpha \in [0, 1) : \text{such that } f \in \mathcal{C}^\alpha(\tau) \big\} \qquad (1.57)$$
$$= \sup \big\{ \alpha \in [0, 1) \cap \mathbb{Q} : \text{such that } f \in \mathcal{C}^\alpha(\tau) \big\},$$

where the last equality follows from (1.54).

The following remark is a straightforward consequence of the Definitions 1.33, 1.39, 1.42 and 1.45

Remark 1.46. For all $\alpha \in [0, 1)$ and compact interval I, one has

$$\mathcal{C}^\alpha(I) \subset \mathcal{C}^\alpha(\tau), \quad \text{for each } \tau \in \mathring{I}; \qquad (1.58)$$

therefore, one also has

$$\tilde{\alpha}_f(\tau) \leq \alpha_f(\tau), \quad \text{for every } \tau \in \mathbb{R}^N. \qquad (1.59)$$

The following lemma, which has been first established [33] in [Daoudi *et al.* (1998)], is an extension to pointwise Hölder exponents of Lemma 1.36.

---

[33]More precisely, in the version of Lemma 1.47 established in [Daoudi *et al.* (1998)], it is assumed that f is defined on the interval $[0, 1]$ of the real line.

Lemma 1.47 ([Daoudi *et al.* (1998)]). *For every $n \in \mathbb{N}$ and $\tau \in \mathbb{R}^N$, let us set*

$$\aleph_{n,f}(\tau) := \sup\left\{|f(t) - f(\tau)| : t \in \mathbb{R}^N \text{ and } 2^{-n} \le |t - \tau| \le 2^{-n+1}\right\} \quad (1.60)$$

and

$$\alpha_{n,f}(\tau) := -n^{-1}\log_2\left(\aleph_{n,f}(\tau) + 2^{-n^2}\right). \quad (1.61)$$

Then, $\aleph_{n,f}$ and $\alpha_{n,f}$ are continuous functions on $\mathbb{R}^N$; moreover, one has

$$\alpha_f(\tau) = \liminf_{n \to +\infty} \alpha_{n,f}(\tau), \quad \text{for all } \tau \in \mathbb{R}^N. \quad (1.62)$$

Before proving Lemma 1.47, let us mention that the proof of (1.62), mainly relies on the following remark, which is a straightforward consequence of (1.60).

Remark 1.48. Assume that ζ is an arbitrary real number such that

$$\sup_{n \in \mathbb{N}}\left\{2^{n\zeta}\aleph_{n,f}(\tau)\right\} < +\infty;$$

then, one has

$$\sup_{t \in \mathcal{B}(\tau,1)} \frac{|f(t) - f(\tau)|}{|t - \tau|^\zeta} < +\infty.$$

Proof of Lemma 1.47. Let us first show that $\aleph_{n,f}$ is a continuous function on $\mathbb{R}^N$ i.e. for each $\tau \in \mathbb{R}^N$, one has

$$\lim_{h \to 0} |\aleph_{n,f}(\tau + h) - \aleph_{n,f}(\tau)| = 0. \quad (1.63)$$

Observe that, in view of (1.60), $\aleph_{n,f}(\tau)$ and $\aleph_{n,f}(\tau + h)$ can be expressed as

$$\aleph_{n,f}(\tau) = \left\|f(\cdot) - f(\tau)\right\|_{\mathcal{R}_n(\tau),\infty}$$

and

$$\aleph_{n,f}(\tau + h) = \left\|f(\cdot + h) - f(\tau + h)\right\|_{\mathcal{R}_n(\tau),\infty},$$

where, for every continuous function g from $\mathbb{R}^N$ to $\mathbb{R}$,

$$\|g\|_{\mathcal{R}_n(\tau),\infty} := \sup_{t \in \mathcal{R}_n(\tau)} |g(t)|,$$

denotes the uniform seminorm of g on the compact ring

$$\mathcal{R}_n(\tau) := \left\{t \in \mathbb{R}^N : 2^{-n} \le |t - \tau| \le 2^{-n+1}\right\}.$$

Then, using the triangle inequality, one gets, for all $h \in \mathbb{R}^N$, that

$$|\aleph_{n,f}(\tau + h) - \aleph_{n,f}(\tau)| \le \left\|f(\cdot + h) - f(\cdot)\right\|_{\mathcal{R}_n(\tau),\infty} + |f(\tau + h) - f(\tau)|.$$

Thus, (1.63) results from the uniform continuity of f on the closed ball $\mathcal{B}(\tau,2)$. Having shown that $\aleph_{n,f}$ is a continuous function on $\mathbb{R}^N$, then (1.61) clearly implies that $\alpha_{n,f}$ is a continuous function on $\mathbb{R}^N$, as well. At last, in view of Remark 1.48, the proof of (1.62) can be done similarly to that of (1.47). □

Remark 1.49. More generally, Lemma 1.47 remains valid when $\aleph_{n,f}(\tau)$ is replaced by

$$\aleph_{n,f}^{b}(\tau) := \sup\left\{|f(t)-f(\tau)| : t \in \mathbb{R}^N \text{ and } 2^{-n}b \leq |t-\tau| \leq 2^{-n+1}b\right\}, \quad (1.64)$$

where b is an arbitrary positive real number.

Remark 1.50. Lemma 1.47 (more particularly the equality (1.62) in it), shows that when f is an arbitrary real-valued continuous nowhere differentiable function on $\mathbb{R}^N$, then α_f, its pointwise Hölder exponent, can be expressed as the lim inf of a sequence of continuous functions. As we have already mentioned, [Daoudi *et al.* (1998)] has established a first version of this lemma, in which $\mathbb{R}^N$ is replaced by the interval $[0,1]$ of the real line; more importantly, a reciprocal of it has also been obtained in [Daoudi *et al.* (1998)], namely, when $H(\cdot)$ is an arbitrary function from $[0,1]$ to itself, which can be written as the lim inf of a sequence of continuous functions, then $H(\cdot)$ is the pointwise Hölder exponent of some continuous functions defined on $[0,1]$. The papers [Jaffard (1995); Andersson (1997)] have generalized Lemma 1.47 as well as the reciprocal of it, by showing that *the class of the pointwise Hölder exponents of all the continuous functions* [34] *from* $\mathbb{R}^N$ *to* $\mathbb{R}$ *is exactly the class of the* lim inf*'s of the sequences of nonnegative continuous functions over* $\mathbb{R}^N$. Several years later, the article [Ayache and Jaffard (2010)] has shown that the pointwise Hölder exponent of an arbitrary function [35] in $L^1_{\text{loc}}(\mathbb{R}^N)$ can also be expressed as a lim inf of a sequence of nonnegative continuous functions. This result is rather surprising since, roughly speaking, it means that the pointwise Hölder exponent of a function having a lot of discontinuities is not "wilder" than that of a continuous function.

From now on we assume that $\{Z(t), t \in \mathbb{R}^N\}$ is a real-valued stochastic field defined on the probability space $(\Omega, \mathcal{A}, \mathbb{P})$. Also, we assume that there is an event of probability 1, denoted by Ω^*, on which the paths of this field are continuous nowhere differentiable functions; the trace σ-field of Ω^* in $\mathcal{A}$ is denoted by $\mathcal{A}_{\Omega^*}$ and defined as $\mathcal{A}_{\Omega^*} := \{A \in \mathcal{A} : \text{such that} A \subseteq \Omega^*\}$. For each $\omega \in \Omega^*$, the global Hölder exponent on an arbitrary compact interval I, for the path $Z(\cdot, \omega) : \mathbb{R}^N \to \mathbb{R}$, $t \mapsto Z(t, \omega)$, is denoted by $\beta_Z(I, \omega)$; moreover $\widetilde{\alpha}_Z(\tau, \omega)$ and $\alpha_Z(\tau, \omega)$ respectively denote its local and

---

[34]Notice that in [Jaffard (1995); Andersson (1997)] the assumption of nowhere differentiability is dropped; thus Hölder exponents may take any nonnegative values.

[35]With the convention that when a function is discontinuous at a point, then the value of its pointwise Hölder exponent in it is assumed to be equal to 0.

pointwise Hölder exponents at an arbitrary point τ. Our next goal is to give some useful properties of the distributions of the random functions $\beta_Z(I)$, $\tilde{\alpha}_Z(\tau)$ and $\alpha_Z(\tau)$. To this end, we need the following proposition whose proof can be found in e.g. [Billingsley (1968)]. Also we need Remark 1.52 below.

Proposition 1.51. *M and D are two arbitrary positive integers. For any compact interval T of $\mathbb{R}^M$, let $\mathcal{C}(T, \mathbb{R}^D)$ be the Banach space of the $\mathbb{R}^D$-valued continuous functions over T; it is equipped with the uniform norm and with the corresponding Borel σ-field $\mathcal{B}(\mathcal{C}(T, \mathbb{R}^D))$.*

(i) *Assume that $\{X(t), t \in \mathbb{R}^M\}$ is an $\mathbb{R}^D$-valued stochastic field with continuous paths, defined on the probability space $(\Omega^*, \mathcal{A}_{\Omega^*}, \mathbb{P})$; we denote by X^T the function from $(\Omega^*, \mathcal{A}_{\Omega^*}, \mathbb{P})$ to $(\mathcal{C}(T, \mathbb{R}^D), \mathcal{B}(\mathcal{C}(T, \mathbb{R}^D)))$, such that, for any $\omega \in \Omega^*$, $X^T(\omega)$ is the restriction to T of the path $t \mapsto X(t, \omega)$. Then X^T is a random variable [36]; the probability measure it induces over $\mathcal{B}(\mathcal{C}(T, \mathbb{R}^D))$ is denoted by P_{X^T} [37] and called the distribution of the field X on $\mathcal{C}(T, \mathbb{R}^D)$.*

(ii) *Assume that $\{X_1(t), t \in \mathbb{R}^M\}$ and $\{X_2(t), t \in \mathbb{R}^M\}$ are two $\mathbb{R}^D$-valued stochastic fields with continuous paths, which have the same finite-dimensional distributions i.e.*

$$\{X_1(t), t \in \mathbb{R}^M\} \overset{d}{=} \{X_2(t), t \in \mathbb{R}^M\}.$$

Then, these two fields also have the same distribution on $\mathcal{C}(T, \mathbb{R}^D)$ i.e.

$$P_{X_1^T} = P_{X_2^T}.$$

The proof of the following remark is very easy.

Remark 1.52. We use the same notations as in Proposition 1.51. Assume that $p \in \mathbb{N}$ is arbitrary, and that $\mathcal{L}_1, \ldots, \mathcal{L}_p$ are arbitrary nonempty compact subsets of T. Then, the function Ξ from $(\mathcal{C}(T, \mathbb{R}^D), \mathcal{B}(\mathcal{C}(T, \mathbb{R}^D)))$ to $(\mathbb{R}_+^{pD}, \mathcal{B}(\mathbb{R}_+^{pD}))$ defined, for all $g = (g_q)_{1 \le q \le D}$, as

$$\Xi(g) := \left(\|g_q\|_{\mathcal{L}_l, \infty}\right)_{1 \le l \le p, 1 \le q \le D} := \left(\sup_{t \in \mathcal{L}_l} |g_q(t)|\right)_{1 \le l \le p, 1 \le q \le D},$$

is continuous; therefore it is measurable.

---

[36]More precisely, a measurable function from $(\Omega^*, \mathcal{A}_{\Omega^*}, \mathbb{P})$ to $(\mathcal{C}(T, \mathbb{R}^D), \mathcal{B}(\mathcal{C}(T, \mathbb{R}^D)))$.
[37]More precisely, P_{X^T} is defined for all $B \in \mathcal{B}(\mathcal{C}(T, \mathbb{R}^D))$, as:

$$P_{X^T}(B) = \mathbb{P}\big(\{\omega \in \Omega^* : X^T(\omega) \in B\}\big).$$

Proposition 1.53. *Let J be an arbitrary fixed compact interval of $\mathbb{R}^N$, for each $\tau \in \mathbb{R}^N$, we denote by $\tau + J$ the compact interval of $\mathbb{R}^N$ defined as*

$$\tau + J := \{\tau + t : t \in J\}. \tag{1.65}$$

Then $\{\beta_Z(\tau + J), \tau \in \mathbb{R}^N\}$ is a measurable [38] *stochastic field; moreover, it is stationary when the field $\{Z(t), t \in \mathbb{R}^N\}$ has stationary increments.*

Before proving Proposition 1.53 recall that:

Definition 1.54. One says that a real-valued stochastic field $\{S(t), t \in \mathbb{R}^N\}$ is stationary, if, for all fixed in $\mathbf{t} \in \mathbb{R}^N$, one has

$$\{S(t + \mathbf{t}), t \in \mathbb{R}^N\} \stackrel{d}{=} \{S(t), t \in \mathbb{R}^N\}. \tag{1.66}$$

Proof of Proposition 1.53. For any fixed $\tau \in \mathbb{R}^N$, we denote by $\{Y_\tau(t^1, t^2), (t^1, t^2) \in \mathbb{R}^N \times \mathbb{R}^N\}$ the real-valued stochastic field with continuous nowhere differentiable paths defined, for each $(t^1, t^2, \omega) \in \mathbb{R}^N \times \mathbb{R}^N \times \Omega^*$, as

$$Y_\tau(t^1, t^2, \omega) := Z(\tau + t^1, \omega) - Z(\tau + t^2, \omega). \tag{1.67}$$

Moreover, for each $n \in \mathbb{N}$, we denote by $\mathcal{K}_n$ the compact subset of $J \times J$ defined as

$$\mathcal{K}_n := \left\{ (t^1, t^2) \in J \times J : \frac{\delta_J}{2^n} \leq |t^1 - t^2| \leq \frac{\delta_J}{2^{n-1}} \right\},$$

where δ_J is the diameter of J (see (1.45)). Next, we assume that $\Theta_{n,Z}(\tau + J, \omega)$ is defined, for each $(\tau, \omega) \in \mathbb{R}^N \times \Omega^*$, through (1.46) in which one takes $I = \tau + J$ and $f = Z(\cdot, \omega)$; then one has

$$\Theta_{n,Z}(\tau + J, \omega) = \left\| Y_\tau(\cdot, \cdot, \omega) \right\|_{\mathcal{K}_n, \infty} := \sup_{(t^1, t^2) \in \mathcal{K}_n} \left| Y_\tau(t^1, t^2, \omega) \right|. \tag{1.68}$$

Thus, it follows from Part (i) of Proposition 1.51 and from Remark 1.52 (in which one takes $M = 2N$, $D = 1$, $T = J \times J$ and $X = Y_\tau$) that $\Theta_{n,Z}(\tau + J)$ is a random variable. Moreover (1.67), (1.68) and the continuity of the paths of $\{Z(t), t \in \mathbb{R}^N\}$ imply that the stochastic field $\{\Theta_{n,Z}(\tau + J), \tau \in \mathbb{R}^N\}$ has continuous paths, which in turn entails [39] that the latter field is measurable. Therefore, the stochastic field $\{\beta_{n,Z}(\tau + J), \tau \in \mathbb{R}^N\}$ defined, for each $(\tau, \omega) \in \mathbb{R}^N \times \Omega^*$, as

$$\beta_{n,Z}(\tau + J, \omega) := -n^{-1} \log_2 \left(\Theta_{n,Z}(\tau + J, \omega) + 2^{-n^2} \right), \tag{1.69}$$

---

[38]That is $(\tau, \omega) \mapsto \beta_Z(\tau + J, \omega)$ is a measurable function from $(\mathbb{R}^N \times \Omega^*, \mathcal{B}(\mathbb{R}^N) \otimes \mathcal{A}_{\Omega^*})$ to $([0, 1], \mathcal{B}([0, 1]))$.

[39]Recall that any stochastic field with continuous paths is measurable (see e.g. [Koralov and Sinai (2007)]).

is measurable. Next, using Lemma 1.36, in which one replaces $\Theta_{n,f}(I,\omega)$ by $\Theta_{n,Z}(\tau + J, \omega)$, one has that

$$\beta_Z(\tau + J, \omega) = \liminf_{n \to +\infty} \beta_{n,Z}(\tau + J, \omega), \quad \text{for all } (\tau, \omega) \in \mathbb{R}^N \times \Omega^*. \quad (1.70)$$

Thus, the measurability of $\{\beta_{n,Z}(\tau + J), \tau \in \mathbb{R}^N\}$ implies that $\{\beta_Z(\tau + J), \tau \in \mathbb{R}^N\}$ is a measurable stochastic field.

From now on, we assume that the field $\{Z(t), t \in \mathbb{R}^N\}$ has stationary increments, and our goal is to show that the field $\{\beta_Z(\tau + J), \tau \in \mathbb{R}^N\}$ is stationary i.e. (see Definition 1.54), for any fixed $\mathbf{t} \in \mathbb{R}^N$, $m \in \mathbb{N}$, and $\tau^1, \ldots, \tau^m \in \mathbb{R}^N$, the random vectors

$$\left(\beta_Z(\mathbf{t} + \tau^q + J)\right)_{1 \le q \le m} \text{ and } \left(\beta_Z(\tau^q + J)\right)_{1 \le q \le m},$$

have the same distribution. For showing the latter fact, in view of (1.69) and (1.70), it is enough to prove that the sequences of real-valued random variables

$$\left\{\Theta_{n,Z}(\mathbf{t} + \tau^q + J), (q,n) \in \{1, \ldots, m\} \times \mathbb{N}\right\}$$

and

$$\left\{\Theta_{n,Z}(\tau^q + J), (q,n) \in \{1, \ldots, m\} \times \mathbb{N}\right\}$$

have the same finite-dimensional distributions i.e. for any fixed $p \in \mathbb{N}$ and $n_1, \ldots, n_p \in \mathbb{N}$, the distribution of the $\mathbb{R}^{pm}$-valued random vector

$$\Upsilon(\mathbf{t}) := \left(\Upsilon_{l,q}(\mathbf{t})\right)_{1 \le l \le p, 1 \le q \le m} := \left(\Theta_{n_l, Z}(\mathbf{t} + \tau^q + J)\right)_{1 \le l \le p, 1 \le q \le m}$$

does not depend on $\mathbf{t}$. In view of (1.68), for each fixed $\mathbf{t} \in \mathbb{R}^N$, let us consider, the $\mathbb{R}^m$-valued stochastic field, with continuous paths, $\mathbb{X}_\mathbf{t} := \left\{\mathbb{X}_\mathbf{t}(t^1, t^2), (t^1, t^2) \in \mathbb{R}^N \times \mathbb{R}^N\right\}$ such that

$$\mathbb{X}_\mathbf{t}(t^1, t^2) := \left(Y_{\mathbf{t}+\tau^q}(t^1, t^2)\right)_{1 \le q \le m}.$$

Next, using (1.67), one has, for all $\mathbf{t}$, q and (t^1, t^2),

$$Y_{\mathbf{t}+\tau^q}(t^1, t^2) := \left(Z(\mathbf{t} + \tau^q + t^1) - Z(\mathbf{t})\right) - \left(Z(\mathbf{t} + \tau^q + t^2) - Z(\mathbf{t})\right).$$

Therefore, the stationarity of the increments of the field $\{Z(t), t \in \mathbb{R}^N\}$ implies that the fields $\mathbb{X}_\mathbf{t}$ and $\mathbb{X}_0$ have the same finite-dimensional distributions. Thus, thanks to Part (ii) of Proposition 1.51, the latter two fields have the same distribution over $\mathcal{C}(J \times J, \mathbb{R}^m)$. Then, taking in Remark 1.52, $\mathcal{L}_l := \mathcal{K}_{n_l}$, for all $1 \le l \le p$, one has

$$\Upsilon(\mathbf{t}) = \Xi\left(\mathbb{X}_\mathbf{t}^{J \times J}\right) \text{ and } \Upsilon(0) = \Xi\left(\mathbb{X}_0^{J \times J}\right),$$

which implies that $\Upsilon(\mathbf{t})$ and $\Upsilon(0)$ have the same distribution. $\qquad \square$

Proposition 1.55. $\{\widetilde{\alpha}_Z(\tau), \tau \in \mathbb{R}^N\}$ is a measurable stochastic field; moreover, it is stationary when the field $\{Z(t), t \in \mathbb{R}^N\}$ has stationary increments.

Proof of Proposition 1.55. Observe that in view of (1.48), for any fixed $\omega \in \Omega^*$, $t \in \mathbb{R}^N$, $m \in \mathbb{N}$, and $\tau^1, \ldots, \tau^m \in \mathbb{R}^N$, one has

$$\lim_{p \to +\infty} \left(\beta_Z(t + \tau^q + J_p, \omega)\right)_{1 \le q \le m} = \left(\widetilde{\alpha}_Z(t + \tau^q, \omega)\right)_{1 \le q \le m}, \qquad (1.71)$$

where $(J_p)_{p \in \mathbb{N}}$ denotes an arbitrary decreasing sequence of compact intervals of $\mathbb{R}^N$ such that $\bigcap_{p \in \mathbb{N}} \mathring{J}_p = \{0\}$. Thus, combining (1.71) with Proposition 1.53 on gets Proposition 1.55. $\qquad \square$

Proposition 1.56. $\{\alpha_Z(\tau), \tau \in \mathbb{R}^N\}$ is a measurable stochastic field; moreover, it is stationary when the field $\{Z(t), t \in \mathbb{R}^N\}$ has stationary increments.

Proof of Proposition 1.56. It is quite similar to that of Proposition 1.53 except that one has to make use of Lemma 1.47 instead of Lemma 1.36. $\qquad \square$

Proposition 1.57. Let a be an arbitrary fixed positive real number and let $\mathcal{I}$ be the class of all the compact intervals of $\mathbb{R}^N$. For any $I \in \mathcal{I}$, the compact interval aI is defined as $aI := \{at : t \in I\}$. Assume that the stochastic field $\{Z(t), t \in \mathbb{R}^N\}$ is globally self-similar. Then, the finite-dimensional distributions of the field $\{\beta_Z(aI), I \in \mathcal{I}\}$ do not depend on a, in other words, one has

$$\{\beta_Z(aI), I \in \mathcal{I}\} \overset{d}{=} \{\beta_Z(I), I \in \mathcal{I}\}. \qquad (1.72)$$

Proof of Proposition 1.57. As usual, we denote by H the global self-similarty exponent of $\{Z(t), t \in \mathbb{R}^N\}$. Observe that, in view of Lemma 1.36, for all $I \in \mathcal{I}$, one has

$$\beta_Z(aI) \overset{\text{a.s.}}{=} \liminf_{n \to +\infty} \left\{ -n^{-1} \log_2 \left(\Theta_{n,Z}(aI) + 2^{-n^2}\right) \right\}$$

$$\overset{\text{a.s.}}{=} \liminf_{n \to +\infty} \left\{ -n^{-1} \log_2 \left(a^{-H} \Theta_{n,Z}(aI) + 2^{-n^2}\right) \right\},$$

where

$$\Theta_{n,Z}(aI)$$
$$:= \sup \left\{ |Z(t^1) - Z(t^2)| : (t^1, t^2) \in aI \times aI \text{ and } 2^{-n} a \delta_I \le |t^1 - t^2| \le 2^{-n+1} a \delta_I \right\}$$
$$= \sup \left\{ |Z(as^1) - Z(as^2)| : (s^1, s^2) \in I \times I \text{ and } 2^{-n} \delta_I \le |s^1 - s^2| \le 2^{-n+1} \delta_I \right\},$$

δ_I being the diameter of I. Therefore, in order to show that the proposition holds, it is enough to derive that

$$\left\{ a^{-H}\,\Theta_{n,Z}(aI),\ (I,n) \in \mathcal{I} \times \mathbb{N} \right\} \overset{d}{=} \left\{ \Theta_{n,Z}(I),\ (I,n) \in \mathcal{I} \times \mathbb{N} \right\}. \qquad (1.73)$$

Assume that $J, K \in \mathbb{N}$, $I_1, \ldots, I_J \in \mathcal{I}$ and $n_1, \ldots, n_K \in \mathbb{N}$ are arbitrary and fixed. Let $V^a = (V_l^a)_{1 \le l \le JK}$ be the $\mathbb{R}^{JK}$-valued random vector defined, for all $l \in \{1, \ldots, JK\}$, as

$$V_l^a := a^{-H}\,\Theta_{n_k,Z}(aI_j),$$

where (j,k) is the unique element of set $\{1, \ldots, J\} \times \{1, \ldots, K\}$ such that

$$l = (j-1)K + k. \qquad (1.74)$$

In order to obtain (1.73), it is sufficient to prove that the random vectors V^a and V^1 have the same distribution. To this end, let us consider the real-valued stochastic field, with continuous paths, $\mathbb{Y}_a := \{\mathbb{Y}_a(s^1, s^2), (s^1, s^2) \in \mathbb{R}^N \times \mathbb{R}^N\}$, such that

$$\mathbb{Y}_a(s^1, s^2) := a^{-H} Z(as^1) - a^{-H} Z(as^2).$$

The self-similarity property of the field $\{Z(t), t \in \mathbb{R}^N\}$ implies that the fields $\mathbb{Y}_a$ and $\mathbb{Y}_1$ have the same finite-dimensional distributions. Thus, thanks to Part (ii) of Proposition 1.51, the latter two fields have the same distribution over $\mathcal{C}(T \times T, \mathbb{R})$, where T is an arbitrary and fixed compact interval of $\mathbb{R}^N$ which contains $I_1, \ldots, I_J$. Then, taking in Remark 1.52, $p = JK$, $D = 1$ and, for all $1 \le l \le JK$,

$$\mathcal{L}_l := \left\{ (s^1, s^2) \in I_j \times I_j : \ 2^{-n_k}\delta_{I_j} \le |s^1 - s^2| \le 2^{-n_k+1}\delta_{I_j} \right\},$$

where j and k are the same as in (1.74), one has $V^a = \Xi\big(\mathbb{Y}_a^{T \times T}\big)$ and $V^1 = \Xi\big(\mathbb{Y}_1^{T \times T}\big)$, which implies that V^a and V^1 have the same distribution. $\square$

Remark 1.58. Propositions 1.53 and 1.57 remain valid when J and I are *closed balls* of $\mathbb{R}^N$ instead of compact intervals.

The following proposition can be proved in a rather similar way to Proposition 1.57.

Proposition 1.59. *Let Q be an arbitrary fixed orthogonal matrix of size N and let $\mathcal{I}_b$ be the class of all the closed balls of $\mathbb{R}^N$. For any $I \in \mathcal{I}_b$, the closed ball QI is defined as $QI := \{Qt : t \in I\}$. Assume that the stochastic field $\{Z(t), t \in \mathbb{R}^N\}$ is isotropic. Then, the finite-dimensional distributions of the field $\{\beta_Z(QI), I \in \mathcal{I}_b\}$ do not depend on Q, in other words, one has*

$$\{\beta_Z(QI), I \in \mathcal{I}_b\} \overset{d}{=} \{\beta_Z(I), I \in \mathcal{I}_b\}. \qquad (1.75)$$

The following proposition easily results from Proposition 1.57 and Definition 1.39.

Proposition 1.60. *Assume that the stochastic field* $\{Z(t), t \in \mathbb{R}^N\}$ *is globally self-similar and that a is an arbitrary fixed positive real number. Then, the finite-dimensional distributions of the field* $\{\widetilde{\alpha}_Z(a\tau), \tau \in \mathbb{R}^N\}$ *do not depend on a, in other words, one has*

$$\{\widetilde{\alpha}_Z(a\tau), \tau \in \mathbb{R}^N\} \overset{d}{=} \{\widetilde{\alpha}_Z(\tau), \tau \in \mathbb{R}^N\}. \tag{1.76}$$

The following proposition easily results from Proposition 1.59 and Definition 1.39.

Proposition 1.61. *Assume that the stochastic field* $\{Z(t), t \in \mathbb{R}^N\}$ *is isotropic and that Q is an arbitrary fixed orthogonal matrix of size N. Then, the finite-dimensional distributions of the field* $\{\widetilde{\alpha}_Z(Q\tau), \tau \in \mathbb{R}^N\}$ *do not depend on Q, in other words, one has*

$$\{\widetilde{\alpha}_Z(Q\tau), \tau \in \mathbb{R}^N\} \overset{d}{=} \{\widetilde{\alpha}_Z(\tau), \tau \in \mathbb{R}^N\}. \tag{1.77}$$

Proposition 1.62. *Assume that the stochastic field* $\{Z(t), t \in \mathbb{R}^N\}$ *is globally self-similar and that a is an arbitrary fixed positive real number. Then, the finite-dimensional distributions of the field* $\{\alpha_Z(a\tau), \tau \in \mathbb{R}^N\}$ *do not depend on a, in other words, one has*

$$\{\alpha_Z(a\tau), \tau \in \mathbb{R}^N\} \overset{d}{=} \{\alpha_Z(\tau), \tau \in \mathbb{R}^N\}. \tag{1.78}$$

Proof of Proposition 1.62. As usual, we denote by H the global self-similarty exponent of $\{Z(t), t \in \mathbb{R}^N\}$. Observe that, in view of Lemma 1.47 and Remark 1.49, one has, for all $\tau \in \mathbb{R}^N$,

$$\alpha_Z(a\tau) \overset{\text{a.s.}}{=} \liminf_{n \to +\infty} \left\{ -n^{-1} \log_2 \left(\aleph_{n,Z}(a\tau) + 2^{-n^2} \right) \right\}$$

$$\overset{\text{a.s.}}{=} \liminf_{n \to +\infty} \left\{ -n^{-1} \log_2 \left(a^{-H} \aleph_{n,Z}(a\tau) + 2^{-n^2} \right) \right\}$$

and

$$\alpha_Z(\tau) \overset{\text{a.s.}}{=} \liminf_{n \to +\infty} \left\{ -n^{-1} \log_2 \left(\aleph_{n,Z}^{a^{-1}}(\tau) + 2^{-n^2} \right) \right\},$$

where

$$\aleph_{n,Z}(a\tau) := \sup \left\{ |Z(t) - Z(a\tau)| : t \in \mathbb{R}^N \text{ and } 2^{-n} \le |t - a\tau| \le 2^{-n+1} \right\}$$

$$= \sup \left\{ |Z(as + a\tau) - Z(a\tau)| : s \in \mathbb{R}^N \text{ and } 2^{-n} a^{-1} \le |s| \le 2^{-n+1} a^{-1} \right\} \tag{1.79}$$

and

$$\aleph_{n,Z}^{a^{-1}}(\tau) := \sup\left\{|Z(t)-Z(\tau)| : t \in \mathbb{R}^N \text{ and } 2^{-n}a^{-1} \leq |t-\tau| \leq 2^{-n+1}a^{-1}\right\}$$
$$= \sup\left\{|Z(s+\tau)-Z(\tau)| : s \in \mathbb{R}^N \text{ and } 2^{-n}a^{-1} \leq |s| \leq 2^{-n+1}a^{-1}\right\}.$$

$$(1.80)$$

Therefore, in order to show that the proposition holds, it is enough to derive that

$$\left\{a^{-H}\aleph_{n,Z}(a\tau),\ (\tau,n) \in \mathbb{R}^N \times \mathbb{N}\right\} \stackrel{\mathrm{d}}{=} \left\{\aleph_{n,Z}^{a^{-1}}(\tau),\ (\tau,n) \in \mathbb{R}^N \times \mathbb{N}\right\}.$$

That is, for all $m,p \in \mathbb{N}$, for every $\tau^1,\ldots,\tau^m \in \mathbb{R}^N$ and for each $n_1,\ldots,n_p \in \mathbb{N}$, the $\mathbb{R}^{pm}$-valued random vectors $\left(a^{-H}\aleph_{n_l,Z}(a\tau^q)\right)_{1\leq l\leq p,1\leq q\leq m}$ and $\left(\aleph_{n_l,Z}^{a^{-1}}(\tau^q)\right)_{1\leq l\leq p,1\leq q\leq m}$ have the same distribution. Let us consider the $\mathbb{R}^m$-valued stochastic field, with continuous paths, $\mathbb{X}_a := \{\mathbb{X}_a(s), s \in \mathbb{R}^N\}$, such that

$$\mathbb{X}_a(s) := \left(a^{-H}Z(as+a\tau^q) - a^{-H}Z(a\tau^q)\right)_{1\leq q\leq m}.$$

The self-similarity property of the field $\{Z(t), t \in \mathbb{R}^N\}$ implies that the fields $\mathbb{X}_a$ and $\mathbb{X}_1$ have the same finite-dimensional distributions. Thus, thanks to Part (ii) of Proposition 1.51, the latter two fields have the same distribution over $\mathcal{C}(T, \mathbb{R}^m)$, where $T := \left[\langle -a^{-1}\rangle, \langle a^{-1}\rangle\right]$. Then, taking in Remark 1.52, for all $1 \leq l \leq p$,

$$\mathcal{L}_l := \left\{s \in \mathbb{R}^N : 2^{-n_l}a^{-1} \leq |s| \leq 2^{-n_l+1}a^{-1}\right\},$$

one has

$$\left(a^{-H}\aleph_{n_l,Z}(a\tau^q)\right)_{1\leq l\leq p,1\leq q\leq m} = \Xi\left(\mathbb{X}_a^T\right)$$

and

$$\left(\aleph_{n_l,Z}^{a^{-1}}\right)_{1\leq l\leq p,1\leq q\leq m} = \Xi\left(\mathbb{X}_1^T\right),$$

which implies that

$$\left(a^{-H}\aleph_{n_l,Z}(a\tau^q)\right)_{1\leq l\leq p,1\leq q\leq m} \quad \text{and} \quad \left(\aleph_{n_l,Z}^{a^{-1}}\right)_{1\leq l\leq p,1\leq q\leq m}$$

have the same distribution. $\qquad\Box$

The following proposition can be proved in a rather similar way to Proposition 1.62.

Proposition 1.63. *Assume that the stochastic field $\{Z(t), t \in \mathbb{R}^N\}$ is isotropic and that Q is an arbitrary orthogonal matrix of size N. Then, the finite-dimensional distributions of the field $\{\alpha_Z(Q\tau), \tau \in \mathbb{R}^N\}$ do not depend on Q, in other words, one has*

$$\{\alpha_Z(Q\tau), \tau \in \mathbb{R}^N\} \stackrel{\mathrm{d}}{=} \{\alpha_Z(\tau), \tau \in \mathbb{R}^N\}. \qquad (1.81)$$

When $N = 1$ and $\{Z(t), t \in \mathbb{R}\}$ is a Brownian Motion, it is well-known (see e.g. Chapter 16 in [Kahane (1968 1st edition, 1985 2nd edition)]) that the exponents $\beta_Z(I, \omega)$, $\widetilde{\alpha}_Z(\tau, \omega)$ and $\alpha_Z(\tau, \omega)$ are almost surely non random and equal to $1/2$. Now, we are going to see that much more generally these three exponents remain almost surely deterministic under the weak additional condition that the field $\{Z(t), t \in \mathbb{R}^N\}$ satisfies some kind of zero-one law. In order to precisely state the kind of zero-one law we shall focus on, it is convenient to identify (see e.g. [Koralov and Sinai (2007)]) the latter field with the functional valued random variable Z from $(\Omega^*, \mathcal{A}_{\Omega^*}, \mathbb{P})$ to $\left(\mathbb{R}^{\mathbb{R}^N}, \mathcal{B}(\mathbb{R})^{\otimes \mathbb{R}^{\mathbb{R}^N}}\right)$ whose value $Z(\omega)$, at an arbitrary $\omega \in \Omega^*$, is the continuous nowhere differentiable path $Z(\cdot, \omega)$. We denote by $\mathbb{R}^{\mathbb{R}^N}$ the space of all the real-valued functions [40] on $\mathbb{R}^N$, and by $\mathcal{B}(\mathbb{R})^{\otimes \mathbb{R}^{\mathbb{R}^N}}$ the tensor product σ-field. Recall that $\mathcal{B}(\mathbb{R})^{\otimes \mathbb{R}^{\mathbb{R}^N}}$ is the smallest σ-field $\mathcal{F}$ on $\mathbb{R}^{\mathbb{R}^N}$ such that, for all fixed $x \in \mathbb{R}^N$, the projection $\pi_x : \left(\mathbb{R}^{\mathbb{R}^N}, \mathcal{F}\right) \to (\mathbb{R}, \mathcal{B}(\mathbb{R}))$,

$$\text{defined, for each } \xi \in \mathbb{R}^{\mathbb{R}^N}, \text{ by } \pi_x(\xi) := \xi(x), \tag{1.82}$$

is a measurable function.

Definition 1.64. We say that the stochastic field $\{Z(t), t \in \mathbb{R}^N\}$ satisfies a zero-one law, if, for each linear subspace V of $\mathbb{R}^{\mathbb{R}^N}$ such that $V \in \mathcal{B}(\mathbb{R})^{\otimes \mathbb{R}^{\mathbb{R}^N}}$, the only possible values for the probability $\mathbb{P}(\{\omega \in \Omega^* : Z(\omega) \in V\})$ are 0 and 1.

It is worth noticing (see for example [Rosinsky and Samorodnitsky (1996)]) that the zero-one law described in Definition 1.64, holds for Gaussian, stable and some infinitely divisible stochastic fields/processes without Gaussian component; their associated finite-order chaos also have this property.

Proposition 1.65. *First recall that for each ω in the probability space $(\Omega^*, \mathcal{A}_{\Omega^*}, \mathbb{P})$, the path $Z(\cdot, \omega)$ of the stochastic field $\{Z(t), t \in \mathbb{R}^N\}$ is a continuous nowhere differentiable function on $\mathbb{R}^N$. Assume that $\{Z(t), t \in \mathbb{R}^N\}$ satisfies the zero-one law described in Definition 1.64. Then:*

(i) for all $\beta \in [0, 1)$ and compact interval $I \subset \mathbb{R}^N$, the set $\{\omega \in \Omega^* : Z(\cdot, \omega) \in \mathcal{C}^\beta(I)\}$ belongs to the σ-field $\mathcal{A}_{\Omega^*}$, moreover its probability is either 0 or 1;

(ii) for all $\alpha \in [0, 1)$ and point $\tau \in \mathbb{R}^N$, the set $\{\omega \in \Omega^* : Z(\cdot, \omega) \in \mathcal{C}^\alpha(\tau)\}$ belongs to the σ-field $\mathcal{A}_{\Omega^*}$, moreover its probability is either 0 or 1.

---

[40]Notice that we do not impose to them to be continuous.

Before proving Proposition 1.65, let us fix some useful notations related to dyadic vectors of $\mathbb{R}^N$.

Definition 1.66. Let j be an arbitrary fixed integer, *the countable subset of $\mathbb{R}^N$ of the dyadic vectors* [41] *(or dyadic points) of order $j \in \mathbb{Z}$ is denoted by $\mathbb{D}_j^N$*, and defined as

$$\mathbb{D}_j^N := 2^{-j}\mathbb{Z}^N := \left\{ 2^{-j}k : k \in \mathbb{Z}^N \right\}. \tag{1.83}$$

Observe that

$$\mathbb{D}_j^N \subset \mathbb{D}_{j+1}^N, \tag{1.84}$$

and that, for any bounded subset S of $\mathbb{R}^N$, $\text{card}\big(\mathbb{D}_j^N \cap S\big)$, the cardinality of $\mathbb{D}_j^N \cap S$, satisfies

$$\text{card}\big(\mathbb{D}_j^N \cap S\big) \le \left(1 + \left[2^{j+1}\text{diam}(S) \right] \right)^N, \tag{1.85}$$

where $[\cdot]$ is the integer part function and $\text{diam}(S) := \sup\left\{ |t^1 - t^2| : (t^1, t^2) \in S^2 \right\}$ the diameter of S. *The countable dense subset of $\mathbb{R}^N$ of all the dyadic vectors* is denoted by $\mathbb{D}^N$, and defined as

$$\mathbb{D}^N := \bigcup_{j=0}^{+\infty} \mathbb{D}_j^N = \bigcup_{j=J}^{+\infty} \mathbb{D}_j^N; \tag{1.86}$$

notice that the latter equality, in which J is an arbitrary integer, easily follows from (1.84). At last, let us mention that by *a dyadic compact interval in $\mathbb{R}^N$ of an arbitrary order $j \in \mathbb{Z}$*, we mean an interval of the form $[2^{-j}k, 2^{-j}(k + \langle 1 \rangle)]$, where $k \in \mathbb{Z}^N$ and $\langle 1 \rangle$ is the vector of $\mathbb{R}^N$ whose coordinates are all equal to 1; such an interval is denoted by $\Lambda_{j,k}$.

Proof of Proposition 1.65. We only give the proof of Part (i) since that of Part (ii) can be done in the same way. Let $V^\beta(I)$ be the linear subspace of $\mathbb{R}^{\mathbb{R}^N}$ defined as

$$V^\beta := \left\{ \xi \in \mathbb{R}^{\mathbb{R}^N} : \sup_{x^1, x^2 \in \mathbb{D}^N(I)} \frac{|\pi_{x^1}(\xi) - \pi_{x^2}(\xi)|}{|x^1 - x^2|^\beta} < +\infty \right\},$$

where $\mathbb{D}^N(I) := \mathbb{D}^N \cap I$ and where for any $x \in \mathbb{R}^N$ the projection π_x has been defined in (1.82). Observe that $V^\beta \in \mathbb{R}^{\mathbb{R}^N}$, since it can be expressed as

$$V^\beta = \bigcup_{m \in \mathbb{N}} \bigcap_{j \in \mathbb{N}} \left\{ \xi \in \mathbb{R}^{\mathbb{R}^N} : \sup_{x^1, x^2 \in \mathbb{D}_j^N(I)} \frac{|\pi_{x^1}(\xi) - \pi_{x^2}(\xi)|}{|x^1 - x^2|^\beta} < m \right\},$$

---

[41] when $N = 1$ a dyadic vector is called a dyadic number.

where $\mathbb{D}_j^N(I) := \mathbb{D}_j^N \cap I$ is either a finite set or the empty set [42]. Thus, in view of the fact that $\{Z(t), t \in \mathbb{R}^N\}$ satisfies the zero-one law described in Definition 1.64, the probability of the event $\{\omega \in \Omega^* : Z(\cdot, \omega) \in V^\beta\}$ is either equal to 0 or to 1. Finally, the fact that for any fixed $\omega \in \Omega^*$, the function $Z(\cdot, \omega)$ is continuous over $\mathbb{R}^N$, implies that the event $\{\omega \in \Omega^* : Z(\cdot, \omega) \in V^\beta\}$ is equal to the set $\{\omega \in \Omega^* : Z(\cdot, \omega) \in \mathcal{C}^\beta(I)\}$. $\square$

Basically, the following proposition is a consequence of Proposition 1.65.

Proposition 1.67. *Assume that $\{Z(t), t \in \mathbb{R}^N\}$ satisfies the zero-one law described in Definition 1.64. Then:*

(i) *for any compact interval $I \subset \mathbb{R}^N$, there exists a deterministic real number $b_Z(I) \in [0, 1]$ such that the global Hölder exponent $\beta_Z(I)$ equals to $b_Z(I)$ on an event of probability 1 a priori depending on I;*

(ii) *for any point $\tau \in \mathbb{R}^N$, there exists a deterministic real number $\widetilde{a}_Z(\tau) \in [0, 1]$ such that the local Hölder exponent $\widetilde{\alpha}_Z(\tau)$ equals to $\widetilde{a}_Z(\tau)$ on an event of probability 1 a priori depending on τ;*

(iii) *for any point $\tau \in \mathbb{R}^N$, there exists a deterministic real number $a_Z(\tau) \in [0, 1]$ such that the pointwise Hölder exponent $\alpha_Z(\tau)$ equals to $a_Z(\tau)$ on an event of probability 1 a priori depending on τ.*

Before proving Proposition 1.67, let us mention that it has been derived in [Ayache and Taqqu (2005)] under a Gaussianity assumption on $\{Z(t), t \in \mathbb{R}^N\}$; few time later, it has been obtained in its present form in [Ayache *et al.* (2005)].

Proof of Proposition 1.67. First observe that Part (*ii*) of the proposition is a straightforward consequence of Part (*i*) and of Definition 1.39. We only give the proof of Part (*i*) since that of Part (*iii*) can be done in the same way. Let $b_Z(I)$ be the deterministic quantity defined as

$$b_Z(I) := \sup\left\{\beta \in [0, 1) : \mathbb{P}\big(\{\omega \in \Omega^* : Z(\cdot, \omega) \in \mathcal{C}^\beta(I)\}\big) = 1\right\}. \quad (1.87)$$

Observe that $b_Z(I)$ is well-defined since the continuity over $\mathbb{R}^N$ of the function $Z(\cdot, \omega)$, for an arbitrary fixed $\omega \in \Omega^*$, implies that at least $\beta = 0$ satisfies the condition $\mathbb{P}\big(\{\omega \in \Omega^* : Z(\cdot, \omega) \in \mathcal{C}^\beta(I)\}\big) = 1$. Also observe that, in view of (1.87) and Part (*i*) of Proposition 1.65, one has

$$\mathbb{P}\big(\{\omega \in \Omega^* : Z(\cdot, \omega) \notin \mathcal{C}^\beta(I)\}\big) = 1, \text{when } \beta \in (b_Z(I), 1). \quad (1.88)$$

---

[42] Notice that when $\mathbb{D}_j^N$ is the empty set, then, by convention, the supremum over $\mathbb{D}_j^N$ is defined as being equal to $-\infty$.

Next, denote by Q_- and Q_+ the sets defined as $Q_- = \mathbb{Q} \cap [0, b_Z(I))$ and $Q_+ = \mathbb{Q} \cap (b_Z(I), 1)$; observe that Q_- is always a nonempty countable set, while Q_+ can be either a nonempty countable set or the empty set. Let then $\widetilde{\Omega}_I$ be the event

$$\widetilde{\Omega}_I := \left(\bigcap_{\beta \in Q_-} \{ \omega \in \Omega^* : Z(\cdot, \omega) \in \mathcal{C}^\beta(I) \} \right) \cap \left(\bigcap_{\beta \in Q_+} \{ \omega \in \Omega^* : Z(\cdot, \omega) \notin \mathcal{C}^\beta(I) \} \right),$$

with the convention that $\bigcap_{\beta \in Q_+} \{ \omega \in \Omega^* : Z(\cdot, \omega) \notin \mathcal{C}^\beta(I) \}$ is Ω^* itself when Q_+ is the empty set. One can easily derive from (1.42), (1.87) and (1.88) that $\mathbb{P}(\widetilde{\Omega}_I) = 1$. Moreover, in view of (1.43), one clearly has that $\beta_Z(I, \omega) = b_Z(I)$, for every $\omega \in \widetilde{\Omega}_I$. □

Proposition 1.68. *Assume that the stochastic field $\{ Z(t), t \in \mathbb{R}^N \}$ satisfies the zero-one law described in Definition 1.64 and that it is globally H-self-similar with stationary increments. Let $a_Z(0)$ be the deterministic value taken almost surely by its pointwise Hölder exponent at 0 (see part (iii) of Proposition 1.67). Then, $\alpha_Z(\tau)$, the pointwise Hölder exponent of $\{ Z(t), t \in \mathbb{R}^N \}$ at an arbitrary $\tau \in \mathbb{R}^N$, satisfies almost surely*

$$\alpha_Z(\tau) \overset{a.s.}{=} a_Z(0) \leq H.$$

Proof of Proposition 1.68. Combinining Proposition 1.56 and part (iii) of Proposition 1.67 it follows that $\alpha_Z(\tau) \overset{a.s.}{=} a_Z(0)$. Next, suppose ad absurdum that $a_Z(0) > H$. Then, putting together (1.3) and Definitions 1.45 and 1.42, one gets that

$$2^{-nH} Z(2^{-n}e) \xrightarrow[n \to +\infty]{a.s.} 0, \qquad (1.89)$$

where e denotes an arbitrary fixed vector belonging to the unit sphere $\mathcal{S}^{N-1}$. On the other hand, one knows from the self-similarity property of the field $\{ Z(t), t \in \mathbb{R}^N \}$ that, for any arbitrary fixed $n \in \mathbb{N}$, the random variables $2^{-nH} Z(2^{-n}e)$ and $Z(e)$ have the same distribution. Therefore, (1.89) implies that $Z(e) \overset{a.s.}{=} 0$. Then, using again the self-similarity property of $\{ Z(t), t \in \mathbb{R}^N \}$ and the fact that e is arbitrary, one obtains, for all $t \in \mathbb{R}^N$, that $Z(t) \overset{a.s.}{=} 0$, which contradicts the assumption that the paths of $\{ Z(t), t \in \mathbb{R}^N \}$ are, almost surely, continuous nowhere differentiable functions. □

Several results, previously obtained in the current section, show that stationarity of increments and/or global self-similarty and/or isotropy for a continuous nowhere differentiable stochastic field impose severe restrictions on its Hölder exponents. In other words, when such a stochastic field has

Hölder exponents which change from place to place almost surely, then it is typically with non-stationary increments; also, in general, it is not globally self-similar. The notion of self-similarity has been weakened (see [Benassi *et al.* (1997); Falconer (2002, 2003)]) in the following way, in order to fit with non-stationary increments frames.

Definition 1.69 (local asymptotic self-similarity). Let $\{Y(t), t \in \mathbb{R}^N\}$ be a real-valued stochastic field, $\tau \in \mathbb{R}^N$ a fixed point and $H(\tau)$ a fixed positive real number. $\{Y(t), t \in \mathbb{R}^N\}$ is said to be at τ locally asymptotically self-similar (lass) with exponent $H(\tau)$, when the following holds: there exists a non-degenerate stochastic field, denoted by $\{T_\tau Y(s), s \in \mathbb{R}^N\}$, such that, for any arbitray sequence $(a_n)_{n \in \mathbb{N}}$ of positive real numbers converging to zero, one has

$$\left\{ \frac{Y(\tau + a_n s) - Y(\tau)}{a_n^{H(\tau)}}, s \in \mathbb{R}^N \right\} \xrightarrow[n \to +\infty]{d} \{T_\tau Y(s), s \in \mathbb{R}^N\}, \qquad (1.90)$$

where the notation $\xrightarrow[n \to +\infty]{d}$ means that: when n goes to infinity, the finite-dimensional distributions of the stochastic field in the left-hand side converge to those of the one in the right-hand side. The field $\{T_\tau Y(s), s \in \mathbb{R}^N\}$ is necessarily unique, it is called *the tangent field of* $\{Y(t), t \in \mathbb{R}^N\}$ *at* τ.

Definition 1.70 (strong local asymptotic self-similarity). When paths of $\{Y(t), t \in \mathbb{R}^N\}$ and $\{T_\tau Y(s), s \in \mathbb{R}^N\}$ are, with probability 1, continuous functions on $\mathbb{R}^N$. The field $\{Y(t), t \in \mathbb{R}^N\}$ is said to be at τ strongly locally asymptotically self-similar (slass) with exponent $H(\tau)$, if the convergence in distribution in (1.90) holds in the space of the continuous functions over an arbitrary compact subset of $\mathbb{R}^N$.

Our next goal is to draw important connections between incremental variance and Hölder exponents in the Gaussian frame.

Theorem 1.71. *Let* $\{Z(t), t \in \mathbb{R}^N\}$ *be a centered real-valued Gaussian field with almost surely continuous nowhere differentiable paths. Then, its global Hölder exponent* $\beta_Z(I)$ *on an arbitrary compact interval* $I \subset \mathbb{R}^N$ *satisfies almost surely*

$$\beta_Z(I) \overset{a.s.}{=} \sup\left\{ \beta \in [0, 1) : \sup_{t^1, t^2 \in I} \frac{\mathbb{E}\big(|Z(t^1) - Z(t^2)|^2\big)}{|t^1 - t^2|^{2\beta}} < +\infty \right\}. \qquad (1.91)$$

Remark 1.72. Let $\{B_H(t), t \in \mathbb{R}^N\}$ be a Fractional Brownian Field (FBF) of Hurst parameter H and let $I \subset \mathbb{R}^N$ be an arbitrary and fixed compact interval. A straightforward consequence of (1.19) and of Theorem 1.71 is that $\beta_{B_H}(I)$, the global Hölder exponent of $\{B_H(t), t \in \mathbb{R}^N\}$ on I, satisfies almost surely

$$\beta_{B_H}(I) \overset{\text{a.s.}}{=} H.$$

Remark 1.73. Let $\tau \in \mathbb{R}^N$ be arbitrary and fixed. A straightforward consequence of Definition 1.39 and of Remark 1.72, is that $\widetilde{\alpha}_{B_H}(\tau)$, the local Hölder exponent of $\{B_H(t), t \in \mathbb{R}^N\}$ at τ, satisfies almost surely

$$\widetilde{\alpha}_{B_H}(\tau) \overset{\text{a.s.}}{=} H.$$

Proof of Theorem 1.71. Using Definitions 1.34 and 1.33 as well as Theorems A.5 and A.10, one can easily get the theorem. □

Theorem 1.74. *Let $\{Z(t), t \in \mathbb{R}^N\}$ be a centered real-valued Gaussian field with almost surely continuous nowhere differentiable paths. Assume that $\tau \in \mathbb{R}^N$ is a point at which the local Hölder exponent $\widetilde{\alpha}_Z(\tau)$ does not vanish almost surely. Then, $\alpha_Z(\tau)$, the pointwise Hölder exponent at $\tau \in \mathbb{R}^N$, satisfies almost surely*

$$\alpha_Z(\tau) \overset{\text{a.s.}}{=} \sup\left\{\alpha \in [0,1) : \sup_{t \in \mathcal{B}(\tau,\delta)} \frac{\mathbb{E}\big(|Z(t) - Z(\tau)|^2\big)}{|t - \tau|^{2\alpha}} < +\infty\right\}, \qquad (1.92)$$

where $\mathcal{B}(\tau, \delta)$ is the closed ball centered at τ and of an arbitrary fixed radius $\delta > 0$.

The following result can easily be derived from (1.19), Remark 1.73 and Theorem 1.74.

Remark 1.75. Let $\tau \in \mathbb{R}^N$ be arbitrary and fixed, $\alpha_{B_H}(\tau)$, the pointwise Hölder exponent of $\{B_H(t), t \in \mathbb{R}^N\}$ at τ, satisfies almost surely

$$\alpha_{B_H}(\tau) \overset{\text{a.s.}}{=} H.$$

Proof of Theorem 1.74. Let us first show that

$$\alpha_Z(\tau) \overset{\text{a.s.}}{\le} \sup\left\{\alpha \in [0,1) : \sup_{t \in \mathcal{B}(\tau,\delta)} \frac{\mathbb{E}\big(|Z(t) - Z(\tau)|^2\big)}{|t - \tau|^{2\alpha}} < +\infty\right\}. \qquad (1.93)$$

To this end, it is sufficient (see Definition 1.45 and part (*ii*) of Proposition 1.65) to prove that, for any $\alpha \in (0,1)$ satisfying

$$\mathbb{P}\big(\{\omega \in \Omega^* : Z(\cdot, \omega) \in \mathcal{C}^\alpha(\tau)\}\big) = 1, \qquad (1.94)$$

one has, for all $\alpha' \in [0, \alpha)$,

$$\mathbb{E}\left(\sup_{t \in \mathcal{B}(\tau,\delta)} \frac{|Z(t) - Z(\tau)|^2}{|t - \tau|^{2\alpha'}}\right) < +\infty. \qquad (1.95)$$

Observe that (1.95) implies that

$$\sup_{t\in\mathcal{B}(\tau,\delta)} \frac{\mathbb{E}\big(|Z(t)-Z(\tau)|^2\big)}{|t-\tau|^{2\alpha'}} < +\infty,$$

since one always has that

$$\sup_{t\in\mathcal{B}(\tau,\delta)} \frac{\mathbb{E}\big(|Z(t)-Z(\tau)|^2\big)}{|t-\tau|^{2\alpha'}} \leq \mathbb{E}\Bigg(\sup_{t\in\mathcal{B}(\tau,\delta)} \frac{|Z(t)-Z(\tau)|^2}{|t-\tau|^{2\alpha'}}\Bigg).$$

Let T be an arbitrary fixed compact interval of $\mathbb{R}^N$ such that $\mathcal{B}(\tau,\delta) \subset T$ and let $\{X(t), t \in T\}$ be the centered real-valued Gaussian field with continuous paths which vanishes on the negligible event $(\Omega \setminus \Omega^*) \cup \big\{\omega \in \Omega^* : Z(\cdot,\omega) \notin \mathcal{C}^\alpha(\tau)\big\}$ and otherwise is defined as $X(t) := 0$ when $t = \tau$ and $X(t) := |t - \tau|^{-\alpha'}(Z(t) - Z(\tau))$ else (recall that $(\Omega, \mathcal{A}, \mathbb{P})$ is the underlying probability space, and $\Omega^* \in \mathcal{A}$ an event of probability 1 such that, for each $\omega \in \Omega^*$, the path $Z(\cdot,\omega)$ is a continuous nowhere differentiable function on $\mathbb{R}^N$). Applying Lemma A.11 to $\{X(t), t \in T\}$, one gets (1.95).

Let us now show that

$$\alpha_Z(\tau) \overset{\text{a.s.}}{\geq} \sup\left\{\alpha \in [0,1) : \sup_{t\in\mathcal{B}(\tau,\delta)} \frac{\mathbb{E}\big(|Z(t)-Z(\tau)|^2\big)}{|t-\tau|^{2\alpha}} < +\infty\right\}. \tag{1.96}$$

The arguments we will use are more or less inspired by the proof of Proposition 13 in [Herbin (2006)]. It is at this stage that the hypothesis that the local Hölder exponent $\widetilde{\alpha}_Z(\tau)$ does not vanish almost surely will play a crucial role. Assume that $\alpha \in (0,1)$ is arbitrary and such that, for all $t \in \mathcal{B}(\tau,\delta)$, the inequality

$$\mathbb{E}\big(|Z(t)-Z(\tau)|^2\big) \leq c_1|t-\tau|^{2\alpha}, \tag{1.97}$$

holds, for some finite positive constant c_1 not depending on t. Thus, in view of Proposition 1.44, in order to get (1.96), it is sufficient to prove that, for all $\alpha' \in [0,\alpha)$, one has

$$\sup_{\rho\in(0,\delta]} \Big(\rho^{-\alpha'} \sup_{t\in\mathcal{B}(\tau,\rho)} |Z(t)-Z(\tau)|\Big) \overset{\text{a.s.}}{<} +\infty. \tag{1.98}$$

Denoting by $[\cdot]$ the integer part function and using the almost sure continuity of paths of $\{Z(t), t \in \mathbb{R}^N\}$ as well as the inequalities, for all $\rho \in (0, +\infty)$,

$$2^{-\alpha'}2^{\alpha'([-\log_2 \rho]+1)} \sup\Big\{|Z(t)-Z(\tau)| : t \in \mathcal{B}\big(\tau, 2^{-([-\log_2 \rho]+1)}\big)\Big\}$$

$$\leq \rho^{-\alpha'} \sup_{t\in\mathcal{B}(\tau,\rho)} |Z(t)-Z(\tau)| \leq$$

$$2^{\alpha'}2^{\alpha'[-\log_2 \rho]} \sup\Big\{|Z(t)-Z(\tau)| : t \in \mathcal{B}\big(\tau, 2^{-[-\log_2 \rho]}\big)\Big\},$$

one can easily show that (1.98) is equivalent to

$$\sup_{n\in\mathbb{N}} \left(2^{n\alpha'} \sup_{t\in\mathcal{B}(\tau,2^{-n})} |Z(t) - Z(\tau)|\right) \overset{\text{a.s.}}{<} +\infty. \tag{1.99}$$

So, from now on our goal is to derive (1.99). According to Part (ii) of Proposition 1.67, one has $\tilde{\alpha}_Z(\tau) \overset{\text{a.s.}}{=} \tilde{a}_Z(\tau)$, for some deterministic quantity $\tilde{a}_Z(\tau) \in [0,1]$. Moreover, the assumption that $\tilde{\alpha}_Z(\tau)$ does not vanish almost surely implies that $\tilde{a}_Z(\tau)$ does not vanish as well. Next, let η be the deterministic positive quantity defined as $\eta := 2^{-1}\tilde{a}_Z(\tau)$. Thus, it follows from Definition 1.39 that there exists $\Omega_0^* \subseteq \Omega^*$ an event of probability 1 satisfying the following property: for each $\omega \in \Omega_0^*$, there are two finite positive constants $C_0^*(\omega)$ and $\nu_0^*(\omega)$, such that the inequality

$$|Z(t^1,\omega) - Z(t^2,\omega)| \leq C_0^*(\omega)|t^1 - t^2|^\eta, \tag{1.100}$$

holds for all $t^1, t^2 \in \mathcal{B}(\tau, \nu_0^*(\omega))$. Next, we set $p_0 = [\alpha'/\eta] + 1 + N$; observe that

$$p_0\eta > \alpha'. \tag{1.101}$$

Let us show that, for all $\omega \in \Omega_0^*$ and positive integer $n \geq -\log_2(\nu_0^*(\omega)) + 1$, one has

$$\sup\left\{|Z(t,\omega) - Z(\tau,\omega)| : t \in \mathcal{B}(\tau, 2^{-n})\right\} \tag{1.102}$$
$$\leq N^{\eta/2}C_0^*(\omega)2^{-np_0\eta} + \sup\left\{|Z(d,\omega) - Z(\tau,\omega)| : d \in \mathbb{D}_{np_0}^N \cap \mathcal{B}(\tau, 2^{-n+1})\right\},$$

where $\mathbb{D}_{np_0}^N$ is the set of the dyadic vectors of $\mathbb{R}^N$ of order np_0 (see Definition 1.66). Assume that $t \in \mathcal{B}(\tau, 2^{-n})$ is arbitrary and fixed; $t_1, \ldots, t_N$ and $\tau_1, \ldots, \tau_N$ respectively denote the coordinates of t and τ. For all $l \in \{1, \ldots, N\}$, let $d_l(t) \in \left\{2^{-np_0}([2^{np_0}t_l]), 2^{-np_0}([2^{np_0}t_l] + 1)\right\}$ be such that

$$\left|d_l(t) - \tau_l\right| = \min\left\{\left|2^{-np_0}([2^{np_0}t_l]) - \tau_l\right|, \left|2^{-np_0}([2^{np_0}t_l] + 1) - \tau_l\right|\right\}.$$

Thus, one has that

$$|t - d(t)| \leq N^{1/2}\, 2^{-np_0}, \tag{1.103}$$

and

$$\left|d_l(t) - \tau_l\right| \leq \max\left\{|t_l - \tau_l|, 2^{-np_0}\right\}, \quad \text{for every } l \in \{1, \ldots, N\}.$$

Then, the fact t belongs to $\mathcal{B}(\tau, 2^{-n})$ and the inequality $N^{1/2}\, 2^{-np_0} \leq 2^{-n}$ imply that the dyadic vector of $\mathbb{D}_{np_0}^N$ $d(t) := \left(d_1(t), \ldots, d_N(t)\right)$ belongs to

$\mathcal{B}(\tau, 2^{-n+1})$. Next, using the triangle inequality (1.103) and (1.100), one gets that

$$|Z(t,\omega) - Z(\tau,\omega)| \leq |Z(t,\omega) - Z(d(t),\omega)| + |Z(d(t),\omega) - Z(\tau,\omega)|$$

$$\leq N^{\eta/2} C_0^*(\omega) 2^{-np_0\eta} + \sup\left\{|Z(d,\omega) - Z(\tau,\omega)| : d \in \mathbb{D}_{np_0}^N \cap \mathcal{B}(\tau, 2^{-n+1})\right\}.$$

This means that (1.102) holds, since $t \in \mathcal{B}(\tau, 2^{-n})$ has been chosen arbitrarily. Next, it follows from (1.101) and (1.102) that, for obtaining (1.99), it is enough to show that

$$\sup_{n \in \mathbb{N}} \left(2^{n\alpha'} \sup\left\{|Z(d) - Z(\tau)| : d \in \mathbb{D}_{np_0}^N \cap \mathcal{B}(\tau, 2^{-n+1})\right\}\right) \overset{\text{a.s.}}{<} +\infty. \quad (1.104)$$

Using Lemma A.25, (1.85) and (1.97), one gets, for each positive integer $n \geq 1 - \log_2 \delta$, that

$$\mathbb{E}\left(\sup\left\{|Z(d) - Z(\tau)| : d \in \mathbb{D}_{np_0}^N \cap \mathcal{B}(\tau, 2^{-n+1})\right\}\right)$$

$$\leq c_2 \sup\left\{\sqrt{\mathbb{E}|Z(d) - Z(\tau)|^2} : d \in \mathbb{D}_{np_0}^N \cap \mathcal{B}(\tau, 2^{-n+1})\right\}$$

$$\times \sqrt{\log_2\left(1 + \operatorname{card}(\mathbb{D}_{np_0}^N \cap \mathcal{B}(\tau, 2^{-n+1}))\right)}$$

$$\leq c_3 2^{-n\alpha}\sqrt{n},$$

where c_2 and c_3 are two finite positive constants not depending on n. Thus, it follows from Markov inequality that

$$\mathbb{P}\left(2^{n\alpha'} \sup\left\{|Z(d) - Z(\tau)| : d \in \mathbb{D}_{np_0}^N \cap \mathcal{B}(\tau, 2^{-n+1})\right\} > 1\right) \leq c_3 2^{-n(\alpha-\alpha')}\sqrt{n}.$$

Next combining the latter inequality with the inequality $\alpha - \alpha' > 0$, one obtains that

$$\sum_{n=1}^{+\infty} \mathbb{P}\left(2^{n\alpha'} \sup\left\{|Z(d) - Z(\tau)| : d \in \mathbb{D}_{np_0}^N \cap \mathcal{B}(\tau, 2^{-n+1})\right\} > 1\right) < +\infty,$$

which in turn implies that (1.104) is satisfied (Borel-Cantelli Lemma). □

Remark 1.76. A careful inspection of the proof of Theorem 1.74 shows that it still be true even if the assumption that $\widetilde{\alpha}_Z(\tau)$ does not vanish almost surely is weakened to

$$\limsup_{\nu \to 0^+} \left(\sup_{t^1, t^2 \in \mathcal{B}(\tau,\nu)} \frac{|Z(t^1) - Z(t^2)|}{\log^n(1/|t^1 - t^2|)}\right) \overset{\text{a.s.}}{<} +\infty, \quad \text{for all fixed } n \in \mathbb{N}.$$

$$(1.105)$$

On the other hand, it is not clear that the theorem remains valid when one drops the assumption (1.105), yet one has that

$$\alpha_Z(\tau) \overset{\text{a.s.}}{\leq} \sup\left\{\alpha \in [0,1) : \sup_{t \in \mathcal{B}(\tau,\delta)} \frac{\mathbb{E}(|Z(t) - Z(\tau)|^2)}{|t - \tau|^{2\alpha}} < +\infty\right\},$$

where $\mathcal{B}(\tau, \delta)$ is the closed ball centered at τ and of an arbitrary fixed radius $\delta > 0$.

The following remark is a straightforward consequence of Definition 1.69, Theorem 1.74 and Remark 1.76.

Remark 1.77. Let $\{Z(t), t \in \mathbb{R}^N\}$ be a centered real-valued Gaussian field with almost surely continuous nowhere differentiable paths. Assume that $\tau \in \mathbb{R}^N$ is a point at which $\{Z(t), t \in \mathbb{R}^N\}$ is locally asymptotically self-similar with exponent $H(\tau)$. Then, the pointwise Hölder exponent $\alpha_Z(\tau)$ satisfies almost surely

$$\alpha_Z(\tau) \overset{a.s.}{\leq} H(\tau). \tag{1.106}$$

Notice that the inequality in (1.106) becomes an equality when $N = 1$ and (1.105) holds.

1.4 Multifractional Brownian Field

Roughly speaking Multifractional Brownian Field (MuBF) is an extension of the Fractional Brownian Field (FBF) $\{B_H(t), t \in \mathbb{R}^N\}$ of the form $\{B_{H(t)}(t), t \in \mathbb{R}^N\}$; in other words, the Hurst parameter becomes dependent on the space variable t. Throughout the section, $H(\cdot)$ denotes a deterministic function from $\mathbb{R}^N$ into a fixed compact interval $[\underline{H}, \overline{H}]$ included in the open interval $(0, 1)$ of the real line. This function will play a role quite similar to that of the Hurst parameter of FBF, therefore it is called the Hurst Multifractional function and also the Hurst functional parameter of MuBF. Similarly to FBF (see Propositions 1.25 and 1.28), MuBF can be constructed in the time domain as well as in the Fourier domain; we will content ourselves with constructing it in the Fourier domain. In order to do so, first we need to introduce the centered Gaussian field $\{X(u, v), (u, v) \in \mathbb{R}^N \times (0, 1)\}$ which generates MuBF.

Definition 1.78 (Generator of MuBF). One calls Generator of MuBF, the real-valued centered Gaussian field $\{X(u, v), (u, v) \in \mathbb{R}^N \times (0, 1)\}$, defined, for each $(u, v) \in \mathbb{R}^N \times (0, 1)$, as:

$$X(u, v) := \int_{\mathbb{R}^N} \frac{e^{iu \cdot \xi} - 1}{|\xi|^{v + N/2}} \, d\widehat{\mathbb{W}}(\xi), \tag{1.107}$$

where $\widehat{\mathbb{W}}$ denotes the orthogonally scattered random measure introduced in (1.28). Observe that, for each fixed $v \in (0, 1)$, $X(\cdot, v) := \{X(u, v), u \in \mathbb{R}^N\}$ is a FBF with Hurst parameter v.

The following proposition provides sharp estimates for incremental standard deviation of $\{X(u,v), (u,v) \in \mathbb{R}^N \times (0,1)\}$ on an arbitrary compact interval $J \times [M_1, M_2] \subset \mathbb{R}^N \times (0,1)$.

Proposition 1.79. *Let J be an arbitrary compact interval of $\mathbb{R}^N$. Let M_1 and M_2 be two arbitrary real numbers satisfying $0 < M_1 < M_2 < 1$. Then, there exist three positive constants c, c' and c'', only depending on J, M_1 and M_2, such that the inequalities*

$$c' \left| u^1 - u^2 \right|^{\min\{v_1, v_2\}} - c'' \left| v_1 - v_2 \right| \tag{1.108}$$
$$\leq \sigma\big(X(u^1, v_1) - X(u^2, v_2)\big) \leq c\big(\left| u^1 - u^2 \right|^{\min\{v_1, v_2\}} + \left| v_1 - v_2 \right|\big),$$

hold for all $(u^1, v_1) \in J \times [M_1, M_2]$ and $(u^2, v_2) \in J \times [M_1, M_2]$.

Proof of Proposition 1.79. There is no restriction to assume that $v_1 \leq v_2$. Using (1.107), the isometry property of the integral $\int_{\mathbb{R}^N} (\,\cdot\,) d\widehat{\mathbb{W}}$ (see (1.30)) and the triangle inequality, one gets that

$$A^{1/2} - B^{1/2} \tag{1.109}$$
$$\leq \sigma\big(X(u^1, v_1) - X(u^2, v_2)\big) = \left(\int_{\mathbb{R}^N} \left| \frac{e^{iu^1 \cdot \xi} - 1}{|\xi|^{v_1 + N/2}} - \frac{e^{iu^2 \cdot \xi} - 1}{|\xi|^{v_2 + N/2}} \right|^2 d\xi \right)^{1/2}$$
$$\leq A^{1/2} + B^{1/2},$$

where

$$A := \int_{\mathbb{R}^N} \frac{\left| e^{i(u^1 - u^2) \cdot \xi} - 1 \right|^2}{|\xi|^{2v_1 + N}} \, d\xi$$

and

$$B := \int_{\mathbb{R}^N} \frac{\left| e^{iu^2 \cdot \xi} - 1 \right|^2}{|\xi|^{2v_1 + N}} \big(|\xi|^{v_1 - v_2} - 1 \big)^2 d\xi.$$

Next, let us show that

$$c_1' \left| u^1 - u^2 \right|^{2v_1} \leq A \leq c_1 \left| u^1 - u^2 \right|^{2v_1}, \tag{1.110}$$

where $0 < c_1' \leq c_1 < +\infty$ are the constants

$$c_1' := \int_{|\eta| \leq 1} \frac{\left| e^{i\eta_1} - 1 \right|^2}{|\eta|^{2M_1 + N}} \, d\eta + \int_{|\eta| > 1} \frac{\left| e^{i\eta_1} - 1 \right|^2}{|\eta|^{2M_2 + N}} \, d\eta$$

and

$$c_1 := \int_{|\eta| \leq 1} \frac{\left| e^{i\eta_1} - 1 \right|^2}{|\eta|^{2M_2 + N}} \, d\eta + \int_{|\eta| > 1} \frac{\left| e^{i\eta_1} - 1 \right|^2}{|\eta|^{2M_1 + N}} \, d\eta.$$

It is clear that (1.110) holds when $|u^1 - u^2| = 0$, so let us assume that $|u^1 - u^2| > 0$ and set $\eta = |u^1 - u^2| Q\xi$ in the integral A, where Q is an arbitrary orthogonal matrix of size N, such that $|u^1 - u^2|^{-1} Q(u^1 - u^2)$ is the vector of $\mathbb{R}^N$ whose first coordinate equals 1 and the others vanish. Thus, one gets

$$A = |u^1 - u^2|^{2v_1} \int_{\mathbb{R}} \frac{\left| e^{i\eta_1} - 1 \right|^2}{|\eta|^{2v_1 + N}} \, d\eta, \qquad (1.111)$$

which clearly implies that (1.110) is satisfied. Let us now show that

$$B \le c_2 |v_1 - v_2|^2, \qquad (1.112)$$

where the finite constant

$$c_2 := \left(4 + \max_{u \in J} |u|^2 \right) \left(\int_{|\xi| \le 1} |\xi|^{2(1-M_2)-N} \log^2 |\xi| \, d\xi + \int_{|\xi| > 1} |\xi|^{-2M_1 - N} \log^2 |\xi| \, d\xi \right).$$

Using the equality $|\xi|^{v_1 - v_2} := e^{(v_1 - v_2) \log |\xi|}$, for all $\xi \in \mathbb{R}^N \setminus \{0\}$, the inequality

$$\left| e^{i\theta} - 1 \right| \le \min\left\{ 2, |\theta| \right\}, \quad \text{for every } \theta \in \mathbb{R}, \qquad (1.113)$$

Cauchy-Schwarz inequality, as well as the inequalities

$$1 - e^{-x} \le \min\{1, x\} \quad \text{and} \quad e^x - 1 \le xe^x, \quad \text{for each } x \in \mathbb{R}_+, \qquad (1.114)$$

one obtains that

$$B \le |v_1 - v_2|^2 \left(|u^2|^2 \int_{|\xi| \le 1} |\xi|^{2(1-v_2)-N} \log^2 |\xi| \, d\xi + 4 \int_{|\xi| > 1} |\xi|^{-2v_1 - N} \log^2 |\xi| \, d\xi \right)$$

$$\le c_2 |v_1 - v_2|^2,$$

which proves (1.112). Finally one can derive (1.108) by putting together (1.109), (1.110) and (1.112). $\qquad\square$

Proposition 1.80 (Hölder continuity of Generator of MuBF).
There exists a modification of $\{X(u,v), (u,v) \in \mathbb{R}^N \times (0,1)\}$, also denoted by $\{X(u,v), (u,v) \in \mathbb{R}^N \times (0,1)\}$, and there exists Ω^* an event of probability 1, such that, for each $\omega \in \Omega^*$, the path $X(\cdot, \cdot, \omega) : (u,v) \mapsto X(u,v,\omega)$ is a continuous functions on $\mathbb{R}^N \times (0,1)$ which further satisfies the following Hölder property: let J be an arbitrary compact interval of $\mathbb{R}^N$, let M_1, M_2 and ε be three arbitrary real numbers such that $0 < \varepsilon < M_1 < M_2 < 1$, then, for all $(u^1, v_1) \in J \times [M_1, M_2]$ and $(u^2, v_2) \in J \times [M_1, M_2]$, one has

$$\left| X(u^1, v_1, \omega) - X(u^2, v_2, \omega) \right| \le C(\omega) \left(|u^1 - u^2| + |v_1 - v_2| \right)^{M_1 - \varepsilon}, \qquad (1.115)$$

where C is a finite nonnegative random variable. Notice that C does not depend on u^1, u^2, v_1, v_2 but it depends on J, M_1, M_2, ε. Also notice that, in view of Lemma A.11, C can be chosen in such a way that all its moments be finite.

Proof of Proposition 1.80. In view of Theorem A.5 and of the fact that J, M_1 and M_2 are arbitrary, in order to get the proposition, it is enough to show that there exists a constant $c_0 > 0$ such that, for all $(u^1, v_1) \in J \times [M_1, M_2]$ and $(u^2, v_2) \in J \times [M_1, M_2]$, one has

$$\mathbb{E}\big|X(u^1, v_1) - X(u^2, v_2)\big|^2 \leq c_0 \big(|u^1 - u^2| + |v_1 - v_2|\big)^{2M_1}. \tag{1.116}$$

The latter inequality easily follows from the second inequality in (1.108).
$\square$

Remark 1.81. Throughout the book, $\{X(u, v), (u, v) \in \mathbb{R}^N \times (0, 1)\}$, the generator of MuBF, defined through (1.107), is identified with its continuous modification introduced in Proposition 1.80.

Definition 1.82 (Multifractional Brownian Field (MuBF)).
The Multifractional Brownian Field (MuBF) of Hurst functional parameter $H(\cdot)$ is the real-valued centered Gaussian field denoted by $\{Y(t), t \in \mathbb{R}^N\}$ and defined, for each $t \in \mathbb{R}^N$, as

$$Y(t) := X(t, H(t)). \tag{1.117}$$

Let us now discuss continuity of paths of MuBF.

Theorem 1.83. *Let Ω^* be the event of probability 1 introduced in Proposition 1.80.*

(i) For all $\omega \in \Omega^*$, the path $Y(\cdot, \omega)$ is continuous at 0.
(ii) The continuity of $H(\cdot)$ on $\mathbb{R}^N$ is a sufficient condition for having, for all $\omega \in \Omega^*$, the continuity on $\mathbb{R}^N$ of the path $Y(\cdot, \omega)$.
(iii) The continuity of $H(\cdot)$ on $\mathbb{R}^N$ is a necessary condition for having, for all $\omega \in \Omega^*$, the continuity on $\mathbb{R}^N$ of the path $Y(\cdot, \omega)$. More precisely, when $H(\cdot)$ is discontinuous at some $\check{t} \in \mathbb{R}^N \setminus \{0\}$, then there exists $\check{\Omega}_{\check{t}} \subseteq \Omega^*$, an event of probability 1 depending on $\check{t}$, such that, for each $\omega \in \check{\Omega}_{\check{t}}$, the path $Y(\cdot, \omega)$ is discontinuous at $\check{t}$.

Proof of Theorem 1.83. First notice that Part *(ii)* easily follows from (1.117) and Proposition 1.80. In order to establish Part *(i)*, one has to prove that, for any sequence $(s^m)_{m \in \mathbb{N}}$ in $\mathbb{R}^N$ such that

$$s^m \xrightarrow[m \to +\infty]{} 0, \tag{1.118}$$

the sequence $\big(Y(s^m, \omega)\big)_{m \in \mathbb{N}}$ converges to $Y(0, \omega) = 0$. In fact, it is enough to show that being given an arbitrary subsequence $\big(Y(s^{m_l}, \omega)\big)_{l \in \mathbb{N}}$, one can always extract from it a subsubsequence $\left(Y(s^{m_{l_n}},\omega)\right)_{n\in\mathbb{N}}$ which converges 0. The fact that $\left(H(s^{m_l})\right)_{l\in\mathbb{N}}$ is a sequence in the compact interval $[\underline{H},\overline{H}]$ implies that there exist $a \in [\underline{H},\overline{H}]$ and a subsequence $\left(H(s^{m_{l_n}})\right)_{n\in\mathbb{N}}$ such that

$$H(s^{m_{l_n}}) \xrightarrow[n\to+\infty]{} a. \tag{1.119}$$

Next, putting together (1.117), Proposition 1.80, (1.118) and (1.119), one gets that

$$Y(s^{m_{l_n}},\omega) := X(s^{m_{l_n}}, H(s^{m_{l_n}}),\omega) \xrightarrow[n\to+\infty]{} X(0,a,\omega) = 0.$$

Let us now establish Part (iii). The fact that $H(\cdot)$ is discontinuous at $\check{t}$ and with values in the compact interval $[\underline{H},\overline{H}]$ entails that there is a sequence $(t^m)_{m\in\mathbb{N}}$ in $\mathbb{R}^N$ such that

$$t^m \xrightarrow[n\to+\infty]{} \check{t} \quad \text{and} \quad H(t^m) \xrightarrow[n\to+\infty]{} b, \tag{1.120}$$

where $b \in [\underline{H},\overline{H}]$ and $b \neq H(\check{t})$. Next combining (1.117) with Proposition 1.80 and (1.120), it follows that, for each $\omega \in \Omega^*$,

$$Y(t^m,\omega) := X(t^m, H(t^m),\omega) \xrightarrow[n\to+\infty]{} X(\check{t},b,\omega).$$

Thus, in order to obtain Part (iii), it is enough to show that almost surely

$$X(\check{t},b) \overset{\text{a.s.}}{\neq} Y(\check{t}) := X(\check{t}, H(\check{t})). \tag{1.121}$$

Using the fact that $X(\check{t},b) - X(\check{t}, H(\check{t}))$ is a centered real-valued Gaussian random variable, it turns out that (1.121) is equivalent to

$$\mathbb{E}\big|X(\check{t},b) - X(\check{t}, H(\check{t}))\big|^2 > 0. \tag{1.122}$$

Next, observe that (1.107) and the isometry property of the stochastic integral $\int_{\mathbb{R}^N} (\,\cdot\,)d\widehat{\mathbb{W}}$, imply that

$$\mathbb{E}\big|X(\check{t},b) - X(\check{t}, H(\check{t}))\big|^2 = \int_{\mathbb{R}^N} \frac{\big|e^{i\check{t}\cdot\xi} - 1\big|^2}{|\xi|^{2b+N}} \left(|\xi|^{b-H(\check{t})} - 1\right)^2 d\xi.$$

Therefore, standard computations show that if (1.122) fails to be true, then one would have $e^{i\check{t}\cdot\xi} = 1$, for all $\xi \in \mathbb{R}^N$. This is impossible, since $\check{t} \neq 0$. $\square$

Remark 1.84. From now on, we assume that the Hurst Multifractional function $H(\cdot)$ is continuous on $\mathbb{R}^N$. Then, we know from Theorem 1.83 that paths of MuBF are, with probability 1, continuous functions on $\mathbb{R}^N$. On the other hand, under some very mild additional assumptions on $H(\cdot)$, it can be shown that they are, with probability 1, nowhere differentiable on $\mathbb{R}^N$. The latter result is difficult to obtain when the hypotheses on $H(\cdot)$ are very weak; it can be derived by using wavelet methods, in the same spirit of those which will be presented in Chapter 6.

The following two propositions, which easily result from Definition 1.82 and Proposition 1.79, provide useful estimates for incremental standard deviation of the MuBF $\{Y(t), t \in \mathbb{R}^N\}$.

Proposition 1.85. *Let $I \subset \mathbb{R}^N$ be an arbitrary compact interval. One sets*

$$\underline{H}(I) := \min\{H(t) : t \in I\} \quad and \quad \overline{H}(I) := \max\{H(t) : t \in I\}. \quad (1.123)$$

Assuming that

$$H(\cdot) \in \mathcal{C}^{\gamma_I}(I), \quad for \ some \ \gamma_I \in (\underline{H}(I), 1), \quad (1.124)$$

where $\mathcal{C}^{\gamma_I}(I)$ denotes the global Hölder space on I of order γ_I (see Definition 1.33). Then, there exists a constant $c > 0$, such that the inequality

$$\sigma\big(Y(t^1) - Y(t^2)\big) \leq c \, |t^1 - t^2|^{\underline{H}(I)}, \quad (1.125)$$

holds, for all $(t^1, t^2) \in I^2$. Also, there exists a constant $c' > 0$, such that the inverse inequality

$$\sigma\big(Y(t) - Y(t_{min})\big) \geq c' \, |t - t_{min}|^{\underline{H}(I)}, \quad (1.126)$$

holds, for each $t \in I$ and every $t_{min} \in I$ satisfying $H(t_{min}) = \underline{H}(I)$. Moreover, when the condition (1.124) is strengthened to

$$H(\cdot) \in \mathcal{C}^{\gamma_I}(I), \quad for \ some \ \gamma_I \in (\overline{H}(I), 1), \quad (1.127)$$

then, one has, for some constants $c_0, c_0', c_0'' > 0$ and all $(t^1, t^2) \in I^2$,

$$c_0 \, |t^1 - t^2|^{\min\{H(t^1), H(t^2)\}} \geq \sigma\big(Y(t^1) - Y(t^2)\big) \geq c_0' \, |t^1 - t^2|^{\min\{H(t^1), H(t^2)\}}$$

$$\geq c_0'' \, |t^1 - t^2|^{\overline{H}(I)}. \quad (1.128)$$

Proposition 1.86. *Assume that $\tau \in \mathbb{R}^N$ satisfies*

$$H(\cdot) \in \mathcal{C}^{\gamma_\tau}(\tau), \quad for \ some \ \gamma_\tau \in (H(\tau), 1), \quad (1.129)$$

where $\mathcal{C}^{\gamma_\tau}(\tau)$ denotes the pointwise Hölder space at τ of order γ_τ (see Definition 1.42). Then for each fixed $\delta > 0$, there are two constants $0 < c' \leq c$, such that the inequalities

$$c' \, |t - \tau|^{H(\tau)} \leq \sigma\big(Y(t) - Y(\tau)\big) \leq c \, |t - \tau|^{H(\tau)}, \quad (1.130)$$

holds for all $t \in \mathcal{B}(\tau, \delta)$ the closed ball centered at τ and of radius δ.

Let us now give three important results concerning global, local and pointwise Hölder regularity of MuBF. The following theorem is a straightforward consequence of Theorem 1.71 and Proposition 1.85.

Theorem 1.87 (global Hölder regularity of MuBF).

Let $I \subset \mathbb{R}^N$ be an arbitrary compact interval such that the condition (1.124) holds. Then, $\beta_Y(I)$, the global Hölder exponent of MuBF $\{Y(t), t \in \mathbb{R}^N\}$ on I, satisfies almost surely

$$\beta_Y(I) \stackrel{a.s.}{=} \min\{H(t) : t \in I\}. \tag{1.131}$$

Theorem 1.88 (local Hölder regularity of MuBF). *Assume that $\tau \in \mathbb{R}^N$ satisfies the following property: there exists a compact interval $J_\tau \subset \mathbb{R}^N$ such that*

$$\tau \in \mathring{J}_\tau \quad and \quad H(\cdot) \in \mathcal{C}^{\gamma_{J_\tau}}(J_\tau), \quad for\ some\ \gamma_{J_\tau} \in (H(\tau), 1), \tag{1.132}$$

where $\mathring{J}_\tau$ denotes the interior of J_τ. Let Ω^* be the event of probability 1 introduced in Proposition 1.80. There exists $\widetilde{\Omega}$ an event of probability 1, not depending on τ and included in Ω^*, such that, for each $\omega \in \widetilde{\Omega}$, $\widetilde{\alpha}_Y(\tau, \omega)$, the local Hölder exponent at τ of the path $Y(\cdot, \omega)$ of MuBF, satisfies

$$\widetilde{\alpha}_Y(\tau, \omega) = H(\tau). \tag{1.133}$$

Proof of Theorem 1.88. We denote by $\widetilde{\mathcal{I}}_\mathbb{Q}$ the class of the compact intervals I of $\mathbb{R}^N$ with vertices having rational coordinates and which satisfy the following property:

$$H(\cdot) \in \mathcal{C}^{\gamma_I}(I), \quad for\ some\ \gamma_I \in (\underline{H}(I), 1), \tag{1.134}$$

where $\underline{H}(I)$ is as in (1.123). It follows from (1.132) that $\widetilde{\mathcal{I}}_\mathbb{Q}$ is a non-empty set, on the other hand it is clearly a countable set. We know from Theorem 1.87 that, for each $I \in \widetilde{\mathcal{I}}_\mathbb{Q}$, there exists an event of probability 1 $\widetilde{\Omega}_I \subseteq \Omega^*$, such that

$$\beta_Y(I, \omega) = \underline{H}(I), \quad for\ all\ \omega \in \widetilde{\Omega}_I. \tag{1.135}$$

Next, let $\widetilde{\Omega}$ be the event of probability 1 defined as $\widetilde{\Omega} := \bigcap_{I \in \widetilde{\mathcal{I}}_\mathbb{Q}} \widetilde{\Omega}_I$. Then, it follows from Definition 1.39, (1.135), (1.123) and the continuity of $H(\cdot)$ that (1.133) holds. □

The following theorem is a straightforward consequence of Theorem 1.74 and Proposition 1.86.

Theorem 1.89 (pointwise Hölder regularity of MuBF). *Assume that $\tau \in \mathbb{R}^N$ satisfies the condition (1.129) and $\widetilde{\alpha}_Y(\tau) \stackrel{a.s.}{>} 0$, where $\widetilde{\alpha}_Y(\tau)$ is the local Hölder exponent of MuBF $\{Y(t), t \in \mathbb{R}^N\}$ at τ. Then, $\alpha_Y(\tau)$, the pointwise Hölder exponent of $\{Y(t), t \in \mathbb{R}^N\}$ at τ, satisfies almost surely*

$$\alpha_Y(\tau) \stackrel{a.s.}{=} H(\tau). \tag{1.136}$$

Remark 1.90. Theorem 1.88 provides a *uniform* result on local Hölder exponent of MuBF $\{Y(t), t \in \mathbb{R}^N\}$; by uniform we mean that the almost sure equality $\widetilde{\alpha}_Y(\tau) \stackrel{\text{a.s.}}{=} H(\tau)$ holds on *an event of probability* 1 *not depending on* τ. It seems natural to wonder whether such a uniformity remains valid in the case of pointwise Hölder exponent of MuBF. The arguments we have just given for deriving Theorem 1.89 do not allow to answer to this question. More precisely, one cannot conclude from Theorem 1.74 whether or not the almost sure equality $\alpha_Y(\tau) \stackrel{\text{a.s.}}{=} H(\tau)$ holds on an event of probability 1 not depending on τ. In fact, uniform results for pointwise Hölder exponents are valid, not only in the setting of MuBF but also in other Gaussian and non-Gaussian settings, as for instance in the case of Multifractional Field with Random Exponent (MuFRE) which is considerably more general than MuBF. Fine path behavior of MuFRE will be studied in detail in Chapter 6 by using a rather new strategy relying on wavelet methods.

Theorem 1.91 (local asymtotic self-similarity of MuBF).
Assume that $\tau \in \mathbb{R}^N$ sastisfies the condition (1.129). Then, the MuBF $\{Y(t), t \in \mathbb{R}^N\}$ is at τ locally asymtotically self-similar (see Definition 1.69) with exponent $H(\tau)$ and the FBF $\{X(s, H(\tau)), s \in \mathbb{R}^N\}$ is the tangent field. Moreover this self-similarity property of MuBF holds in the strong sense (see Definition 1.70) when the condition (1.129) is replaced by the stronger condition (1.132).

Proof of Theorem 1.91. In view of (1.117) and the Gaussianity of the field $\{X(u,v), (u,v) \in \mathbb{R}^N \times (0,1)\}$, in order to get the desired local asymptotic self-similarity property, it is enough to show that, for each fixed $(s^1, s^2) \in \mathbb{R}^N \times \mathbb{R}^N$ and for any sequence $(a_n)_{n \in \mathbb{N}}$ of real numbers in the interval $(0, 1/2]$ converging to zero, one has

$$G_{\tau,n}(s^1, s^2)$$
$$:= a_n^{-H(\tau)} \sigma \big(X(\tau + a_n s^1, H(\tau + a_n s^1)) - X(\tau + a_n s^2, H(\tau + a_n s^2)) \big)$$
$$\xrightarrow[n \to +\infty]{} G_\tau(s^1, s^2) := \sigma \big(X(s^1, H(\tau)) - X(s^2, H(\tau)) \big). \quad (1.137)$$

Let $A_{\tau,n}(s^1, s^2)$ and $B_{\tau,n}(s^1, s^2)$ be defined as:

$$A_{\tau,n}(s^1, s^2) := a_n^{-H(\tau)} \sigma \big(X(\tau + a_n s^1, H(\tau + a_n s^1)) - X(\tau + a_n s^2, H(\tau + a_n s^1)) \big) \quad (1.138)$$

and

$$B_{\tau,n}(s^1, s^2) := G_{\tau,n}(s^1, s^2) - A_{\tau,n}(s^1, s^2). \quad (1.139)$$

Thus, in order to get (1.137), we will prove that

$$A^2_{\tau,n}(s^1, s^2) \xrightarrow[n\to+\infty]{} G^2_\tau(s^1, s^2), \tag{1.140}$$

where $A^2_{\tau,n}(s^1, s^2)$ (resp. $G^2_\tau(s^1, s^2)$) denotes the square of $A_{\tau,n}(s^1, s^2)$ (resp. $G_\tau(s^1, s^2)$), and

$$B_{\tau,n}(s^1, s^2) \xrightarrow[n\to+\infty]{} 0. \tag{1.141}$$

Let us now focus on (1.140). Similarly to (1.111), one can show that

$$A^2_{\tau,n}(s^1, s^2) = a_n^{2(H(\tau+a_n s^1) - H(\tau))} |s^1 - s^2|^{2H(\tau + a_n s^1)} \int_{\mathbb{R}^N} \frac{\left| e^{i\eta_1} - 1 \right|^2}{|\eta|^{2H(\tau + a_n s^1) + N}} \, d\eta \tag{1.142}$$

and

$$G^2_\tau(s^1, s^2) = |s^1 - s^2|^{2H(\tau)} \int_{\mathbb{R}^N} \frac{\left| e^{i\eta_1} - 1 \right|^2}{|\eta|^{2H(\tau) + N}} \, d\eta. \tag{1.143}$$

Next, observe that the condition (1.129) and the fact $H(\cdot)$ is a bounded function imply that there exists a constant $c_1 > 0$, not depending on n, s^1 and s^2, such that, for all $t \in \mathbb{R}^N$, one has

$$|H(t) - H(\tau)| \le c_1 |t - \tau|^{\gamma_\tau}. \tag{1.144}$$

Also, observe that one has

$$|a_n^{2(H(\tau + a_n s^1) - H(\tau))} - 1| \le 2c_1 |s^1|^{\gamma_\tau} a_n^{\gamma_\tau} \log(a_n^{-1}) \exp\left\{ 2c_1 |s^1|^{\gamma_\tau} a_n^{\gamma_\tau} \log(a_n^{-1}) \right\}; \tag{1.145}$$

the inequality (1.145) easily results from

$$|a_n^{2(H(\tau + a_n s^1) - H(\tau))} - 1|$$
$$\le \max\Big\{ \exp\left\{ 2|H(\tau + a_n s^1) - H(\tau)| \log(a_n^{-1}) \right\} - 1,$$
$$1 - \exp\left\{ -2|H(\tau + a_n s^1) - H(\tau)| \log(a_n^{-1}) \right\} \Big\},$$

(1.114) and (1.144). Next, using (1.142), (1.143), (1.145), the continuity of $H(\cdot)$ at τ, and the dominated convergence Theorem, one obtains (1.140).

Let us now prove that (1.141) is satisfied. It follows from (1.139), (1.138), the definition of $G_{\tau,n}(s^1, s^2)$ (see the first equality in (1.137)) and the triangle inequality, that

$$|B_{\tau,n}(s^1, s^2)|$$
$$\le a_n^{-H(\tau)} \sigma\big(X(\tau + a_n s^2, H(\tau + a_n s^1)) - X(\tau + a_n s^2, H(\tau + a_n s^2))\big).$$

Therefore, (1.107) and the isometry property of the integral $\int_{\mathbb{R}^N}(\,\cdot\,)d\widehat{\mathbb{W}}$ imply that

$$B^2_{\tau,n}(s^1, s^2)$$

$$\leq a_n^{-2H(\tau)} \int_{\mathbb{R}^N} \frac{\left|e^{i(\tau+a_n s^2)\cdot\xi} - 1\right|^2}{|\xi|^{2H(\tau+a_n s^1)+N}} \left(|\xi|^{H(\tau+a_n s^1)-H(\tau+a_n s^2)} - 1\right)^2 d\xi.$$

Next, using the latter inequality, similarly to (1.112), one can derive that

$$B^2_{\tau,n}(s^1, s^2) \leq c_2\left(1 + |\tau| + |s^2|\right)^2 a_n^{-2H(\tau)} \left|H(\tau + a_n s^1) - H(\tau + a_n s^2)\right|^2,$$
(1.146)

where the constant

$$c_2 := 4\left(\int_{|\xi|\leq 1} |\xi|^{2(1-\overline{H})-N} \log^2 |\xi| \, d\xi + \int_{|\xi|>1} |\xi|^{-2\underline{H}-N} \log^2 |\xi| \, d\xi\right).$$

Thus, combining (1.146) with the triangle inequality, (1.144) and the inequality $\gamma_\tau > H(\tau)$, one gets (1.141).

Let us now show that the local asymptotic self-similarity property also holds in the strong sense as soon as one has (1.132). More precisely, one has to prove that the sequence $\left(\{Z_n(s), s \in \mathbb{R}^N\}\right)_{n\in\mathbb{N}}$ of the centered real-valued Gaussian fields with almost surely continuous paths, defined as

$$Z_n(s) := a_n^{-H(\tau)}\left(Y(\tau + a_n s) - Y(\tau)\right)$$
(1.147)
$$:= a_n^{-H(\tau)}\left(X(\tau + a_n s, H(\tau + a_n s)) - X(\tau, H(\tau))\right),$$

satisfies the conditions (i) and (ii) in Lemma A.21. In view of (1.147), the first one of these two conditions is clearly satisfied. For the second one, the issue can be reduced to showing that, for any fixed positive real number δ, there are two constants $c_3 > 0$ and $\nu > 0$, only depending on δ, such that the inequality

$$G_{\tau,n}(s^1, s^2) \leq c_3|s^1 - s^2|^\nu$$
(1.148)

holds, for every $n \in \mathbb{N}$ and for all $s^1, s^2 \in \mathbb{R}^N$ satisfying $|s^1| \leq \delta$, $|s^2| \leq \delta$. Assume that the finite constants c_4, c_5 and c_6 are defined as

$$c_4 := 2(1 + \delta)^{\overline{H}-\underline{H}}\left(\int_{|\xi|\leq 1} |\xi|^{2(1-\overline{H})-N} \, d\xi + \int_{|\xi|>1} |\xi|^{-2\underline{H}-N} \, d\xi\right)^{1/2},$$

$$c_5 := 1 + \left(2c_1\delta^{\gamma_\tau}\right)^{1/2} \sup_{n\in\mathbb{N}} \left\{a_n^{\gamma_\tau/2} \log^{1/2}(a_n^{-1}) \exp\left\{c_1\delta^{\gamma_\tau} a_n^{\gamma_\tau} \log(a_n^{-1})\right\}\right\}$$

and $c_6 := c_4 c_5$. Then, using (1.142), (1.145), and standard computations, one can derive that

$$A_{\tau,n}(s^1, s^2) \leq c_6|s^1 - s^2|^{\underline{H}}.$$
(1.149)

Next, observe that (1.132) implies that there are three constants $c_7 > 0$, $\varepsilon > 0$ and $\eta > 0$, such that the inequality

$$|H(\tau + h^1) - H(\tau + h^2)| \leq c_7 |h^1 - h^2|^{H(\tau)+\varepsilon} \qquad (1.150)$$

holds, for each $h^1, h^2 \in \mathbb{R}^N$ satisfying $|h^1| \leq \eta$, $|h^2| \leq \eta$. Also observe that the fact that the sequence of positive numbers $(a_n)_{n \in \mathbb{N}}$ converges to zero, allows, without any restriction, to assume that $a_n \delta \leq \eta$, for all $n \in \mathbb{N}$. Thus, combining (1.146) with (1.150), it follows that

$$\left| B_{\tau,n}(s^1, s^2) \right| \leq c_8 |s^1 - s^2|^{H(\tau)+\varepsilon}, \qquad (1.151)$$

where the constant $c_8 := c_2^{1/2} c_7 (1 + |\tau| + \delta)$. Finally, putting together (1.149), (1.151) and (1.139), one gets (1.148).

□

Chapter 2

Erraticism, local times and local nondeterminism

2.1 Introduction

Let $\{Y(t), t \in \mathbb{R}^N\}$ be a Multifractional Brownian Field (MuBF) having a continuous Hurst functional parameter $H(\cdot)$ with values in a compact interval $[\underline{H}, \overline{H}] \subset (0, 1)$. We assume in addition that, for each $\tau \in \mathbb{R}^N$, the condition (1.132) holds, which implies that the condition (1.129) holds as well. Then, we know from Theorem 1.89 that $\alpha_Y(\tau)$, the pointwise Hölder exponent of $\{Y(t), t \in \mathbb{R}^N\}$ at an arbitrary $\tau \in \mathbb{R}^N$, satisfies almost surely $\alpha_Y(\tau) \overset{\text{a.s.}}{=} H(\tau)$. Yet, so far, we have not shown the uniform validity of the latter equality, that is on an event of probability 1 not depending on τ. In fact, in order to derive such a uniform result, it remains to us to prove that the inequality $\alpha_Y(\tau) \overset{\text{a.s.}}{\leq} H(\tau)$ holds uniformly, since we already know from Theorem 1.88 and (1.59) that the inverse inequality $\alpha_Y(\tau) \overset{\text{a.s.}}{\geq} H(\tau)$ is satisfied uniformly.

That kind of issue is one of the main motivation behind this book. Perhaps, one might assert that in real analysis showing that a function has everywhere some irregularity is in general more difficult than to show that it has everywhere some regularity; indeed, lower bounds for oscillations of functions are usually harder to obtain than upper bounds. Brownian paths seems to be a good illustration of this assertion: their Hölder continuity easily follows from a well-known Kolmogorov's Theorem (see Theorems A.6 and A.5), while deriving their nowhere differentiability is considerably more tricky (see e.g. [Paley *et al.* (1933); Kahane (1968 1st edition, 1985 2nd edition); Karatzas and Shreve (1987); Revuz and Yor (1991)]). It is worth mentioning that in the proofs of the latter result independence of increments of Brownian Motion plays a crucial role.

In order to deal with the problem of showing that the pointwise Hölder exponent of MuBF satisfies uniformly, for each $\tau \in \mathbb{R}^N$, $\alpha_Y(\tau) \overset{\text{a.s.}}{\leq} H(\tau)$, it seems natural to try to adapt the classical strategy which, long time ago, allowed to solve this problem in the case of Fractional Brownian Field (FBF) and many other stochastic fields having some stationarity of increments (see e.g. [Berman (1970, 1972, 1973); Dozzi (2003); Ehm (1981); Geman and Horowitz (1980); Nolan (1989); Pitt (1978); Xiao (1997, 2006, 2009, 2013)]). The main goal of the present chapter is to outline this strategy and then to explain what is the main new difficulty which appears in the frame of MuBF. Let us emphasize that the very important concept of *local time* is fundamental in the latter strategy. This concept was first introduced by Paul Lévy in [Lévy (1948 1st edition, 1965 2nd edition)] for study of behavior of Brownian paths.

   Throughout the present chapter, $\mathcal{B}(I)$ denotes the Borel σ-field over an arbitrary fixed compact interval $I \subset \mathbb{R}^N$, and $\{Z(t), t \in I\}$ denotes an arbitrary real-valued measurable [1] stochastic field defined on the probability space $(\Omega, \mathcal{A}, \mathbb{P})$; also, one frequently assumes that $\{Z(t), t \in I\}$ is Gaussian centered. Roughly speaking, the local time associated with $\{Z(t), t \in I\}$ is a nonnegative random field $\{L(x, T), (x, T) \in \mathbb{R} \times \mathcal{B}(I)\}$ defined on $(\Omega, \mathcal{A}, \mathbb{P})$, providing, for each (x, T), a measure of the "random amout of time t spent by a path of $\{Z(t), t \in I\}$ at the point x during the time period T". In this setting, though it is multidimensional, the index t is viewed as *the time variable* [2], x is called *the space variable* and T *the set variable*.

   The remaining of the present chapter is organized as follows.

   Section 2.2 is devoted to the definitions of the notions of occupation measure and local time as well as to the presentation of some of their basic properties. Also, it presents a classical criterion for showing the existence of a local time and its square integrability as a function of $(x, \omega) \in \mathbb{R} \times \Omega$. At the end of the section, two important fomulas allowing to determine in the Gaussian frame moments of local times and their increments are given.

   Section 2.3 defines the concept of *jointly continuous local time* and provides, thanks to this concept, some formalization of an important principle due to [Berman (1972)] according to which: *the more regular (smooth) is a path of the local time $\{L(x, T), (x, T) \in \mathbb{R} \times \mathcal{B}(I)\}$ in the set variable T uniformly in the space variable x, the more irregular (erratic) is the corresponding path of the stochastic field $\{Z(t), t \in I\}$.*

---

[1]That is $(t, \omega) \mapsto Z(t, \omega)$ is a measurable function from $(I \times \Omega, \mathcal{B}(I) \otimes \mathcal{A})$ into $(\mathbb{R}, \mathcal{B}(\mathbb{R}))$.
[2]Notice that, in contrast with all the other chapters of the book, in the present chapter the index t is never called *a space variable*, for avoiding to confuse it with x.

The goal of Section 2.4 is to present, in a general Gaussian frame, the main ideas of very classical Fourier analytic methods, initially introduced in [Berman (1973)], which allow to derive joint continuity for local time; these methods are presented with a view to be applied to Fractional Brownian Field (FBF). First, it is explained how the latter methodology mainly relies on appropriate estimates for moments of local time and moments of its increments in the space variable; this leads to the fundamental concept of *local nondeterminism*, initially due to [Berman (1973)], which plays a crucial role in getting such estimates. Then, it is shown that FBF is strongly locally nondeterministic which allows to derive joint continuity of its local time. Thus, it turns out that the main difficulty for deriving through, such Fourier analytic methods, joint continuity and other regularity properties for local time of Multifractional Brownian Field (MuBF), is to show local nondeterminism for the latter field. It is worth mentioning that recently this problem has been partially solved in [Ayache *et al.* (2011)] thanks to wavelet methods reminiscent of those which will be presented in the book; the solution is partial since the strong version of local nondeterminism has not yet been obtained.

2.2 Local time: definition and basic properties

Definition 2.1 (occupation measure). Let $T \in \mathcal{B}(I)$ be arbitrary and fixed. Assume that $Z(\cdot, \omega) : t \mapsto Z(t, \omega)$ is an arbitrary path of the stochastic field $\{Z(t), \, t \in I\}$; we recall that $Z(\cdot, \omega)$ is a Borel measurable function in the variable t. The *occupation measure* associated with $Z(\cdot, \omega)$ is the deterministic positive finite measure $\mu_T(\bullet, \omega)$ on $\mathcal{B}(\mathbb{R})$, the Borel σ-field over $\mathbb{R}$, defined as

$$\mu_T(A, \omega) := \lambda_N\big(\{t \in T : Z(t, \omega) \in A\}\big) = \int_T \mathbb{1}_A\big(Z(t, \omega)\big)\, dt, \text{ for all } A \in \mathcal{B}(\mathbb{R}),$$

(2.1)

where λ_N denotes [3] the Lebesgue measure on $\mathbb{R}^N$. Observe that the quantity $\mu_T(A, \omega)$ can be viewed as a measure of "the amout of time t spent by the path $Z(\cdot, \omega)$ in the Borel set A during the time period T".

Remark 2.2.

(i) The measure $\mu_T(\bullet, \omega)$ is supported on $\overline{Z(T, \omega)}$ the (topological) closure of the set $Z(T, \omega) := \big\{Z(t, \omega) : t \in T\big\}$.

---

[3]The classical Lebesgue integral associated with λ_N is denoted by $\int(\cdot)\, dt$ rather than $\int(\cdot)\, d\lambda(t)$.

(ii) For each fixed $A \in \mathcal{B}(\mathbb{R})$, the set function, from $\mathcal{B}(I)$ into $[0, \lambda_N(I)]$, $T \mapsto \mu_T(A, \omega)$ is a measure; that is, for any sequence $(T_m)_{m \geq 1}$ of disjoint Borel subsets of I, one has

$$\mu_{\bigcup_{m=1}^{+\infty} T_m}(A, \omega) = \sum_{m=1}^{+\infty} \mu_{T_m}(A, \omega).$$

(iii) Assume that f is either a Borel measurable nonnegative function [4] on $\mathbb{R}$, or a Borel measurable bounded complex-valued function on $\mathbb{R}$. Then $\int_{\mathbb{R}} f(x) \, d\mu_T(x, \omega)$, the integral of f with respect to the measure $\mu_T(\bullet, \omega)$, satisfies

$$\int_{\mathbb{R}} f(x) \, d\mu_T(x, \omega) = \int_T f\big(Z(t, \omega)\big) \, dt. \qquad (2.2)$$

(iv) It follows from (2.2) that the function $\omega \mapsto \int_{\mathbb{R}} f(x) \, d\mu_T(x, \omega)$ from $(\Omega, \mathcal{A})$ into $(\mathbb{R}_+, \mathcal{B}(\mathbb{R}_+))$ (or into $(\mathbb{C}, \mathcal{B}(\mathbb{C}))$ when f is complex-valued) is a random variable (i.e. a measurable function). In particular, for any fixed $A \in \mathcal{B}(\mathbb{R})$, $\omega \mapsto \mu_T(A, \omega)$ is a nonnegative random variable.

(v) The occupation measure $\mu_T(\bullet, \omega)$ is completely determined by its characteristic function $\chi_{\mu_T}(\bullet, \omega)$ defined as

$$\chi_{\mu_T}(\xi, \omega) := \int_{\mathbb{R}} e^{i\xi x} \, d\mu_T(x, \omega), \quad \text{for each } \xi \in \mathbb{R}; \qquad (2.3)$$

also, in view of (2.2), one has

$$\chi_{\mu_T}(\xi, \omega) = \int_T e^{i\xi Z(t, \omega)} \, dt. \qquad (2.4)$$

Definition 2.3 (local time). *For having the existence of $L(\bullet, T)$, the local time on T of the stochastic field $\{Z(t), \, t \in I\}$, the occupation measure $\mu_T(\bullet, \omega)$ needs to be with probability 1 (that is for $\mathbb{P}$-almost all ω) absolutely continuous with respect to λ, the Lebesgue measure on $\mathbb{R}$.* When this condition is fulfilled, then $L(\bullet, T, \omega)$ is defined as the Radon-Nikodým derivative of $\mu_T(\bullet, \omega)$ with respect to λ. In other words, $L(\bullet, T, \omega)$ is defined as the unique (up to a Lebesgue negligible set) nonnegative function in the Lebesgue space $L^1(\mathbb{R}, \mathcal{B}(\mathbb{R}), \lambda)$, such that the equality

$$\int_{\mathbb{R}} f(x) \, d\mu_T(x, \omega) = \int_{\mathbb{R}} f(x) L(x, T, \omega) \, dx, \qquad (2.5)$$

in which dx denotes the Lebesgue measure λ, holds for any function f satisfying the same assumptions as in part *(iii)* of Remark 2.2.

Notice that when $L(\bullet, T)$ exists then $L(\bullet, S)$ also exists for any Borel subset S of T (this is a consequence of the Radon-Nikodým Theorem (see e.g. [Rudin (1986)]).

---

[4] Notice that the integrals in (2.2) and (2.5) may then be equal to $+\infty$.

Definition 2.4 (a nice version of local time). When it exists, the local time $L(\bullet, T)$ is always identified with its version defined as:

$$L(x, T, \omega) \tag{2.6}$$

$$:= \liminf_{n \to +\infty, n \in \mathbb{N}} \left\{ 2^{-1} n \, \mu_T \big([x - n^{-1}, x + n^{-1}], \omega \big) \right\}, \quad \text{for all } (x, \omega) \in \mathbb{R} \times \Omega.$$

We refer to [Geman and Horowitz (1980)] for a presentation of several nice properties of this version of the local time.

Remark 2.5.

(i) It easily results from part (i) of Remark 2.2 that one has $L(x, T, \omega) = 0$, for λ-almost all $x \notin \overline{Z(T, \omega)}$.

(ii) Assuming that $(T_m)_{m \geq 1}$ is an arbitrary sequence of disjoint Borel subsets of I such that $L\big(\bullet, \bigcup_{m=1}^{+\infty} T_m\big)$ exists; then, it easily results from part (ii) of Remark 2.2 that

$$L\Big(x, \bigcup_{m=1}^{+\infty} T_m, \omega\Big) = \sum_{m=1}^{+\infty} L(x, T_m, \omega), \quad \text{for } \lambda \otimes \mathbb{P}\text{-almost all } (x, \omega).$$

(iii) Assume that f is a function satisfying the same assumptions as in part (iii) of Remark 2.2; then, combining (2.2) with (2.5), one gets the so called *occupation density formula*:

$$\int_T f\big(Z(t, \omega)\big) \, dt = \int_{\mathbb{R}} f(x) L(x, T, \omega) \, dx, \tag{2.7}$$

which holds for $\mathbb{P}$-almost all ω, namely for all ω such that $L(\bullet, T, \omega)$ exists.

(iv) The Fourier transform of the function $x \mapsto L(x, T, \omega)$ is denoted by $\mathcal{F}\big(L(\bullet, T, \omega)\big)$ and defined as

$$\mathcal{F}\big(L(\bullet, T, \omega)\big)(\xi) := (2\pi)^{-1/2} \int_{\mathbb{R}} e^{-i\xi x} L(x, T, \omega) \, dx, \quad \text{for each } \xi \in \mathbb{R}. \tag{2.8}$$

Thus, it follows from (2.5), (2.3) and (2.4) that

$$(2\pi)^{1/2} \mathcal{F}\big(L(\bullet, T, \omega)\big)(-\xi) = \chi_{\mu_T}(\xi, \omega) = \int_T e^{i\xi Z(t, \omega)} \, dt, \quad \text{for all } \xi \in \mathbb{R}. \tag{2.9}$$

(v) In view of Definition 2.4 and parts (iii) and (iv) of Remark 2.2, $\{L(x, T), \, x \in \mathbb{R}\}$ is a measurable nonnegative stochastic process, which may take the value $+\infty$, only on a $\lambda \otimes \mathbb{P}$ negligible subset of $\mathbb{R} \times \Omega$. Also, notice that the sequence $\big\{ 2^{-1} n \, \mu_T \big([x - n^{-1}, x + n^{-1}], \omega \big) \big\}_{n \in \mathbb{N}}$, which was introduced in Definition 2.4, converges to a finite limit, for $\lambda \otimes \mathbb{P}$-almost all fixed $(x, \omega) \in \mathbb{R} \times \Omega$.

(vi) In view of (2.6) and (2.1), for each fixed $(x, \omega) \in \mathbb{R} \times \Omega$, the set function, from $\mathcal{B}(I)$ into $\mathbb{R}_+ \cup \{+\infty\}$, $T \mapsto L(x, T, \omega)$ is non-decreasing: one has $L(x, T_1, \omega) \leq L(x, T_2, \omega)$ for all $T_1, T_2 \in \mathcal{B}(I)$ satisfying $T_1 \subseteq T_2$.

Proposition 2.6 (square integrability of local time). *One has, for all fixed $\xi \in \mathbb{R}$,*

$$\mathbb{E}\big(|\chi_{\mu_T}(\xi)|^2\big) := \int_\Omega |\chi_{\mu_T}(\xi, \omega)|^2 \, d\mathbb{P}(\omega) = \int_{T^2} \chi_{Z(t^1)-Z(t^2)}(\xi) \, d\bar{t}, \quad (2.10)$$

where $\bar{t} := (t^1, t^2)$ and $\chi_{Z(t^1)-Z(t^2)}$ is the characteristic function of the real-valued random variable $Z(t^1) - Z(t^2)$. Moreover, when

$$\int_\mathbb{R} \mathbb{E}\big(|\chi_{\mu_T}(\xi)|^2\big) \, d\xi < +\infty, \quad (2.11)$$

then $L(\bullet, T)$ the local time on T of $\{Z(t),\ t \in I\}$ exists; also, the function $(x, \omega) \mapsto L(x, T, \omega)$ belongs to the space $L^2(\mathbb{R} \times \Omega, \mathcal{B}(\mathbb{R}) \otimes \mathcal{A}, \lambda \otimes \mathbb{P})$.

Proof of Proposition 2.6. It easily follows from (2.9) and Fubini's Theorem that, for $\mathbb{P}$-almost all fixed $\omega \in \Omega$ and for all fixed $\xi \in \mathbb{R}$,

$$|\chi_{\mu_T}(\xi, \omega)|^2 = \int_{T^2} e^{i\xi(Z(t^1, \omega) - Z(t^2, \omega))} \, d\bar{t}; \quad (2.12)$$

thus, using again Fubini's Theorem and the fact that

$$\chi_{Z(t^1)-Z(t^2)}(\xi) := \mathbb{E}\big(e^{i\xi(Z(t^1)-Z(t^2))}\big), \quad \text{for each } \xi \in \mathbb{R},$$

one gets (2.10). In view of (2.9), the condition (2.11) is equivalent to

$$\int_\mathbb{R} \left(\int_\Omega \big|\mathcal{F}(L(\bullet, T, \omega))(\xi)\big|^2 \, d\mathbb{P}(\omega) \right) d\xi < +\infty. \quad (2.13)$$

On the other hand, thanks to Tonelli's Theorem, we know that

$$\int_\mathbb{R} \left(\int_\Omega \big|\mathcal{F}(L(\bullet, T, \omega))(\xi)\big|^2 \, d\mathbb{P}(\omega) \right) d\xi = \int_\Omega \left(\int_\mathbb{R} \big|\mathcal{F}(L(\bullet, T, \omega))(\xi)\big|^2 \, d\xi \right) d\mathbb{P}(\omega). \quad (2.14)$$

Then (2.13) and (2.14) imply that, for $\mathbb{P}$-almost all ω, the function $\xi \mapsto \mathcal{F}(L(\bullet, T, \omega))(\xi)$ belongs to $L^2(\mathbb{R}, \mathcal{B}(\mathbb{R}), \lambda)$. Thus, for $\mathbb{P}$-almost all ω, the function $x \mapsto L(x, T, \omega)$ belongs to $L^2(\mathbb{R}, \mathcal{B}(\mathbb{R}), \lambda)$ as well. Moreover, one knows from the isometry property of the Fourier transform that

$$\int_\mathbb{R} \big|\mathcal{F}(L(\bullet, T, \omega))(\xi)\big|^2 \, d\xi = \int_\mathbb{R} L(x, T, \omega)^2 \, dx. \quad (2.15)$$

Finally, putting together (2.13), (2.14) and (2.15), one gets that

$$\int_\Omega \left(\int_\mathbb{R} L(x, T, \omega)^2 \, dx \right) d\mathbb{P}(\omega) < +\infty.$$

$\square$

Proposition 2.7. *Assume that* $\{Z(t), t \in I\}$ *is a Gaussian centered field. Then, a sufficient condition for having the existence of* $L(\bullet, T)$ *and its square integrability in* (x, ω) *is that*

$$\int_{T^2} \sigma\big(Z(t^1) - Z(t^2)\big)^{-1} d\bar{t} < +\infty, \qquad (2.16)$$

where $\sigma\big(Z(t^1) - Z(t^2)\big)$ *denotes the standard deviation of the centered Gaussian random variable* $Z(t^1) - Z(t^2)$.

Proof of Proposition 2.7. The characteristic function $\chi_{Z(t^1)-Z(t^2)}$ of $Z(t^1) - Z(t^2)$ can be expressed as

$$\chi_{Z(t^1)-Z(t^2)}(\xi) = \exp\Big\{ -2^{-1}\big(\sigma\big(Z(t^1) - Z(t^2)\big)\xi\big)^2 \Big\}.$$

Thus, one has that

$$\int_{\mathbb{R}} \chi_{Z(t^1)-Z(t^2)}(\xi)\, d\xi = (2\pi)^{1/2}\, \sigma\big(Z(t^1) - Z(t^2)\big)^{-1}, \qquad (2.17)$$

with the convention that

$$\sigma\big(Z(t^1) - Z(t^2)\big)^{-1} = +\infty, \quad \text{when } \sigma\big(Z(t^1) - Z(t^2)\big) = 0.$$

Then using (2.16), (2.17), the Fubini Theorem and Proposition 2.6, one gets Proposition 2.7. $\qquad \square$

Corollary 2.8. *Assume that the condition (1.127) holds. Then* $L_Y(\bullet, I)$ *the local time on the compact interval* I *of the MuBF* $\{Y(t), t \in I\}$ *exists; moreover,* $L_Y(\bullet, I)$ *is square integrable in* (x, ω).

Proof of Corollary 2.8. Let $\overline{H}(I)$ be as in (1.123). In view of Proposition 2.7, the last inequality in (1.128) and Tonelli's Theorem, for deriving the corollary, it is enough to show that

$$A := \int_I \Big(\int_I |t^1 - t^2|^{-\overline{H}(I)}\, dt^1 \Big)\, dt^2 < +\infty.$$

Let $\mathcal{B}(0, \nu_I)$ be the closed ball of $\mathbb{R}^N$ centered at 0 and of radius $\nu_I := 2\max_{u \in I} |u|$. For each arbitrary and fixed $t^2 \in I$, let $I - t^2$ be the compact interval of $\mathbb{R}^N$ defined as $I - t^2 := \{t^1 - t^2 : t^1 \in I\}$. The change of variable $s = t^1 - t^2$, the inclusion $I - t^2 \subset \mathcal{B}(0, \nu_I)$, and the fact that $\overline{H}(I)$ belongs to the open interval $(0, 1)$ imply that

$$\int_I |t^1 - t^2|^{-\overline{H}(I)}\, dt^1 = \int_{I-t^2} |s|^{-\overline{H}(I)}\, ds \leq \int_{\mathcal{B}(0,\nu_I)} |s|^{-\overline{H}(I)}\, ds < +\infty.$$

Therefore, one has that

$$A \leq \lambda_N(I) \int_{\mathcal{B}(0,\nu_I)} |s|^{-\overline{H}(I)}\, ds < +\infty.$$

$$\square$$

Definition 2.9. Let k be an arbitrary positive integer and let J^k be the cartesian product of J with itself k times, where J is a closed subinterval of I. An arbitrary element $(t^1, \ldots, t^k) \in J^k$ is denoted by $\bar{t}$. Moreover we denote by $J^k_{\preceq}$ the Lebesgue measurable subset of J^k defined as

$$J^k_{\preceq} := \left\{ \bar{t} := (t^1, \ldots, t^k) \in J^k : \forall\, m \in \{2, \ldots, k\},\ |t^m - t^{m-1}| = \min_{1 \le n < m} |t^m - t^n| \right\},$$
(2.18)

with the convention that $J^1_{\preceq} := J$ and $J^2_{\preceq} := J^2$.

We refer to Subsection 25 in [Geman and Horowitz (1980)] for a proof of the following proposition which will play a crucial role in the sequel.

Proposition 2.10 (Further integrability properties). *Assume that $k \ge 2$ is an arbitrary fixed even integer and that $\{Z(t),\ t \in I\}$ is a Gaussian centered field. For each $\bar{t} := (t^1, \ldots, t^k) \in I^k$, denote by $\mathbb{V}_{Z,k}(\bar{t})$ the covariance matrix of the centered Gaussian column [5] vector $\big(Z(t^1), \ldots, Z(t^k)\big)^*$. Let $\det\big(\mathbb{V}_{Z,k}(\bar{t})\big)$ be the determinant of $\mathbb{V}_{Z,k}(\bar{t})$. If the condition*

$$\int_{I^k_{\preceq}} \det\big(\mathbb{V}_{Z,k}(\bar{t})\big)^{-1/2}\, d\bar{t} < +\infty \tag{2.19}$$

holds (with the convention that $\det\big(\mathbb{V}_{Z,k}(\bar{t})\big)^{-1/2} = +\infty$ when $\det\big(\mathbb{V}_{Z,k}(\bar{t})\big) = 0$); then $L(\bullet, I)$ exists, moreover its satisfies the following two integrability properties.

(i) *For $\mathbb{P}$-almost all fixed $\omega \in \Omega$, the function $x \mapsto L(x, I, \omega)$ belongs to the Lebesgue space $L^k(\mathbb{R}, \mathcal{B}(\mathbb{R}), \lambda)$.*

(ii) *For every fixed $x \in \mathbb{R}$, the nonnegative random variable $\omega \mapsto L(x, I, \omega)$ belongs to the space $L^k(\Omega, \mathcal{A}, \mathbb{P})$. Moreover, for any closed subinterval $J \subseteq I$, one has*

$$\mathbb{E}\Big(L(x, J)^k \Big) \tag{2.20}$$

$$= \frac{1}{(2\pi)^k} \int_{J^k} \int_{\mathbb{R}^k} \exp\Big(-ix \sum_{l=1}^{k} u_l \Big) \exp\Big(-2^{-1} u^* \mathbb{V}_{Z,k}(\bar{t}) u \Big)\, du\, d\bar{t}$$

$$= \frac{(k-1)!}{(2\pi)^k} \int_{J^k_{\preceq}} \int_{\mathbb{R}^k} \exp\Big(-ix \sum_{l=1}^{k} u_l \Big) \exp\Big(-2^{-1} u^* \mathbb{V}_{Z,k}(\bar{t}) u \Big)\, du\, d\bar{t},$$

---

[5] As, usual the symbol $*$ means the transposition operation on matrices.

where u is the column matrix with coefficients $u_1, \ldots, u_k$ and u^* its transpose. Also, for all fixed $y \in \mathbb{R}$, one has

$$\mathbb{E}\Big(\big(L(x, J) - L(y, J) \big)^k \Big) \qquad (2.21)$$

$$= \frac{1}{(2\pi)^k} \int_{J^k} \int_{\mathbb{R}^k} \left(\prod_{l=1}^{k} \big(e^{-iu_l x} - e^{-iu_l y} \big) \right) \exp\Big(- 2^{-1} u^* \mathbb{V}_{Z,k}(\bar{t}) u \Big) \, du \, d\bar{t}$$

$$= \frac{(k-1)!}{(2\pi)^k} \int_{J^k_{\preceq}} \int_{\mathbb{R}^k} \left(\prod_{l=1}^{k} \big(e^{-iu_l x} - e^{-iu_l y} \big) \right) \exp\Big(- 2^{-1} u^* \mathbb{V}_{Z,k}(\bar{t}) u \Big) \, du \, d\bar{t}.$$

Remark 2.11.

(i) The two equalities in (2.20) remain valid when k is a positive odd integer.

(ii) The quantity $\exp\Big(- 2^{-1} u^* \mathbb{V}_{Z,k}(\bar{t}) u \Big)$ in (2.20) and (2.21) is in fact the value at 1 of the characteristic function of the centered real-valued Gaussian random variable $\sum_{l=1}^{k} u_l Z(t^l)$.

(iii) Proposition 2.10 can be extended to non-Gaussian settings (see [Geman and Horowitz (1980)]), as for instance the setting of heavy-tailed stable distributions.

2.3   Joint continuity and the Berman's principle

Definition 2.12 (jointly continuous local time). Assume that one has the existence of $L(\bullet, I)$, the local time on the whole interval I of the stochastic field $\{Z(t), t \in I\}$. Thus, as we have already mentioned, we know from the Radon-Nikodým Theorem that $L(\bullet, T)$ also exists, for any $T \in \mathcal{B}(I)$. This is in particular the case when $T = I(s)$, where $s = (s_1, \ldots, s_N) \in I$ is an arbitrary fixed point and $I(s)$ the closed subinterval of I defined as

$$I(s) := \big\{ t = (t_1, \ldots, t_N) \in I : \text{such that } t_l \leq s_l, \text{ for all } l = 1, \ldots, N \big\}. \qquad (2.22)$$

Let then $\big\{ L(x, I(s)), (x, s) \in \mathbb{R} \times I \big\}$ be the stochastic field of local times of $\{Z(t), t \in I\}$ defined through (2.6) where $T = I(s)$. $\{Z(t), t \in I\}$ *is said to have a jointly continuous local time on (the compact interval) I, if* $\big\{ L(x, I(s)), (x, s) \in \mathbb{R} \times I \big\}$ *has a continuous modification.* When it exists, we denote this continuous modification by $\big\{ \mathcal{L}(x, I(s)), (x, s) \in \mathbb{R} \times I \big\}$.

Jointly continuous local times have several nice properties (see e.g. [Adler (1981); Berman (1972); Ehm (1981); Geman and Horowitz (1980); Xiao (1997, 2006, 2009, 2013)]). Let us mention that:

Remark 2.13. For $\mathbb{P}$-almost all ω, and each fixed $x \in \mathbb{R}$, the set function $I(s) \mapsto \mathcal{L}(x, I(s), \omega)$ can be extended in a canonical and unique way (see for instance [Adler (1981)]) to be a finite measure $\mathcal{L}(x, \bullet, \omega) : T \mapsto \mathcal{L}(x, T, \omega)$ defined on the Borel subsets T of I, that is on the σ-field $\mathcal{B}(I)$. Moreover, when $\{Z(t), t \in I\}$ has continuous paths, this measure $\mathcal{L}(x, \bullet, \omega)$ is supported on the level set $\{t \in I : Z(t, \omega) = x\}$. Thus, among other things, $\mathcal{L}(x, \bullet, \omega)$ plays a crucial role in determining the Hausdorff dimension/measure of the latter set.

Let us also mention the following result which provides a first illustation of the Berman's principle.

Proposition 2.14 (nowhere differentiablity of paths). *Assume that I is a compact interval in the real line (i.e. $N = 1$) and that the stochastic process $\{Z(t), t \in I\}$ has continuous paths as well as a jointly continuous local time on I. Then, with probability 1, the paths of $\{Z(t), t \in I\}$ are nowhere differentiable functions on $\overset{\circ}{I}$ the (topological) interior of I.*

Proof of Proposition 2.14. Let $\{\mathcal{L}(x, I(s)), (x, s) \in \mathbb{R} \times I\}$ be the continuous modification of the stochastic field of local times of $\{Z(t), t \in I\}$. Also, let $\omega \in \Omega$ be fixed and such that:

- the function, from I into $\mathbb{R}$, $t \mapsto Z(t, \omega)$ is continuous;
- the function, from $\mathbb{R} \times I$ into $\mathbb{R}_+$, $(x, s) \mapsto \mathcal{L}(x, I(s), \omega)$ is continuous.

Observe that part (i) of Remark 2.5 and the inclusion $I(s) \subseteq I$ imply that the function $(x, s) \mapsto \mathcal{L}(x, I(s), \omega)$ has a compact support included in the compact subset $Z(I, \omega) \times I$ of $\mathbb{R} \times I$. We denote by $\tau \in \overset{\circ}{I}$ an arbitrary fixed point and we assume that ρ is an arbitrary positive real number small enough so that $[\tau - \rho, \tau + \rho] \subseteq I$. Then taking in (2.1), $T = [\tau - \rho, \tau + \rho]$ and $A = \mathbb{R}$, one obtains that

$$\mu_{[\tau-\rho,\tau+\rho]}(\mathbb{R}, \omega) = 2\rho. \tag{2.23}$$

On the other hand, taking in (2.5) $T = [\tau - \rho, \tau + \rho]$ and $f = \mathbb{1}_{\mathbb{R}}$, one gets that

$$\mu_{[\tau-\rho,\tau+\rho]}(\mathbb{R}, \omega) = \int_{\mathbb{R}} \mathcal{L}\big(x, [\tau - \rho, \tau + \rho], \omega\big) \, dx \tag{2.24}$$

$$= \int_{R_Z(\tau,\rho,\omega)} \mathcal{L}\big(x, [\tau - \rho, \tau + \rho], \omega\big) \, dx,$$

where $R_Z(\tau, \rho, \omega)$ is the compact interval of $\mathbb{R}$ defined as

$$R_Z(\tau, \rho, \omega) := \{Z(t, \omega) : t \in [\tau - \rho, \tau + \rho]\} \tag{2.25}$$
$$= \left[\inf_{t \in [\tau - \rho, \tau + \rho]} Z(t, \omega), \sup_{t \in [\tau - \rho, \tau + \rho]} Z(t, \omega) \right].$$

Observe that the last equality in (2.24) follows from part (i) of Remark 2.5. Also observe that, in view of Definition 1.43 and (2.25), $\lambda\big(R_Z(\tau, \rho, \omega)\big)$, the Lebesgue measure of $R_Z(\tau, \rho, \omega)$, satisfies

$$\lambda\big(R_Z(\tau, \rho, \omega)\big) = \mathrm{osc}_Z\big([\tau - \rho, \tau + \rho], \omega\big), \tag{2.26}$$

where $\mathrm{osc}_Z\big([\tau - \rho, \tau + \rho], \omega\big)$ denotes the oscillation of the continuous function $t \mapsto Z(t, \omega)$ on $[\tau - \rho, \tau + \rho]$. Next, putting together equation (2.23) to (2.26) one gets, for each $\rho > 0$ small enough, that

$$2 \leq \left(\frac{\mathrm{osc}_Z\big([\tau - \rho, \tau + \rho], \omega\big)}{\rho} \right) \times \mathcal{L}^*\big([\tau - \rho, \tau + \rho], \omega\big), \tag{2.27}$$

where $\mathcal{L}^*\big([\tau - \rho, \tau + \rho], \omega\big)$ is the quantity defined as

$$\mathcal{L}^*\big([\tau - \rho, \tau + \rho], \omega\big) := \sup\Big\{ \mathcal{L}\big(x, [\tau - \rho, \tau + \rho], \omega\big) : x \in \mathbb{R} \Big\}. \tag{2.28}$$

Notice that the equality, for all $x \in \mathbb{R}$,

$$\mathcal{L}\big(x, [\tau - \rho, \tau + \rho], \omega\big) = \mathcal{L}(x, I(\tau + \rho)) - \mathcal{L}(x, I(\tau - \rho)),$$

and the continuity on $\mathbb{R} \times I$ of the compactly supported nonnegative function $(x, s) \mapsto \mathcal{L}(x, I(s), \omega)$, entail that $\mathcal{L}^*\big([\tau - \rho, \tau + \rho], \omega\big)$ is finite and satisfies

$$\lim_{\rho \to 0, \, \rho > 0} \mathcal{L}^*\big([\tau - \rho, \tau + \rho], \omega\big) = 0. \tag{2.29}$$

Finally, in view of (2.29), it turns out that a necessary condition for the inequality (2.27) to hold, for all $\rho > 0$ small enough, is that

$$\liminf_{\rho \to 0, \, \rho > 0} \left\{ \frac{\mathrm{osc}_Z\big([\tau - \rho, \tau + \rho], \omega\big)}{\rho} \right\} = +\infty.$$

The latter equality implies that the function $t \mapsto Z(t, \omega)$ is not differentiable at τ. $\qquad\square$

The following theorem is a quite natural generalization of Proposition 2.14.

Theorem 2.15 (some formulation of the Berman's principle).
Assume that I is a compact interval of $\mathbb{R}^N$ and that the stochastic field $\{Z(t),\, t \in I\}$ has continuous paths as well as $\{\mathcal{L}(x, I(s)),\, (x,s) \in \mathbb{R} \times I\}$ a jointly continuous local time on I. Also, assume that there exists an event Ω_0 of probability 1 on which the paths of $\{\mathcal{L}(x, I(s)),\, (x,s) \in \mathbb{R} \times I\}$ are continuous functions which further satisfy, for all fixed $\omega \in \Omega_0$ and $\tau \in \overset{\circ}{I}$,

$$\limsup_{\rho \to 0,\, \rho > 0} \left\{ \frac{\mathcal{L}^*\big([\tau - \langle \rho \rangle, \tau + \langle \rho \rangle], \omega\big)}{\rho^N\, \Theta_\tau(\rho)^{-1}} \right\} < +\infty, \qquad (2.30)$$

where:

- *$[\tau - \langle \rho \rangle, \tau + \langle \rho \rangle]$ is the closed cube of $\mathbb{R}^N$ centered at τ and of edge size 2ρ;*
- *$\mathcal{L}^*\big([\tau - \langle \rho \rangle, \tau + \langle \rho \rangle], \omega\big) := \sup \Big\{ \mathcal{L}\big(x, [\tau - \langle \rho \rangle, \tau + \langle \rho \rangle], \omega\big) : x \in \mathbb{R} \Big\}$;*
- *Θ_τ is a continuous increasing nonnegative function vanishing at 0 and defined on $[0, \delta_\tau] \subset \mathbb{R}_+$, for some $\delta_\tau > 0$.*

Then, for each fixed $\omega \in \Omega_0$, $\tau \in \overset{\circ}{I}$ and $\varepsilon > 0$, $\mathrm{osc}_Z\big([\tau - \langle \rho \rangle, \tau + \langle \rho \rangle], \omega\big)$, the oscillation of the continuous function $t \mapsto Z(t, \omega)$ on $[\tau - \langle \rho \rangle, \tau + \langle \rho \rangle]$ (see Definition 1.43), satisfies

$$\liminf_{\rho \to 0,\, \rho > 0} \left\{ \frac{\mathrm{osc}_Z\big([\tau - \langle \rho \rangle, \tau + \langle \rho \rangle], \omega\big)}{\Theta_\tau(\rho)^{1+\varepsilon}} \right\} = +\infty.$$

The proof of Theorem 2.15 has been omitted since it is very similar to that of Proposition 2.14.

Remark 2.16. Assume that, for every fixed $\tau \in \overset{\circ}{I}$, the function Θ_τ in Theorem 2.15 is such that $\Theta_\tau(\rho) = \rho^{H(\tau)}$, for some constant $H(\tau) \in (0,1)$ and for all $\rho \in [0, \delta_\tau]$. Then combining the theorem with Proposition 1.44 and Definition 1.45, it follows that, for any $\tau \in \overset{\circ}{I}$, $\alpha_Z(\tau)$, the pointwise Hölder exponent of $\{Z(t),\, t \in I\}$ at τ, satisfies the inequality $\alpha_Z(\tau) \overset{\text{a.s.}}{\leq} H(\tau)$ uniformly, that is on a event of probability 1 not depending on τ, more precisely the event Ω_0 introduced in Theorem 2.15.

2.4   The concept of local nondeterminism

Having explained in the previous section the importance of the joint continuity property of local times in studying erraticism of paths of stochastic fields, let us now present, in a general Gaussian frame, the main ideas of classical Fourier analytic methods, initially introduced in [Berman (1973)], which allow to derive this continuity property of local times. The reader should always keep in his mind that our final goal is to prove joint continuity of local time of Fractional Brownian Field.

As usual I denotes a compact interval of $\mathbb{R}^N$. Moreover, for the sake of simplicity, in this section, we always suppose that $\{Z(t), t \in I\}$ is a centered Gaussian field satisfying the condition (2.19); thus we know from Proposition 2.10 that $L(\bullet, I)$, the local time on I of $\{Z(t), t \in I\}$, exists and satisfies the properties described in the proposition.

In order to show that $\{L(x, I(s)), (x, s) \in \mathbb{R} \times I\}$, the stochastic field of local times defined through (2.6) where $T = I(s)$ (see (2.22)), has a continuous modification, it seems natural to try to make use of the Kolmogorov's continuity Theorem, that is of Theorem A.6. To this end, it is useful to have, for any fixed even positive integer k, a control on $\Delta_{I,k}\big(x_1, x_2; s^1, s^2\big)$ the moment of increment of order k defined, for all $(x_1, x_2) \in \mathbb{R}^2$ and $(s^1, s^2) \in I^2$, as

$$\Delta_{I,k}\big(x_1, x_2; s^1, s^2\big) := \mathbb{E}\Big(\big|L\big(x_1, I(s^1)\big) - L\big(x_2, I(s^2)\big)\big|^k\Big). \qquad (2.31)$$

Notice that one has to assume that k is an even positive integer, in order to be allowed to use the formula (2.21); the formula (2.20) can be used even when k is odd.

In order to get control on $\Delta_{I,k}\big(x_1, x_2; s^1, s^2\big)$, first we need to introduce some notations.

- For any fixed $\rho \in \big[0, \mathrm{diam}(I)\big]$, we denote by $\mathcal{U}_{I,\rho}$ the class of the closed subintervals U of I, such that each U can be expressed as the intersection of I with a closed cube in $\mathbb{R}^N$ of edge size lying in the interval $[\rho, 2\rho]$. Moreover we set

$$\tilde{\Delta}_{I,k}(\rho) := \sup\Big\{\mathbb{E}\big(L(x, U)^k\big) : (x, U) \in \mathbb{R} \times \mathcal{U}_{I,\rho}\Big\}. \qquad (2.32)$$

- We denote by $\mathcal{J}_I$ the class of all the closed subintervals of I. Moreover, for any fixed $(x_1, x_2) \in \mathbb{R}^2$, we set

$$\check{\Delta}_{I,k}(x_1, x_2) := \sup\Big\{\mathbb{E}\Big(\big|L\big(x_1, J\big) - L\big(x_2, J\big)\big|^k\Big) : J \in \mathcal{J}_I\Big\}. \qquad (2.33)$$

The following lemma provides a first useful upper bound for $\Delta_{I,k}\big(x_1, x_2; s^1, s^2\big)$.

Lemma 2.17. *There exists a finite constant $c > 0$, only depending on N, $\mathrm{diam}(I)$ and k, such that the inequality*

$$\Delta_{I,k}\big(x_1, x_2; s^1, s^2\big) \le c\Big(\big|s^1 - s^2\big|^{k(1-N)} \tilde{\Delta}_{I,k}\big(|s^1 - s^2|\big) + \check{\Delta}_{I,k}(x_1, x_2)\Big), \qquad (2.34)$$

holds, for all $(x_1, x_2) \in \mathbb{R}^2$ and $(s^1, s^2) \in I^2$, with the convention that $(+\infty) \times 0 = 0$.

Proof of Lemma 2.17. First observe that using the equalities

$$L\big(x_1, I(s^1)\big) - L\big(x_2, I(s^2)\big)$$
$$= L\big(x_1, I(s^1)\big) - L\big(x_1, I(s^2)\big) + L\big(x_1, I(s^2)\big) - L\big(x_2, I(s^2)\big)$$
$$= L\big(x_1, I(s^1) \setminus I(s^2)\big) + L\big(x_1, I(s^2)\big) - L\big(x_2, I(s^2)\big),$$

the convexity property of the function, from $\mathbb{R}$ to $\mathbb{R}$, $z \mapsto z^k$, and (2.33), one gets that

$$\Delta_{I,k}\big(x_1, x_2; s^1, s^2\big) \leq 2^{k-1} \bigg(\mathbb{E}\Big(L\big(x_1, I(s^1) \setminus I(s^2)\big)^k\Big) + \check{\Delta}_{I,k}(x_1, x_2) \bigg).$$
(2.35)

Next, we set

$$c_1 := N \operatorname{diam}(I)^{N-1}.$$

Let us now show that there exists a positive integer Q satisfying

$$Q \leq c_1 \big|s^1 - s^2\big|^{1-N}, \tag{2.36}$$

and there are $U_1, U_2, \ldots, U_Q$ belonging to $\mathcal{U}_{I,|s^1-s^2|}$ such that

$$I(s^1) \setminus I(s^2) \subset \bigcup_{q=1}^{Q} U_q. \tag{2.37}$$

In the sequel, we assume without any restriction that $s^1 \neq s^2$ since the left-hand side of (2.37) reduces to the empty set when $s^1 = s^2$. For each fixed arbitrary $m \in \{1, \ldots, N\}$, $z \in \mathbb{R}$ and $\delta \in \big[0, 2^{-1}\operatorname{diam}(I)\big]$, we denote by $J_m(z, \delta)$ the closed subinterval of I defined as

$$J_m(z, \delta) := \big\{t = (t_1, \ldots, t_N) \in I : \text{such that } |z - t_m| \leq \delta\big\}. \tag{2.38}$$

Notice that $J_m(z, \delta)$ can be the empty set. Also notice that there exists a positive integer P_m satisfying

$$P_m \leq \big(2^{-1}\delta^{-1}\operatorname{diam}(I)\big)^{N-1}, \tag{2.39}$$

and there are $U_{m,1}, U_{m,2}, \ldots, U_{m,P_m}$ belonging to $\mathcal{U}_{I,2\delta}$ such that

$$J_m(z, \delta) \subset \bigcup_{p=1}^{P_m} U_{m,p}. \tag{2.40}$$

Next observe that, in view of (2.22) and (2.38), one has

$$I(s^1) \setminus I(s^2) \subset \bigcup_{m=1}^{N} J_m\big(2^{-1}(s_m^1 + s_m^2), 2^{-1}|s^1 - s^2|\big), \tag{2.41}$$

where s_m^1 (resp. s_m^2) is the m-th coordinate of s^1 (resp. s^2). Thus, (2.39) to (2.41) imply that (2.36) and (2.37) hold. Next, using (2.37), part (ii) of Remark 2.5, the convexity property of the function $z \mapsto z^k$, (2.32) and (2.36), it follows that

$$\mathbb{E}\Big(L\big(x_1, I(s^1) \setminus I(s^2)\big)^k\Big) \leq \mathbb{E}\Big(\Big(\sum_{q=1}^{Q} L(x_1, U_q)\Big)^k\Big)$$

$$\leq Q^{k-1} \sum_{q=1}^{Q} \mathbb{E}\big(L(x_1, U_q)^k\big)$$

$$\leq Q^k \tilde{\Delta}_{I,k}\big(|s^1 - s^2|\big)$$

$$\leq c_1^k |s^1 - s^2|^{k(1-N)} \tilde{\Delta}_{I,k}\big(|s^1 - s^2|\big).$$

Finally, combining the latter inequality with (2.35), one obtains (2.34). $\square$

Let us now focus on the issue of finding appropriate upper bounds for $\tilde{\Delta}_{I,k}(\rho)$ and $\check{\Delta}_{I,k}(x_1, x_2)$ defined through (2.32) and (2.33). As we will show it, the strategy allowing to get such upper bounds, mainly requires the stochastic field $\{Z(t), t \in I\}$ to be *locally nondeterministic*. Roughly speaking, this means that $\{Z(t), t \in I\}$ is "locally unpredictable", in other words, its local evolution is everywhere perturbed by some unremovable element of "noise". More precisely, the fundamental concept of local nondeterminism, which was first introduced by [Berman (1973)] in the setting of Gaussian processes and later was extended by [Pitt (1978)] to the setting of Gaussian fields, can be defined in the following way:

Definition 2.18. A centered real-valued Gaussian field $\{Z(t), t \in I\}$ is said to be *locally nondeterministic* on the compact interval I of $\mathbb{R}^N$, when

$$\inf_{t \in I} \mathrm{Var}\big(Z(t)\big) > 0, \tag{2.42}$$

and there exists $\delta > 0$ such that $\{Z(t), t \in I\}$ has the following two properties:

(i) $\mathrm{Var}\big(Z(t') - Z(t'')\big) \neq 0$, for all $(t', t'') \in I^2$ satisfying $0 < |t' - t''| \leq \delta$;
(ii) for each fixed integer $n \geq 2$, there is a constant $c \in (0, 1)$, such that the inequality

$$\mathrm{Var}\big(Z(t^n) | Z(t^1), \ldots, Z(t^{n-1})\big) \geq c \, \mathrm{Var}\big(Z(t^n) - Z(t^{n-1})\big), \tag{2.43}$$

holds, for every $(t^1, \ldots, t^n) \in I_{\underset{\sim}{\leq}}^n$ satisfying

$$\max\big\{|t^l - t^m| : (l, m) \in \{1, \ldots, n\}^2\big\} \leq \delta, \tag{2.44}$$

the set $I_{\underset{\sim}{\leq}}^n$ being defined through Definition 2.9 with $J = I$ and $k = n$.

$\{Z(t),\ t \in I\}$ is said to be *strongly locally nondeterministic* on the compact interval I, if its locally nondeterministic on this interval and the constant c in (2.43) does not depend on n.

Lemma 2.19. *Assume that* $\{Z(t),\ t \in I\}$ *is locally nondeterministic on* I *and that*

$$\sup_{t \in I} \operatorname{Var}\big(Z(t)\big) < +\infty. \tag{2.45}$$

Then, for each fixed integer $n \geq 3$*, there is a constant* $c \in (0,1)$*, such that the inequality*

$$\operatorname{Var}\big(Z(t^q)|Z(t^l) : 1 \leq l \leq n,\ l \neq q\big) \tag{2.46}$$
$$\geq c\operatorname{Var}\big(Z(t^q) - Z(t^{q-1})\big) \times \operatorname{Var}\big(Z(t^q)|Z(t^l) : q < l \leq n\big),$$

holds, for all $(t^1, \ldots, t^n) \in I_{\preccurlyeq}^n$ *satisfying the condition (2.44) and for every* $q \in \{2, \ldots, n-1\}$*.*

Notice that when $\{Z(t),\ t \in I\}$ is strongly locally nondeterministic, then the constant c in (2.46) does not depend on n.

Proof of Lemma 2.19. Let us first show that, for every $j \in \{1, \ldots, n\}$ satisfying

$$q \neq j \text{ and } |t^q - t^j| = \min\big\{|t^q - t^l| : l \in \{1, \ldots, n\} \setminus \{q\}\big\}, \tag{2.47}$$

one has

$$\operatorname{Var}\big(Z(t^q)|Z(t^l) : 1 \leq l \leq n,\ l \neq q\big) \geq c_1\operatorname{Var}\big(Z(t^q) - Z(t^j)\big), \tag{2.48}$$

where c_1 is the constant c in (2.43). Let π be a permutation of $\{1, \ldots, n\}$ such that $\pi(n) := q$, $\pi(n-1) := j$ and, for all $m \in \{1, \ldots, n-2\}$, $\pi(m)$ is chosen such that

$$|t^{\pi(m+1)} - t^{\pi(m)}| \tag{2.49}$$
$$= \min\big\{|t^{\pi(m+1)} - t^l| : l \in \{1, \ldots, n\} \setminus \{\pi(m+1), \pi(m+2), \ldots, \pi(n)\}\big\}.$$

Then, (2.47) and (2.49) imply that $(t^{\pi(1)}, \ldots, t^{\pi(n)})$ is an element of $I_{\preccurlyeq}^n$. Thus, (2.48) is obtained by replacing in (2.43), $(t^1, \ldots, t^n)$ by $(t^{\pi(1)}, \ldots, t^{\pi(n)})$. Having derived (2.48), from now on "we forget" $(t^{\pi(1)}, \ldots, t^{\pi(n)})$ and we focus on the initial $(t^1, \ldots, t^n)$ which is assumed to be an element of $I_{\preccurlyeq}^n$ according to the hypotheses of the lemma. The later fact, entails that (see (2.18))

$$|t^q - t^{q-1}| = \min_{1 \leq n \leq q-1} |t^q - t^n|,$$

and consequently that

$$\min\left\{|t^q - t^l| : l \in \{1, \ldots, n\} \backslash \{q\}\right\} = \min\left\{|t^q - t^l| : l \in \{q-1, \ldots, n\} \backslash \{q\}\right\}. \tag{2.50}$$

Next, let us consider an arbitrary $j' \in \{q - 1, \ldots, n\} \backslash \{q\}$ such that

$$|t^q - t^{j'}| = \min\left\{|t^q - t^l| : l \in \{q-1, \ldots, n\} \backslash \{q\}\right\}. \tag{2.51}$$

Thus, it follows from (2.48), (2.50) and (2.51) that

$$\mathrm{Var}\big(Z(t^q)|Z(t^l) : 1 \le l \le n, \, l \ne q\big) \ge c_1 \mathrm{Var}\big(Z(t^q) - Z(t^{j'})\big). \tag{2.52}$$

Also, observe that

$$\mathrm{Var}\big(Z(t^q) - Z(t^{j'})\big) \tag{2.53}$$
$$\ge \min\left\{\mathrm{Var}\big(Z(t^q) - Z(t^{q-1})\big), \mathrm{Var}\big(Z(t^q)|Z(t^l) : q < l \le n\big)\right\};$$

it is clear that (2.53) is true when $j' = q - 1$, on the other hand, when $q < j' \le n$, using the definition of a conditional variance one gets that

$$\mathrm{Var}\big(Z(t^q) - Z(t^{j'})\big) \ge \mathrm{Var}\big(Z(t^q)|Z(t^l) : q < l \le n\big)$$
$$\ge \min\left\{\mathrm{Var}\big(Z(t^q) - Z(t^{q-1})\big), \mathrm{Var}\big(Z(t^q)|Z(t^l) : q < l \le n\big)\right\}.$$

Next, let c_2 be the finite postive constant defined as

$$c_2 := \left(1 + 4\sup_{t \in I} \mathrm{Var}\big(Z(t)\big)\right)^{-1};$$

then it is clear that

$$0 \le c_2 \mathrm{Var}\big(Z(t^q) - Z(t^{q-1})\big) \le 1, \tag{2.54}$$

$$0 \le c_2 \mathrm{Var}\big(Z(t^q)|Z(t^l) : q < l \le n\big) \le 1, \tag{2.55}$$

and

$$\min\left\{\mathrm{Var}\big(Z(t^q) - Z(t^{q-1})\big), \mathrm{Var}\big(Z(t^q)|Z(t^l) : q < l \le n\big)\right\} \tag{2.56}$$
$$\ge \min\left\{c_2\mathrm{Var}\big(Z(t^q) - Z(t^{q-1})\big), c_2\mathrm{Var}\big(Z(t^q)|Z(t^l) : q < l \le n\big)\right\}.$$

Moreover, using (2.54), (2.55) and the inequality $\min\{a, b\} \ge ab$, for all $(a, b) \in [0, 1]^2$, one has that

$$\min\left\{c_2\mathrm{Var}\big(Z(t^q) - Z(t^{q-1})\big), c_2\mathrm{Var}\big(Z(t^q)|Z(t^l) : q < l \le n\big)\right\}$$
$$\ge c_2^2 \mathrm{Var}\big(Z(t^q) - Z(t^{q-1})\big) \times \mathrm{Var}\big(Z(t^q)|Z(t^l) : q < l \le n\big). \tag{2.57}$$

Finally, let c be the constant defined as $c := c_1 c_2^2$; putting together (2.52), (2.53), (2.56) and (2.57), it follows that (2.46) holds. $\qquad\square$

The following result was obtained in [Pitt (1978)].

Proposition 2.20. *Fractional Brownian Field (FBF)* $\{B_H(t), t \in \mathbb{R}^N\}$
*(see Definition 1.17 as well as Propositions 1.25 and 1.28), of an arbitrary
Hurst parameter $H \in (0,1)$, is strongly locally nondeterministic on any
compact interval I of $\mathbb{R}^N$ such that $0 \notin I$; moreover one has $\delta = \mathrm{diam}(I)$.*

Proposition 2.20 can easily be derived by using (1.19), the fact that
$B_H(0) \overset{\mathrm{a.s.}}{=} 0$ and the following lemma borrowed from [Pitt (1978)].

Lemma 2.21. *Let I be a compact interval of $\mathbb{R}^N$ which does not contain 0.
Then, there exists a constant $c > 0$ such that the inequality*

$$\mathrm{Var}\big(B_H(\tau)|B_H(t) : t \in I \text{ and } |\tau - t| \geq r\big) \geq c\, r^{2H}, \qquad (2.58)$$

holds, for all $r \in \big[0, \mathrm{diam}(I)\big]$ and for each $\tau \in I$.

Proof of Lemma 2.21. Using the definition of a conditional variance in
the Gaussian frame, it turns out that the lemma can be equivalently stated
as follows: *there exists a constant $c > 0$ such that, for all $r \in \big[0, \mathrm{diam}(I)\big]$,
for each $\tau \in I$, for any integer $n \geq 1$, for each $a_1, \ldots, a_n \in \mathbb{R}$ and for every
$t^1, \ldots, t^n \in I$ satisfying*

$$\min_{1 \leq l \leq n} |t^l - \tau| \geq r, \qquad (2.59)$$

one has

$$\mathbb{E}\Big| B_H(\tau) - \sum_{l=1}^{n} a_l B(t^l) \Big|^2 \geq c\, r^{2H}. \qquad (2.60)$$

The lemma clearly holds when $r = 0$, so from now on we assume that $r > 0$
and we set

$$\rho = \rho(r, |\tau|) := \min\big(r, |\tau|\big). \qquad (2.61)$$

Observe that

$$\rho(r, |\tau|) \geq c_1\, r, \qquad (2.62)$$

where the positive constant $c_1 := \min\big\{1, \mathrm{diam}(I)^{-1} \min_{t \in I} |t|\big\}$. Also, ob-
serve that Proposition 1.28 and the isometry property of the stochastic
integral $\int_{\mathbb{R}^N} (\,\cdot\,)\, d\widehat{W}$ imply that

$$\mathbb{E}\Big| B_H(\tau) - \sum_{l=1}^{n} a_l B(t^l) \Big|^2 = c_{H,N}^2 \int_{\mathbb{R}^N} \Big| e^{i\tau \cdot \xi} - 1 - \sum_{l=1}^{n} a_l\big(e^{it^l \cdot \xi} - 1\big) \Big|^2 \frac{d\xi}{|\xi|^{2H+N}}, \qquad (2.63)$$

where $c_{H,N}$ is a non-vanishing constant only depending on H and N. Next, let φ be an arbitrary infinitely differentiable bump function, from $\mathbb{R}^N$ into the compact interval $[0,1]$ of the real line, such that

$$\varphi(0) = 1 \quad \text{and} \quad \operatorname{Supp} \varphi \subseteq \{\xi \in \mathbb{R}^N : |\xi| \leq 1\}. \tag{2.64}$$

Observe that φ belongs to the Schwartz class $\mathcal{S}(\mathbb{R}^N)$ (see Definition 4.21) and consequently (see Proposition 4.22) that its Fourier transform $\widehat{\varphi}$ belongs to $\mathcal{S}(\mathbb{R}^N)$ as well. Next, we denote by φ_ρ the function defined as

$$\varphi_\rho(s) = \rho^{-N} \varphi(\rho^{-1} s), \quad \text{for all } s \in \mathbb{R}^N. \tag{2.65}$$

It is clear that φ_ρ belongs to $\mathcal{S}(\mathbb{R}^N)$. Also, observe that (2.64) and (2.65) entail that

$$\varphi_\rho(0) = \rho^{-N} \quad \text{and} \quad \operatorname{Supp} \varphi_\rho \subseteq \{\xi \in \mathbb{R}^N : |\xi| \leq \rho\}. \tag{2.66}$$

Next, let $\Lambda = \Lambda(\tau; a_1, \ldots, a_n; t^1, \ldots, t^n; \rho)$ be the integral defined as

$$\Lambda := \int_{\mathbb{R}^N} e^{-i\tau \cdot \xi} \widehat{\varphi}_\rho(\xi) \left(e^{i\tau \cdot \xi} - 1 - \sum_{l=1}^{n} a_l \left(e^{it^l \cdot \xi} - 1 \right) \right) d\xi, \tag{2.67}$$

where $\widehat{\varphi}_\rho$ denotes the Fourier transform of φ_ρ. We mention in passing that, in view of (2.65) and of part (iv) of Proposition 4.5, the function $\widehat{\varphi}_\rho$ can be expressed as

$$\widehat{\varphi}_\rho(\xi) = \widehat{\varphi}(\rho\xi), \quad \text{for every } \xi \in \mathbb{R}^N. \tag{2.68}$$

Next, using (2.67) and the equality

$$\varphi_\rho(s) := (2\pi)^{-N/2} \int_{\mathbb{R}^N} e^{is \cdot \xi} \widehat{\varphi}_\rho(\xi) \, ds, \quad \text{for all } s \in \mathbb{R}^N,$$

one gets that

$$\Lambda = \int_{\mathbb{R}^N} \widehat{\varphi}_\rho(\xi) \, d\xi + \left(-1 + \sum_{l=1}^{n} a_l \right) \int_{\mathbb{R}^N} e^{-i\tau \cdot \xi} \widehat{\varphi}_\rho(\xi) \, d\xi$$

$$- \sum_{l=1}^{n} a_l \int_{\mathbb{R}^N} e^{i(t^l - \tau) \cdot \xi} \widehat{\varphi}_\rho(\xi) \, d\xi$$

$$= (2\pi)^{N/2} \varphi_\rho(0) + (2\pi)^{N/2} \left(-1 + \sum_{l=1}^{n} a_l \right) \varphi_\rho(-\tau) - (2\pi)^{N/2} \sum_{l=1}^{n} a_l \varphi_\rho(t^l - \tau).$$

Thus, (2.66), (2.59) and (2.61) imply that

$$\Lambda = (2\pi)^{N/2} \rho^{-N}. \tag{2.69}$$

On the other hand, using (2.67) and Cauchy-Schwarz inequality, one obtains that

$$|\Lambda|^2 = \left| \int_{\mathbb{R}^N} e^{-i\tau\cdot\xi}\, \widehat{\varphi}_\rho(\xi)|\xi|^{H+N/2}\left(e^{i\tau\cdot\xi} - 1 - \sum_{l=1}^{n} a_l\left(e^{it^l\cdot\xi} - 1\right)\right)\frac{d\xi}{|\xi|^{H+N/2}}\right|^2$$

$$\leq \int_{\mathbb{R}^N}\left|\widehat{\varphi}_\rho(\xi)\right|^2 |\xi|^{2H+N}\, d\xi \times \int_{\mathbb{R}^N}\left|e^{i\tau\cdot\xi} - 1 - \sum_{l=1}^{n} a_l\left(e^{it^l\cdot\xi} - 1\right)\right|^2 \frac{d\xi}{|\xi|^{2H+N}}.$$

Therefore, it follows from (2.69), (2.68) and the change of variable $\eta = \rho\xi$ that

$$(2\pi)^N \rho^{-2N}$$

$$\leq \int_{\mathbb{R}^N}\left|\widehat{\varphi}(\rho\xi)\right|^2 |\xi|^{2H+N}\, d\xi \times \int_{\mathbb{R}^N}\left|e^{i\tau\cdot\xi} - 1 - \sum_{l=1}^{n} a_l\left(e^{it^l\cdot\xi} - 1\right)\right|^2 \frac{d\xi}{|\xi|^{2H+N}}$$

$$= \rho^{-2(H+N)}\int_{\mathbb{R}^N}\left|\widehat{\varphi}(\eta)\right|^2 |\eta|^{2H+N}\, d\eta$$

$$\times \int_{\mathbb{R}^N}\left|e^{i\tau\cdot\xi} - 1 - \sum_{l=1}^{n} a_l\left(e^{it^l\cdot\xi} - 1\right)\right|^2 \frac{d\xi}{|\xi|^{2H+N}}.$$

Thus, using (2.61) and (2.62), one gets that

$$\int_{\mathbb{R}^N}\left|e^{i\tau\cdot\xi} - 1 - \sum_{l=1}^{n} a_l\left(e^{it^l\cdot\xi} - 1\right)\right|^2 \frac{d\xi}{|\xi|^{2H+N}} \geq c_2\rho^{2H} \geq c_3 r^{2H}, \qquad (2.70)$$

where the finite constant

$$c_2 := (2\pi)^N \left(\int_{\mathbb{R}^N}\left|\widehat{\varphi}(\eta)\right|^2 |\eta|^{2H+N}\, d\eta\right)^{-1}$$

and the finite constant $c_3 := c_1^{2H} c_2$. Finally, combining (2.70) with (2.63), it follows that (2.60) holds.

$$\square$$

The following general lemma on centered Gaussian vectors will allow to draw important connections between the concept of local nondeterminism and the issue of finding appropriate upper bounds for $\tilde{\Delta}_{I,k}(\rho)$ and $\check{\Delta}_{I,k}(x_1, x_2)$.

Lemma 2.22. *Assume that $p \geq 2$ is an arbitrary integer and $\mathcal{Z} := (\mathcal{Z}_1, \ldots, \mathcal{Z}_{p-1}, \mathcal{Z}_p)^*$ is an $\mathbb{R}^p$-valued centered Gaussian column* [6] *random vector; its covariance matrix is denoted* $\mathrm{Var}(\mathcal{Z})$. *Then, one has*

$$\det\left(\mathrm{Var}(\mathcal{Z})\right) = \mathrm{Var}(\mathcal{Z}_1)\prod_{n=2}^{p} \mathrm{Var}\left(\mathcal{Z}_n\middle|\mathcal{Z}_1, \ldots, \mathcal{Z}_{n-1}\right); \qquad (2.71)$$

---

[6]As, usual the symbol $*$ means the transposition operation on matrices.

moreover, when $\det(\mathrm{Var}(\mathcal{Z})) > 0$, *for any fixed positive real number* θ *and* $q \in \{1, \dots, p\}$, *one has*

$$\frac{1}{(2\pi)^{p/2}} \int_{\mathbb{R}^p} |u_q|^\theta \exp\left(-2^{-1}u^*\mathrm{Var}(\mathcal{Z})u\right) du \qquad (2.72)$$

$$= \frac{2^{(\theta+1)/2}}{(2\pi)^{1/2}} \Gamma\left(\frac{\theta+1}{2}\right) \det(\mathrm{Var}(\mathcal{Z}))^{-1/2} \mathrm{Var}\left(\mathcal{Z}_q \,\middle|\, all\ the\ \mathcal{Z}_l\ with\ l \neq q\right)^{-\theta/2};$$

where u_q *denotes the q-th coordinate of the column vector u and Γ is the Gamma function (see (A.7)).*

In order to show that Lemma 2.22 holds, one needs the following lemma.

Lemma 2.23. *Assume that $p \geq 2$ is an arbitrary integer and $\mathcal{Z} := (\mathcal{Z}_1, \dots, \mathcal{Z}_{p-1}, \mathcal{Z}_p)^*$ is an $\mathbb{R}^p$-valued centered Gaussian column random vector. Let $\breve{\mathcal{Z}}_p$ be the $\mathbb{R}^{p-1}$-valued centered Gaussian column random vector defined as $\breve{\mathcal{Z}}_p := (\mathcal{Z}_1, \dots, \mathcal{Z}_{p-1})^*$, and let R be a deterministic row matrix of size $p-1$ such that $\mathbb{E}(\mathcal{Z}_p|\breve{\mathcal{Z}}_p) = R\breve{\mathcal{Z}}_p$, where $\mathbb{E}(\mathcal{Z}_p|\breve{\mathcal{Z}}_p)$ is the conditional expectation of $\mathcal{Z}_p$ given $\breve{\mathcal{Z}}_p$ [7]. We denote Q the lower triangular matrix of size p defined as*

$$Q := \begin{pmatrix} I_{p-1} & 0 \\ -R & 1 \end{pmatrix}, \qquad (2.73)$$

where I_{p-1} is the identity matrix of size $p-1$. We respectively denote $\mathrm{Var}(\mathcal{Z})$ and $\mathrm{Var}(\breve{\mathcal{Z}}_p)$ the covariance matrices of $\mathcal{Z}$ and $\breve{\mathcal{Z}}_p$. Moreover, we denote $\mathrm{Var}(\mathcal{Z}_p|\breve{\mathcal{Z}}_p) := \mathbb{E}(|\mathcal{Z}_p - \mathbb{E}(\mathcal{Z}_p|\breve{\mathcal{Z}}_p)|^2)$ the conditional variance of $\mathcal{Z}_p$ given $\breve{\mathcal{Z}}_p$ [8]. Then, $\mathrm{Var}(\mathcal{Z})$ can be expressed as the matrix product:

$$\mathrm{Var}(\mathcal{Z}) = S \begin{pmatrix} \mathrm{Var}(\breve{\mathcal{Z}}_p) & 0 \\ 0 & \mathrm{Var}(\mathcal{Z}_p|\breve{\mathcal{Z}}_p) \end{pmatrix} S^*, \quad where\ S := Q^{-1}. \qquad (2.74)$$

Proof of Lemma 2.23. It follows from the definition of $\mathbb{E}(\mathcal{Z}_p|\breve{\mathcal{Z}}_p)$ that the centered Gaussian column random vector $\breve{\mathcal{Z}}_p$ and the centered Gaussian random variable $\mathcal{Z}_p - \mathbb{E}(\mathcal{Z}_p|\breve{\mathcal{Z}}_p)$ are independent. Therefore

$$\mathrm{Var}(Q\mathcal{Z}) := \mathbb{E}(Q\mathcal{Z}(Q\mathcal{Z})^*) = Q(\mathbb{E}(ZZ^*))Q^* = Q(\mathrm{Var}(\mathcal{Z}))Q^*, \qquad (2.75)$$

the covariance matrix of the centered Gaussian column random vector

$$Q\mathcal{Z} = (\mathcal{Z}_1, \dots, \mathcal{Z}_{p-1}, \mathcal{Z}_p - \mathbb{E}(\mathcal{Z}_p|\breve{\mathcal{Z}}_p))^* = (\breve{\mathcal{Z}}_p^*, \mathcal{Z}_p - \mathbb{E}(\mathcal{Z}_p|\breve{\mathcal{Z}}_p))^*,$$

---

[7]In other words, $\mathbb{E}(\mathcal{Z}_p|\breve{\mathcal{Z}}_p)$ is $\mathbb{E}(\mathcal{Z}_p|\mathcal{Z}_1, \dots, \mathcal{Z}_{p-1})$, the conditional expectation of $\mathcal{Z}_p$ given all the other random variables $\mathcal{Z}_1, \dots, \mathcal{Z}_{p-1}$

[8]In other words, $\mathrm{Var}(\mathcal{Z}_p|\breve{\mathcal{Z}}_p)$ is $\mathrm{Var}(\mathcal{Z}_p|\mathcal{Z}_1, \dots, \mathcal{Z}_{p-1})$, the conditional variance of $\mathcal{Z}_p$ given all the other random variables $\mathcal{Z}_1, \dots, \mathcal{Z}_{p-1}$

can also be expressed as:

$$\mathrm{Var}(Q\mathcal{Z}) = \begin{pmatrix} \mathrm{Var}(\breve{\mathcal{Z}}_p) & 0 \\ 0 & \mathrm{Var}(\mathcal{Z}_p|\breve{\mathcal{Z}}_p) \end{pmatrix}. \tag{2.76}$$

Thus, (2.74) results from (2.75) and (2.76). $\qquad\qquad\qquad\qquad\square$

We are now in position to prove Lemma 2.22.

Proof of Lemma 2.22. We use the same notations as in Lemma 2.23. It easily follows (2.74) and (2.73) that

$$\det\big(\mathrm{Var}(\mathcal{Z})\big) = \det\big(\mathrm{Var}(\breve{\mathcal{Z}}_p)\big)\,\mathrm{Var}\Big(\mathcal{Z}_n\Big|\mathcal{Z}_1,\ldots,\mathcal{Z}_{p-1}\Big);$$

thus, (2.71) can be obtained by induction. Let us now show that (2.72) holds, for the sake of simplicity, we assume that $q = p$. Setting $w = (Q^*)^{-1}u$ i.e. $u = Q^*w$, and using again (2.74) as well as (2.73), one gets that

$$\frac{1}{(2\pi)^{p/2}} \int_{\mathbb{R}^p} |u_p|^\theta \exp\big(-2^{-1}u^*\mathrm{Var}(\mathcal{Z})u\big)\,du$$

$$= \frac{1}{(2\pi)^{(p-1)/2}} \int_{\mathbb{R}^{p-1}} \exp\big(-2^{-1}\breve{w}_p^*\mathrm{Var}(\breve{\mathcal{Z}}_p)\breve{w}_p\big)\,d\breve{w}_p \tag{2.77}$$

$$\times \frac{1}{(2\pi)^{1/2}} \int_{\mathbb{R}} |w_p|^\theta \exp\big(-2^{-1}\mathrm{Var}(\mathcal{Z}_p|\breve{\mathcal{Z}}_p)w_p^2\big)\,dw_p,$$

where w_p is the last coordinate of w and $\breve{w}_p$ is the vector constituted by the other coordinates. Next, let $\breve{\Gamma}$ be the square matrix such that $\breve{\Gamma}^*\breve{\Gamma} = \mathrm{Var}(\breve{\mathcal{Z}}_p)$. Setting in (2.77),

$$\breve{z}_p = \breve{\Gamma}\breve{w}_p \quad \text{and} \quad z_p = \mathrm{Var}(\mathcal{Z}_p|\breve{\mathcal{Z}}_p)^{1/2}\,w_p,$$

it follows from standard computations and (2.71) that (2.72) is satisfied. $\quad\square$

Remark 2.24. Assume that $p \geq 2$ is an arbitrary integer and $\mathcal{Z} := (\mathcal{Z}_1,\ldots,\mathcal{Z}_{p-1},\mathcal{Z}_p)^*$ is an $\mathbb{R}^p$-valued centered Gaussian column random vector with covariance matrix $\mathrm{Var}(\mathcal{Z})$. By using (2.71) and the fact that the value of $\det\big(\mathrm{Var}(\mathcal{Z})\big)$ does not depend on the way on which the coordinates of $\mathcal{Z}$ are labelled, one gets that

$$\det\big(\mathrm{Var}(\mathcal{Z})\big) = \mathrm{Var}(\mathcal{Z}_{\pi(1)}) \prod_{n=2}^{p} \mathrm{Var}\Big(\mathcal{Z}_{\pi(n)}\Big|\mathcal{Z}_{\pi(1)},\ldots,\mathcal{Z}_{\pi(n-1)}\Big), \tag{2.78}$$

where π denotes an arbitrary permutation of $\{1,\ldots,p\}$. In particular, one has

$$\det\big(\mathrm{Var}(\mathcal{Z})\big) = \mathrm{Var}(\mathcal{Z}_p) \prod_{n=2}^{p} \mathrm{Var}\Big(\mathcal{Z}_{p+1-n}\Big|\mathcal{Z}_p,\ldots,\mathcal{Z}_{p+2-n}\Big). \tag{2.79}$$

Lemma 2.25. *Assume that the condition (2.19) holds for some integer* $k \geq 2$. *Then, for any fixed* $\rho \in [0, \mathrm{diam}(I)]$, *the quantity* $\tilde{\Delta}_{I,k}(\rho)$, *defined through (2.32), can be bounded from above in the following way:*

$$\tilde{\Delta}_{I,k}(\rho) \leq \frac{(k-1)!}{(2\pi)^{k/2}} \times \tag{2.80}$$

$$\sup_{U \in \mathcal{U}_{I,\rho}} \left\{ \int_{U_{\preccurlyeq}^k} \left(\mathrm{Var}(Z(t^1)) \prod_{n=2}^k \mathrm{Var}(Z(t^n)|Z(t^l) : 1 \leq l < n) \right)^{-1/2} d\bar{t} \right\}.$$

Recall that $\mathcal{U}_{I,\rho}$ *denotes the class of the closed subintervals* U *of* I, *such that each* U *can be expressed as the intersection of* I *with a closed cube in* $\mathbb{R}^N$ *of edge size lying in the interval* $[\rho, 2\rho]$. *Also recall that the set* $U_{\preccurlyeq}^k$ *is defined through (2.18) with* $J = U$.

Proof of Lemma 2.25. Thanks to the second equality in (2.20), for every $x \in \mathbb{R}$ and $U \in \mathcal{U}_{I,\rho}$, the quantity $\mathbb{E}\left(L(x, U)^k\right)$, can be bound from above, independently on x, in following way:

$$\mathbb{E}\left(L(x, U)^k\right) \leq \mathbb{E}\left(L(0, U)^k\right) = \frac{(k-1)!}{(2\pi)^{k/2}} \int_{U_{\preccurlyeq}^k} \det\left(\mathbb{V}_{Z,k}(\bar{t})\right)^{-1/2} d\bar{t}, \tag{2.81}$$

where $\bar{t} := (t^1, \ldots, t^k)$ and $\mathbb{V}_{Z,k}(\bar{t})$ denotes the covariance matrix of the $\mathbb{R}^k$-valued centered Gaussian column random vector $\left(Z(t^1), \ldots, Z(t^k)\right)^*$. Notice that the equality in (2.81) results from the fact that, for λ_{kN}-almost all $\bar{t} \in I_{\preccurlyeq}^k$, one has

$$\frac{\det\left(\mathbb{V}_{Z,k}(\bar{t})\right)^{1/2}}{(2\pi)^{k/2}} \int_{\mathbb{R}^k} \exp\left(-2^{-1} u^* \mathbb{V}_{Z,k}(\bar{t}) u\right) du = 1, \tag{2.82}$$

since (2.19) implies that the matrix $\mathbb{V}_{Z,k}(\bar{t})$ is nonsingular for λ_{kN}-almost all $\bar{t} \in I_{\preccurlyeq}^k$. Next, using (2.81) and (2.71) one obtains (2.80). □

The following lemma is a straightforward consequence of Definition 2.18 and of Lemma 2.25.

Lemma 2.26. *Let an arbitrary integer* $k \geq 2$. *Assume that the centered Gaussian field* $\{Z(t), \ t \in I\}$ *is locally nondeterministic on the compact interval* I *and that*

$$\int_{I_{\preccurlyeq}^k} \prod_{n=2}^k \sigma\left(Z(t^n) - Z(t^{n-1})\right)^{-1} d\bar{t} < +\infty. \tag{2.83}$$

Then, there exists a finite constant $c > 0$, such that, for all $\rho > 0$ small enough, one has

$$\tilde{\Delta}_{I,k}(\rho) \leq c^k (k-1)! \sup_{U \in \mathcal{U}_{I,\rho}} \left\{ \int_{U_\preccurlyeq^k} \prod_{n=2}^{k} \sigma \big(Z(t^n) - Z(t^{n-1}) \big)^{-1} d\bar{t} \right\}; \quad (2.84)$$

moreover, the constant c does not depend on k when the local nondeterminism property holds in the strong sense.

Lemma 2.27. *Let $\{B_H(t), t \in \mathbb{R}^N\}$ be the FBF of an arbitrary Hurst parameter $H \in (0,1)$. Assume that I is an arbitrary compact interval of $\mathbb{R}^N$ satisfying $0 \notin I$. Then, there exists a finite constant $c > 0$ such that, for all integer $k \geq 2$ and $\rho \in \big[0, \operatorname{diam}(I)\big]$, one has*

$$\tilde{\Delta}_{I,k}(\rho) \leq c^k (k-1)! \, \rho^{k(N-H)}. \quad (2.85)$$

Proof of Lemma 2.27. For deriving the lemma one will make use of Lemma 2.26. In order to be allowed to do so, one needs to check that $\{B_H(t), t \in \mathbb{R}^N\}$ satisfies the condition (2.83); recall that one knows from Proposition 2.20 that $\{B_H(t), t \in \mathbb{R}^N\}$ is strongly locally nondeterministic on I with $\delta = \operatorname{diam}(I)$. In view (1.19), in the case of FBF, the condition (2.83) can be expressed as

$$\int_{I_\preccurlyeq^k} \prod_{n=2}^{k} |t^n - t^{n-1}|^{-H} \, d\bar{t} < +\infty.$$

Thus, because of the inclusion $I_\preccurlyeq^k \subseteq I^k$, it is enough to show that

$$\int_{I^k} \prod_{n=2}^{k} |t^n - t^{n-1}|^{-H} \, d\bar{t} < +\infty. \quad (2.86)$$

One denotes by $(f_m)_{m \geq 2}$ the sequence of the positive Borel functions defined on I, by induction, in the following way:

$$f_2(t) := \int_I |t - s|^{-H} \, ds, \quad \text{for each } t \in I, \quad (2.87)$$

and

$$f_m(t) := \int_I |t - s|^{-H} f_{m-1}(s) \, ds, \quad \text{for all } m \geq 3 \text{ and } t \in I. \quad (2.88)$$

Observe that using Tonelli's Theorem, one has

$$\int_{I^k} \prod_{n=2}^{k} |t^n - t^{n-1}|^{-H} \, d\bar{t} = \int_I f_k(t) \, dt. \quad (2.89)$$

Thus, for obtaining (2.86), it is enough to show that the functions f_m are bounded. In order to do so, one will proceed by induction. Let $\mathcal{B}(0, \nu_I)$ be the closed ball of $\mathbb{R}^N$ centered at 0 and of radius $\nu_I := 2\max_{u \in I} |u|$. For each arbitrary and fixed $t \in I$, let $t - I$ be the compact interval of $\mathbb{R}^N$, defined as $t - I := \{t - s : s \in I\}$. The change of variable $x = t - s$ in (2.87), the inclusion $t - I \subset \mathcal{B}(0, \nu_I)$, and the fact that H belongs to the open interval $(0, 1)$ imply that

$$\sup_{t \in I} f_2(t) = \int_{t-I} |x|^{-H}\, ds \leq \int_{\mathcal{B}(0, \nu_I)} |x|^{-H}\, ds < +\infty.$$

Next, assuming that, for an arbitrary $m \geq 3$, one has

$$\sup_{t \in I} f_{m-1}(t) \leq \left(\int_{\mathcal{B}(0, \nu_I)} |x|^{-H}\, ds \right)^{m-2},$$

then, the change of variable $x = t - s$ in (2.88) and the same arguments as before, allow to show that

$$\sup_{t \in I} f_m(t) \leq \left(\int_{\mathcal{B}(0, \nu_I)} |x|^{-H}\, ds \right)^{m-1}. \tag{2.90}$$

Notice that, not only (2.89) and (2.90) with $m = k$, imply that (2.86) holds, but also that

$$\int_{I^k} \prod_{n=2}^{k} |t^n - t^{n-1}|^{-H}\, d\bar{t} \leq \lambda_N(I) \left(\int_{\mathcal{B}(0, \nu_I)} |x|^{-H}\, ds \right)^{k-1}. \tag{2.91}$$

For later purposes, let us mention, that similarly to (2.91), one can show that there exists a positive finite constant c_1, not depending on k, such that

$$\int_{[-\langle 1 \rangle, \langle 1 \rangle]^k} \prod_{n=2}^{k} |y^n - y^{n-1}|^{-H}\, d\bar{y} \leq c_1^k, \tag{2.92}$$

where $\bar{y} := (y^1, \ldots, y^k)$ and $[-\langle 1 \rangle, \langle 1 \rangle]$ is the closed cube of $\mathbb{R}^N$ centered at 0 and of edge size 2. Let us now prove (2.85), one can restrict to $\rho \in (0, \operatorname{diam}(I)]$ without any loss of generality. In view of (1.19) and (2.84), it is enough to show that the inequality

$$\int_{U^k} \prod_{n=2}^{k} |t^n - t^{n-1}|^{-H}\, d\bar{t} \leq c_1^k \rho^{k(N-H)}, \tag{2.93}$$

holds, for all $\rho \in (0, \operatorname{diam}(I)]$ and $U \in \mathcal{U}_{I,\rho}$. One knows from the definition of the class $\mathcal{U}_{I,\rho}$, that there exist $\tau \in \mathbb{R}^N$ and $r \in [\rho/2, \rho]$ such that $U = I \cap [\tau - \langle r \rangle, \tau + \langle r \rangle]$, where $[\tau - \langle r \rangle, \tau + \langle r \rangle]$ denotes the closed cube of $\mathbb{R}^N$

centered at τ and of edge size $2r$. The inclusion $U \subseteq [\tau - \langle r \rangle, \tau + \langle r \rangle]$ and the change of variable $t^n = \tau + r y^n$, for all $n = 1, \ldots, k$, allow to show that

$$
\int_{U^k} \prod_{n=2}^{k} |t^n - t^{n-1}|^{-H} \, d\bar{t} \leq \int_{[\tau - \langle r \rangle, \tau + \langle r \rangle]^k} \prod_{n=2}^{k} |t^n - t^{n-1}|^{-H} \, d\bar{t}
$$

$$
= r^{kN} \int_{[-\langle 1 \rangle, \langle 1 \rangle]^k} \prod_{n=2}^{k} |\tau + r y^n - \tau - r y^{n-1}|^{-H} \, d\bar{y}
$$

$$
= r^{k(N-H)} \int_{[-\langle 1 \rangle, \langle 1 \rangle]^k} \prod_{n=2}^{k} |y^n - y^{n-1}|^{-H} \, d\bar{y}.
$$

Thus, using (2.92) and the inequality $r \leq \rho$, one gets (2.93). $\qquad \square$

Lemma 2.28. *Assume that $\gamma \in (0,1]$ is arbitrary and fixed and that the condition (2.19) holds. Then, for all fixed even integer $k \geq 2$ and for any fixed $(x_1, x_2) \in \mathbb{R}^2$, the quantity $\check{\Delta}_{I,k}(x_1, x_2)$, defined through (2.33), can be bounded from above in the following way:*

$$
\check{\Delta}_{I,k}(x_1, x_2) \leq \left(\frac{2}{\pi}\right)^{k/2} (k-1)! \, \Gamma\left(\frac{k\gamma + 1}{2}\right) |x_1 - x_2|^{k\gamma} \tag{2.94}
$$

$$
\times \int_{I_{\preccurlyeq}^k} \left\{ \left(\mathrm{Var}(Z(t^1)) \prod_{n=2}^{k} \mathrm{Var}(Z(t^n)|Z(t^l) : 1 \leq l < n) \right)^{-1/2} \right.
$$

$$
\left. \times \left(\prod_{q=1}^{k} \mathrm{Var}(Z(t^q)|Z(t^l) : 1 \leq l \leq k, \, l \neq q) \right)^{-\gamma/2} \right\} d\bar{t},
$$

where Γ is the Gamma function (see (A.7)).

Proof of Lemma 2.28. Let J be an arbitrary closed subinterval of I. Using (2.21) with $(x,y) = (x_1, x_2)$, the inclusion $J_{\preccurlyeq}^k \subseteq I_{\preccurlyeq}^k$, the elementary inequality

$$
|e^{i\theta} - 1| \leq 2^{1-\gamma} |\theta|^\gamma \quad \text{for all } \theta \in \mathbb{R}, \tag{2.95}
$$

and Proposition A.19 (the generalized Hölder inequality) with $p_1 = \ldots = p_k = k$, one gets that

$$\mathbb{E}\left(\left| L(x_1, J) - L(x_2, J) \right|^k \right)$$

$$\leq \frac{(k-1)!}{(2\pi)^k} \int_{I_{\preceq}^k} \int_{\mathbb{R}^k} \left(\prod_{q=1}^{k} \left| e^{-iu_q x_1} - e^{-iu_q x_2} \right| \right) \exp\left(-2^{-1} u^* \mathbb{V}_{Z,k}(\bar{t}) u \right) du \, d\bar{t}$$

$$\leq \frac{2^{k(1-\gamma)}(k-1)!}{(2\pi)^k} \left| x_1 - x_2 \right|^{k\gamma} \int_{I_{\preceq}^k} \int_{\mathbb{R}^k} \left(\prod_{q=1}^{k} |u_q|^{\gamma} \exp\left(-2^{-1} k^{-1} u^* \mathbb{V}_{Z,k}(\bar{t}) u \right) \right) du \, d\bar{t}$$

$$\leq \frac{2^{k(1-\gamma)}(k-1)!}{(2\pi)^k} \left| x_1 - x_2 \right|^{k\gamma} \int_{I_{\preceq}^k} \left(\prod_{q=1}^{k} \int_{\mathbb{R}^k} |u_q|^{k\gamma} \exp\left(-2^{-1} u^* \mathbb{V}_{Z,k}(\bar{t}) u \right) du \right)^{1/k} d\bar{t}.$$

$$(2.96)$$

Next, recall that (2.19) implies that the matrix $\mathbb{V}_{Z,k}(\bar{t})$ is nonsingular for λ_{kN}-almost all $\bar{t} \in I_{\preceq}^k$. Thus, it follows from Lemma 2.22 that, for λ_{kN}-almost all $\bar{t} \in I_{\preceq}^k$ and each $q \in \{1, \ldots, k\}$, one has

$$\frac{1}{(2\pi)^{k/2}} \int_{\mathbb{R}^k} |u_q|^{k\gamma} \exp\left(-2^{-1} u^* \mathbb{V}_{Z,k}(\bar{t}) u \right) du \qquad (2.97)$$

$$= \frac{2^{(k\gamma+1)/2}}{(2\pi)^{1/2}} \Gamma\left(\frac{k\gamma+1}{2} \right) \left(\mathrm{Var}(Z(t^1)) \prod_{n=2}^{k} \mathrm{Var}(Z(t^n)|Z(t^l) : 1 \leq l < n) \right)^{-1/2}$$

$$\times \left(\mathrm{Var}(Z(t^q)|Z(t^l) : 1 \leq l \leq k, l \neq q) \right)^{-k\gamma/2}$$

Finally combining (2.96) and (2.97), one obtains (2.94). □

Lemma 2.29. *Let an arbitrary even integer $k \geq 2$. Assume that the centered Gaussian field $\{Z(t), t \in I\}$ is locally nondeterministic on the compact interval I with $\delta = \mathrm{diam}(I)$. Also, assume that (2.83) and (2.45) hold. Then, there exists a finite constant $c > 0$, such that, for any fixed $\gamma \in (0,1]$ and for all $(x_1, x_2) \in \mathbb{R}^2$, the quantity $\check{\Delta}_{I,k}(x_1, x_2)$ can be bounded from above in the following way:*

$$\check{\Delta}_{I,k}(x_1, x_2) \leq c^k (k-1)! \, \Gamma\left(\frac{k\gamma+1}{2} \right) \left| x_1 - x_2 \right|^{k\gamma} \qquad (2.98)$$

$$\times \int_{I_{\preceq}^k} \left(\prod_{n=2}^{k} \sigma(Z(t^n) - Z(t^{n-1})) \right)^{-1-2\gamma} d\bar{t};$$

moreover, the constant c does not depend on k when the local nondeterminism property holds in the strong sense.

Proof of Lemma 2.29. Let us set

$$c_1^{-1} := \min\left\{1, \inf_{t\in I} \sigma\big(Z(t)\big)\right\} > 0; \qquad (2.99)$$

notice that the inequality in (2.99) results from (2.42). First we prove that the lemma holds when $k = 2$. Using (2.99), the fact that $I_{\preccurlyeq}^2 = I^2$, the equality $\delta = \mathrm{diam}(I)$, (2.43) and the inequality $\gamma \leq 1$, one gets, for all $\bar{t} := (t^1, t^2) \in I_{\preccurlyeq}^2$, that

$$\Big(\mathrm{Var}\big(Z(t^1)\big)\mathrm{Var}\big(Z(t^2)|Z(t^1)\big)\Big)^{-1/2} \times \Big(\mathrm{Var}\big(Z(t^1)|Z(t^2)\big)\mathrm{Var}\big(Z(t^2)|Z(t^1)\big)\Big)^{-\gamma/2}$$
$$\leq c_1\, c_2^{1+2\gamma}\, \sigma\big(Z(t^2) - Z(t^1)\big)^{-1-2\gamma} \leq c_1\, c_2^3\, \sigma\big(Z(t^2) - Z(t^1)\big)^{-1-2\gamma}, \qquad (2.100)$$

where $c_2^{-1} \in (0,1)$ denotes the square root of the constant c in (2.43). Thus, combining (2.100) and (2.94), one obtains (2.98) when $k = 2$. Let us now assume that the even integer $k \geq 4$ and $\bar{t} := (t^1, \ldots, t^k) \in I_{\preccurlyeq}^k$ are arbitrary. Standard computations, the equality $\delta = \mathrm{diam}(I)$, (2.43), (2.46) and the inequality $\gamma \leq 1$, led to

$$\Big(\textstyle\prod_{q=1}^{k} \mathrm{Var}\big(Z(t^q)|Z(t^l) : 1 \leq l \leq k,\, l \neq q\big) \Big)^{-\gamma/2}$$
$$= \Big(\mathrm{Var}\big(Z(t^k)|Z(t^l) : 1 \leq l < k\big)\mathrm{Var}\big(Z(t^1)|Z(t^l) : 1 < l \leq k\big)$$
$$\times \Big(\textstyle\prod_{q=2}^{k-1} \mathrm{Var}\big(Z(t^q)|Z(t^l) : 1 \leq l \leq k,\, l \neq q\big) \Big) \Big)^{-\gamma/2}$$
$$\leq c_2^\gamma\, c_3^{(k-2)\gamma} \Big(\mathrm{Var}\big(Z(t^k) - Z(t^{k-1})\big)\mathrm{Var}\big(Z(t^1)|Z(t^l) : 1 < l \leq k\big)$$
$$\times \textstyle\prod_{q=2}^{k-1} \Big(\mathrm{Var}\big(Z(t^q) - Z(t^{q-1})\big)\mathrm{Var}\big(Z(t^q)|Z(t^l) : q < l \leq k\big) \Big) \Big)^{-\gamma/2}$$
$$\leq c_2\, c_3^{k-2} \Big(\textstyle\prod_{q=2}^{k} \sigma\big(Z(t^q) - Z(t^{q-1})\big) \Big)^{-\gamma} \times \Big(\textstyle\prod_{q=1}^{k-1} \mathrm{Var}\big(Z(t^q)|Z(t^l) : q < l \leq k\big) \Big)^{-\gamma/2},$$
$$\qquad (2.101)$$

where $c_3^{-1} \in (0,1)$ is the square root of the constant c in (2.46). Moreover setting

$$c_4 := 1 + \sup_{t\in I} \sigma\big(Z(t)\big),$$

and using (2.79) and (2.71), one gets that

$$\prod_{q=1}^{k-1} \mathrm{Var}\big(Z(t^q)|Z(t^l) : q < l \le k\big)$$

$$\ge c_4^{-2}\,\mathrm{Var}\big(Z(t^k)\big)\Big(\prod_{q=1}^{k-1}\mathrm{Var}\big(Z(t^q)|Z(t^l) : q < l \le k\big)\Big) = c_4^{-2}\det\big(\mathbb{V}_{Z,k}(\bar{t})\big)$$

$$= c_4^{-2}\,\mathrm{Var}\big(Z(t^1)\big)\Big(\prod_{n=2}^{k}\mathrm{Var}\big(Z(t^n)|Z(t^l) : 1 \le l < n\big)\Big), \tag{2.102}$$

where $\mathbb{V}_{Z,k}(\bar{t})$ denotes the covariance matrix of the centered Gaussian column vector $\big(Z(t^1), \ldots, Z(t^k)\big)^*$. On the other hand, it follows from (2.99) and (2.43) that

$$\Big(\mathrm{Var}\big(Z(t^1)\big)\Big(\prod_{n=2}^{k}\mathrm{Var}\big(Z(t^n)|Z(t^l) : 1 \le l < n\big)\Big)\Big)^{-1/2}$$

$$\le c_1\, c_2^{k-1}\prod_{n=2}^{k}\sigma\big(Z(t^n) - Z(t^{n-1})\big). \tag{2.103}$$

Next, putting together (2.101), (2.102), (2.103) and the inequality $\gamma \le 1$, one obtains that

$$\Big(\mathrm{Var}\big(Z(t^1)\big)\prod_{n=2}^{k}\mathrm{Var}\big(Z(t^n)\big|Z(t^l) : 1 \le l < n\big)\Big)^{-1/2}$$

$$\times\Big(\prod_{q=1}^{k}\mathrm{Var}\big(Z(t^q)\big|Z(t^l) : 1 \le l \le k,\, l \ne q\big)\Big)^{-\gamma/2}$$

$$\le c_5^k\Big(\prod_{n=2}^{k}\sigma\big(Z(t^n) - Z(t^{n-1})\big)\Big)^{-1-2\gamma}, \tag{2.104}$$

where the constant $c_5 := c_1\, c_2^2\, c_3\, c_4$. Finally, combining (2.104) with (2.94), it follows (2.98) holds. $\qquad\square$

Lemma 2.30. *Let $\{B_H(t), t \in \mathbb{R}^N\}$ be the FBF of an arbitrary Hurst parameter $H \in (0,1)$. Assume that I is an arbitrary compact interval of $\mathbb{R}^N$ satisfying $0 \notin I$. Then, for any real number γ, satisfying*

$$0 < \gamma < \min\big\{1, 2^{-1}(H^{-1} - 1)\big\}, \tag{2.105}$$

there exists a finite constant $c > 0$ such that, for all integer $k \ge 2$ and $(x_1, x_2) \in \mathbb{R}^2$, one has

$$\check{\Delta}_{I,k}(x_1, x_2) \le c^k(k-1)!\,\Gamma\Big(\frac{k\gamma + 1}{2}\Big)\,|x_1 - x_2|^{k\gamma}. \tag{2.106}$$

Proof of Lemma 2.30. The proof relies on Lemma 2.29. The conditions allowing to use the latter lemma are fulfilled. More precisely, one knows from Proposition 2.20 that $\{B_H(t), t \in \mathbb{R}^N\}$ is locally nondeterministic on I with $\delta = \mathrm{diam}(I)$, and one knows from the proof of Lemma 2.27 that $\{B_H(t), t \in \mathbb{R}^N\}$ satisfies the condition (2.83); moreover, it easily follows from (1.19) that the condition (2.45) holds in the case of the FBF.

Next, combining Lemma 2.29 with (1.19) and the inclusion $I_{\preccurlyeq}^k \subseteq I^k$, one gets that

$$\check{\Delta}_{I,k}(x_1, x_2) \le c_1^k (k-1)! \, \Gamma\left(\frac{k\gamma + 1}{2}\right) |x_1 - x_2|^{k\gamma} \qquad (2.107)$$

$$\times \int_{I^k} \prod_{n=2}^{k} |t^n - t^{n-1}|^{-H(1+2\gamma)} \, d\bar{t},$$

where c_1 is a finite positive constant not depending on γ, k and (x_1, x_2). Moreover, using the condition (2.105), similarly to (2.91), one can show that

$$\int_{I^k} \prod_{n=2}^{k} |t^n - t^{n-1}|^{-H(1+2\gamma)} \, d\bar{t} \le c_2^k, \qquad (2.108)$$

where c_2 is a finite positive constant only depending on $\lambda_N(I)$, H and γ. Finally, it follows from (2.107) and (2.108) that (2.106) holds. $\square$

Theorem 2.31 (joint continuity of local time of FBF). *Let $\{B_H(t), t \in \mathbb{R}^N\}$ be the FBF of an arbitrary Hurst parameter $H \in (0,1)$. Assume that I is an arbitrary compact interval of $\mathbb{R}^N$ satisfying $0 \notin I$. Then $\{L(x, I(s)), (x, s) \in \mathbb{R} \times I\}$ the stochastic field of local times defined through (2.6) with $\{Z(t), t \in I\} = \{B_H(t), t \in I\}$ and $T = I(s)$ (see (2.22)) has a continuous modification.*

Proof of Theorem 2.31. Let γ be a fixed positive real number satisfying (2.105). Combining (2.31) with Lemmas 2.17, 2.27 and 2.30, it follows that for any arbitrary even integer $k \ge 2$, there exists a finite constant $c > 0$, only depending on N, I, H, γ and k, such that the inequality

$$\mathbb{E}\left(\left|L\big(x_1, I(s^1)\big) - L\big(x_2, I(s^2)\big)\right|^k\right) \le c\left(\left|s^1 - s^2\right|^{k(1-H)} + |x_1 - x_2|^{k\gamma}\right),$$
$$(2.109)$$

holds, for all $(s^1, s^2) \in I^2$ and $(x_1, x_2) \in \mathbb{R}^2$. Thus, choosing k such that $k(1-H) > N+1$ and $k\gamma > N+1$, the theorem can be obtained thanks to (2.109) and Theorem A.6 (the Kolmogorov's continuity Theorem). $\square$

Chapter 3

Brownian Motion and the Faber-Schauder system

3.1 Introduction

The wavelet methods, employed in Chapter 6 of the book, for finely study-ing path behavior of Multifractional Field with Random Exponent, are, to a certain extent, reminiscent of methods which, long time ago, were used to derive sharp results on roughness of Brownian paths (see for instance the book [Kahane (1968 1st edition, 1985 2nd edition)]) through their ex-pansion in the Faber-Schauder system due to Paul Lévy (see for instance the book [Lévy (1948 1st edition, 1965 2nd edition)]). Thus, we believe it important to describe in the present chapter these pioneering methods. Also, we describe the important recursive midpoint displacement technique for simulating Brownian Motion, which in fact relies on its representation in the Faber-Schauder system.

3.2 The two variants of the Haar basis of $L^2(\mathbb{R})$

First we recall the fundamental notion of an orthonormal basis of a Hilbert space.

Definition 3.1 (orthonormal basis). Let $\mathbb{H}$ be a Hilbert space over $\mathbb{C}$ (the field of complex numbers) equipped with the inner product $\langle \cdot, \cdot \rangle$ and the associated norm $\| \cdot \|$. Let Λ be an arbitrary countable set. One says that a sequence $\{e_\lambda : \lambda \in \Lambda\}$ of elements of $\mathbb{H}$ is an orthonormal basis of this space if and only if $\{e_\lambda : \lambda \in \Lambda\}$ satisfies the following two properties.

(i) For all $\lambda', \lambda'' \in \Lambda$ one has: $\langle e_{\lambda'}, e_{\lambda''} \rangle = 1$ if $\lambda' = \lambda''$ and $\langle e_{\lambda'}, e_{\lambda''} \rangle = 0$ else.
(ii) The (finite) linear combinations of the e_λ's are dense in $\mathbb{H}$.

Remark 3.2. A necessary and sufficient condition for a Hilbert space to have an orthonormal basis is that it be separable which means that it contains a countable dense subset.

The sequence $\{e_\lambda : \lambda \in \Lambda\}$ is said to be an orthonormal basis of $\mathbb{H}$ because of the following fundamental theorem.

Theorem 3.3. *Let a be an arbitrary element of $\mathbb{H}$. It can be expanded as*

$$a = \sum_{\lambda \in \Lambda} \langle a, e_\lambda \rangle e_\lambda,$$

where the series is unconditionally convergent in the sense of the norm $\|\cdot\|$. That is, one has

$$\lim_{n \to +\infty} \left\| a - \sum_{\lambda \in \Lambda_n} \langle a, e_\lambda \rangle e_\lambda \right\| = 0,$$

where $(\Lambda_n)_{n \in \mathbb{Z}_+}$ denotes any arbitrary nondecreasing sequence of nonempty finite subsets of Λ which satisfies $\Lambda = \bigcup_{n \in \mathbb{Z}_+} \Lambda_n$.

Proof of Theorem 3.3. For all $n \in \mathbb{Z}_+$, one denotes by $\mathrm{Proj}_{\Lambda_n}(\cdot)$ the projection operator from $\mathbb{H}$ into $\mathrm{span}\{e_\lambda : \lambda \in \Lambda_n\}$, the finite-dimensional subspace of $\mathbb{H}$ spanned by the finite orthonormal sequence $\{e_\lambda\}_{\lambda \in \Lambda_n}$. Observe that, for all $a \in \mathbb{H}$, one has

$$\mathrm{Proj}_{\Lambda_n}(a) = \sum_{n \in \Lambda_n} \langle a, e_\lambda \rangle e_\lambda.$$

Therefore, in order to derive the theorem, it is enough to prove that

$$\rho_n := \left\| a - \mathrm{Proj}_{\Lambda_n}(a) \right\| \xrightarrow[n \to +\infty]{} 0. \qquad (3.1)$$

The classical equality

$$\rho_n = \min \left\{ \|a - b\| : b \in \mathrm{span}\{e_\lambda : \lambda \in \Lambda_n\} \right\}, \quad \text{for all } n \in \mathbb{Z}_+, \qquad (3.2)$$

and the inclusion

$$\Lambda_n \subseteq \Lambda_{n+1}, \quad \text{for every } n \in \mathbb{Z}_+, \qquad (3.3)$$

imply that the sequence of the nonnegative real numbers $(\rho_n)_{n \in \mathbb{Z}_+}$ is nonincreasing. Thus, for showing (3.1), it is enough to prove that, for any arbitrarily small fixed positive real number ϵ, there exists $n_0(\epsilon) \in \mathbb{Z}_+$ such that $\rho_{n_0(\epsilon)} \le \epsilon$. This can be easily obtained by using (ii) in Definition 3.1, the equality $\mathbb{Z}_+ = \bigcup_{n \in \mathbb{Z}_+} \Lambda_n$, (3.3), and (3.2). $\qquad \square$

The classical L^2 functional spaces are the most common Hilbert spaces. A very important one of them is $L^2(\mathbb{T})$ (see Definition A.29 and Remark A.31 (ii)), where $\mathbb{T}$ is the 1-torus i.e. the unit circle. Let us point out that the first and most famous example of an orthonormal basis is the trigonometric system $\{(2\pi)^{-1/2}e^{il\xi} : l \in \mathbb{Z}\}$ for the space $L^2(\mathbb{T})$. It was initially introduced in 1807 by the celebrated mathematician Joseph Fourier. Despite its great utility the trigonometric system suffers from some drawbacks. One of them is the fail of pointwise convergence of partial sums for some continuous functions. This curious phenomenon was first revealed in [Du Bois-Reymond (1875)] which provides the first example of a continuous 2π-periodic function whose Fourier series diverges at a given point. The primary motivation for which the Haar orthonormal basis was introduced in [Haar (1910)] was to avoid this drawback of the trigonometric system: we will soon see that the series expansion in it of an arbitrary compactly supported continuous function is uniformly convergent in each point. The Haar basis was initially constructed in the frame of the space $L^2([0,1])$, also, in the literature it appears quite often in this same frame. Yet, in this book we prefer to present it in the more general frame of the space $L^2(\mathbb{R})$, since the transition from it to the orthonormal wavelet bases, presented in the next chapter, becomes more natural. We recall, in passing, that the inner product on $L^2(\mathbb{R})$ is defined as:

$$\langle f_1, f_2 \rangle_{L^2(\mathbb{R})} := \int_{\mathbb{R}} f_1(s)\overline{f_2(s)}\, ds, \quad \text{for all } f_1, f_2 \in L^2(\mathbb{R}). \tag{3.4}$$

The construction of the Haar basis relies on the following classical result: *any arbitrary compactly supported continuous function $g : \mathbb{R} \to \mathbb{C}$ can be approximated, in uniform norm, by a step function g_J which is constant on each one of the dyadic intervals of order J of the type $[2^{-J}l, 2^{-J}(l+1))$, $l \in \mathbb{Z}$, where J is an arbitrary big enough fixed integer. More precisely g_J is defined as:*

$$g_J := \sum_{l \in \mathbb{Z}} \overline{g}_{J,l} \mathbb{1}_{[2^{-J}l, 2^{-J}(l+1))}, \tag{3.5}$$

where, for all $(J, l) \in \mathbb{Z}^2$,

$$\overline{g}_{J,l} := 2^J \int_{l/2^J}^{(l+1)/2^J} g(s)\, ds \tag{3.6}$$

is the average value of g on the dyadic interval $[2^{-J}l, 2^{-J}(l+1))$.

In order to connect the step function g_J, in (3.5), with the Haar basis, one needs to introduce some notations.

Definition 3.4 (Haar functions). One denotes by $\mathcal{K}$ and $\mathcal{H}$ the two very simple step functions defined, for each $s \in \mathbb{R}$, as

$$\mathcal{K}(s) := \mathbb{1}_{[0,1)}(s) \tag{3.7}$$

and

$$\mathcal{H}(s) := \mathbb{1}_{[0,1/2)}(s) - \mathbb{1}_{[1/2,1)}(s). \tag{3.8}$$

Notice that in the frame of the wavelet theory and its applications $\mathcal{K}$ is called the Haar scaling function, and $\mathcal{H}$ the Haar mother wavelet.

Moreover, for every $(J, l) \in \mathbb{Z}^2$ and $(j, k) \in \mathbb{Z}^2$, one denotes by $\mathcal{K}_{J,l}$ and $\mathcal{H}_{j,k}$ the translated and dilated copies of $\mathcal{K}$ and $\mathcal{H}$, defined, for all $s \in \mathbb{R}$, as

$$\mathcal{K}_{J,l}(s) := 2^{J/2}\mathcal{K}(2^J s - l) = 2^{J/2}\mathbb{1}_{[2^{-J}l, 2^{-J}(l+1))}(s) \tag{3.9}$$

and

$$\mathcal{H}_{j,k}(s) := 2^{j/2}\mathcal{H}(2^j s - k) \tag{3.10}$$

$$= 2^{j/2}\left(\mathbb{1}_{[2^{-j}k, 2^{-j}(k+1/2))}(s) - \mathbb{1}_{[2^{-j}(k+1/2), 2^{-j}(k+1))}(s)\right).$$

Remark 3.5. Easy computations allow to derive that the sequence of functions

$$\left\{\mathcal{H}_{j,k} : (j, k) \in \mathbb{Z}^2\right\}$$

satisfies the orthonormality condition (i) in Definition 3.1. Also, it is clear that, for each fixed $J \in \mathbb{Z}$, the sequence of functions

$$\left\{\mathcal{K}_{J,l} : l \in \mathbb{Z}\right\} \cup \left\{\mathcal{H}_{j,k} : (j, k) \in \mathbb{Z}^2 \text{ and } j \geq J\right\}$$

satisfies this condition. We mention that later we will see that these two sequences of functions are the two variants of the Haar basis of $L^2(\mathbb{R})$.

Notice that in view of (3.5), (3.6) and (3.9), for every $J \in \mathbb{Z}$, one has

$$g_{J+1} = \sum_{l \in \mathbb{Z}} \langle g, \mathcal{K}_{J+1,l} \rangle_{L^2(\mathbb{R})} \mathcal{K}_{J+1,l}. \tag{3.11}$$

Now, we are going to introduce the functions $\mathcal{H}_{J,k}$, $k \in \mathbb{Z}$, in (3.11). To this end, we need the following remark which can be easily derived from (3.9) and (3.10).

Remark 3.6. Let $J \in \mathbb{Z}$ be arbitrary. In view of (3.9) and (3.10), for all $k \in \mathbb{Z}$, one has

$$\mathcal{K}_{J+1,2k} = 2^{-1/2}(\mathcal{K}_{J,k} + \mathcal{H}_{J,k}) \tag{3.12}$$

and

$$\mathcal{K}_{J+1,2k+1} = 2^{-1/2}(\mathcal{K}_{J,k} - \mathcal{H}_{J,k}). \tag{3.13}$$

The following important lemma easily results from (3.11) and Remark 3.6.

Lemma 3.7. *Let $J \in \mathbb{Z}$ be arbitrary. One has*

$$g_{J+1} = g_J + \sum_{k \in \mathbb{Z}} \langle g, \mathcal{H}_{J,k} \rangle_{L^2(\mathbb{R})} \mathcal{H}_{J,k} \tag{3.14}$$

$$= \sum_{l \in \mathbb{Z}} \langle g, \mathcal{K}_{J,l} \rangle_{L^2(\mathbb{R})} \mathcal{K}_{J,l} + \sum_{k \in \mathbb{Z}} \langle g, \mathcal{H}_{J,k} \rangle_{L^2(\mathbb{R})} \mathcal{H}_{J,k} \,.$$

More generally, for all $J' \in \mathbb{Z}$ satisfying $J' > J$, one has

$$g_{J'} = g_J + \sum_{j=J}^{J'-1} \sum_{k \in \mathbb{Z}} \langle g, \mathcal{H}_{j,k} \rangle_{L^2(\mathbb{R})} \mathcal{H}_{j,k} \tag{3.15}$$

$$= \sum_{l \in \mathbb{Z}} \langle g, \mathcal{K}_{J,l} \rangle_{L^2(\mathbb{R})} \mathcal{K}_{J,l} + \sum_{j=J}^{J'-1} \sum_{k \in \mathbb{Z}} \langle g, \mathcal{H}_{j,k} \rangle_{L^2(\mathbb{R})} \mathcal{H}_{j,k} \,.$$

Lemma 3.7 is the main ingredient of the proof of the following theorem.

Theorem 3.8. *Let $g : \mathbb{R} \to \mathbb{C}$ be an arbitrary compactly supported continuous function. The following two results hold*

(i) For each fixed $J \in \mathbb{Z}$, one has

$$\lim_{J' \to +\infty} \left\| g - \sum_{l \in \mathbb{Z}} \langle g, \mathcal{K}_{J,l} \rangle_{L^2(\mathbb{R})} \mathcal{K}_{J,l} - \sum_{j=J}^{J'-1} \sum_{k \in \mathbb{Z}} \langle g, \mathcal{H}_{j,k} \rangle_{L^2(\mathbb{R})} \mathcal{H}_{j,k} \right\|_\infty = 0 \,.$$
$$\tag{3.16}$$

(ii) One has

$$\lim_{M \to +\infty} \left\| g - \sum_{j=-M}^{M} \sum_{k \in \mathbb{Z}} \langle g, \mathcal{H}_{j,k} \rangle_{L^2(\mathbb{R})} \mathcal{H}_{j,k} \right\|_\infty = 0 \,. \tag{3.17}$$

We recall that the uniform norm $\|f\|_\infty$ of an arbitrary bounded function $f : \mathbb{R} \to \mathbb{C}$ is defined as $\|f\|_\infty := \sup_{s \in \mathbb{R}} |f(s)|$; in addition, we stress that the supremum is taken on all the real numbers and not only on (Lebesgue) almost all of them.

Proof of Theorem 3.8. The assumptions on g imply that it is a uniformly continuous function on the real line. Therefore, part (i) of the theorem can be easily obtained by using (3.15) and (3.5). From now on, we focus on its part (ii). It follows from (3.15) and the triangle inequality that, for all $M \in \mathbb{N}$, one has

$$\left\| g - \sum_{j=-M}^{M} \sum_{k \in \mathbb{Z}} \langle g, \mathcal{H}_{j,k} \rangle_{L^2(\mathbb{R})} \mathcal{H}_{j,k} \right\|_\infty$$

$$\leq \left\| g - \sum_{l \in \mathbb{Z}} \langle g, \mathcal{K}_{-M,l} \rangle_{L^2(\mathbb{R})} \mathcal{K}_{-M,l} - \sum_{j=-M}^{M} \sum_{k \in \mathbb{Z}} \langle g, \mathcal{H}_{j,k} \rangle_{L^2(\mathbb{R})} \mathcal{H}_{j,k} \right\|_\infty$$

$$+ \left\| g_{-M} \right\|_\infty$$

$$= \left\| g - g_{M+1} \right\|_\infty + \left\| g_{-M} \right\|_\infty. \tag{3.18}$$

As we have already mentioned, (3.5) and the uniform continuity of g on the real line entail that

$$\lim_{M \to +\infty} \left\| g - g_{M+1} \right\|_\infty = 0. \tag{3.19}$$

Moreover, it easily follows from (3.5) and (3.6) that

$$\left\| g_{-M} \right\|_\infty \leq 2^{-M} \int_\mathbb{R} |g(s)| \, ds,$$

and consequently that

$$\lim_{M \to +\infty} \left\| g_{-M} \right\|_\infty = 0. \tag{3.20}$$

Finally, combining (3.18) with (3.19) and (3.20), one obtains (3.17). □

Lemma 3.9. *Theorem 3.8 remains valid when one replaces in it the uniform norm $\| \cdot \|_\infty$ by the $L^2(\mathbb{R})$-norm.*

Proof of Lemma 3.9. First notice that, in view of the compacity of the support of g, one knows that there is $M_0 \in \mathbb{N}$ such that

$$\mathrm{Supp}\, g \subseteq [-2^{M_0}, 2^{M_0}]. \tag{3.21}$$

On the other hand, one knows from (3.15) and (3.16) that, for all integers J and J', satisfying $J' > J$, one has

$$g(s) - g_{J'}(s) = g(s) - \sum_{l \in \mathbb{Z}} \langle g, \mathcal{K}_{J,l} \rangle_{L^2(\mathbb{R})} \mathcal{K}_{J,l}(s) - \sum_{j=J}^{J'-1} \sum_{k \in \mathbb{Z}} \langle g, \mathcal{H}_{j,k} \rangle_{L^2(\mathbb{R})} \mathcal{H}_{j,k}(s)$$

$$= \sum_{j=J'}^{+\infty} \sum_{k \in \mathbb{Z}} \langle g, \mathcal{H}_{j,k} \rangle_{L^2(\mathbb{R})} \mathcal{H}_{j,k}(s), \tag{3.22}$$

where the last numerical series is uniformly convergent in $s \in \mathbb{R}$. Thus, it follows from (3.4), (3.10), (3.21) and (3.22) that

$$\mathrm{Supp}\, (g - g_{J'}) \subseteq [-2^{M_0}, 2^{M_0}], \quad \text{when } J' \geq M_0.$$

This implies in turn that $\|g - g_{J'}\|_{L^2(\mathbb{R})} \leq 2^{(M_0+1)/2}\|g - g_{J'}\|_\infty$. Therefore, it results from (3.16) and the first equality in (3.22) that

$$\|g - g_{J'}\|_{L^2(\mathbb{R})} \tag{3.23}$$

$$= \left\|g - \sum_{l \in \mathbb{Z}} \langle g, \mathcal{K}_{J,l} \rangle_{L^2(\mathbb{R})} \mathcal{K}_{J,l} - \sum_{j=J}^{J'-1} \sum_{k \in \mathbb{Z}} \langle g, \mathcal{H}_{j,k} \rangle_{L^2(\mathbb{R})} \mathcal{H}_{j,k}\right\|_{L^2(\mathbb{R})} \xrightarrow[J' \to +\infty]{} 0.$$

One just has shown that (3.16) remains valid when the norm $\|\cdot\|_\infty$ is replaced by the norm $\|\cdot\|_{L^2(\mathbb{R})}$. Let us now prove that a similar result is true in the case of (3.17). Similarly to (3.18), it can be shown that, for all $M \in \mathbb{N}$, one has

$$\left\|g - \sum_{j=-M}^{M} \sum_{k \in \mathbb{Z}} \langle g, \mathcal{H}_{j,k} \rangle_{L^2(\mathbb{R})} \mathcal{H}_{j,k}\right\|_{L^2(\mathbb{R})} \leq \left\|g - g_{M+1}\right\|_{L^2(\mathbb{R})} + \left\|g_{-M}\right\|_{L^2(\mathbb{R})}.$$

Thus, in view of (3.23), in order to prove that

$$\left\|g - \sum_{j=-M}^{M} \sum_{k \in \mathbb{Z}} \langle g, \mathcal{H}_{j,k} \rangle_{L^2(\mathbb{R})} \mathcal{H}_{j,k}\right\|_{L^2(\mathbb{R})} \xrightarrow[M \to +\infty]{} 0,$$

it is enough to derive that

$$\left\|g_{-M}\right\|_{L^2(\mathbb{R})} \xrightarrow[M \to +\infty]{} 0. \tag{3.24}$$

Combining (3.21) with (3.5) and (3.6), one gets, for all $M \geq M_0$, that

$$g_{-M} = \overline{g}_{-M,-1} \mathbb{1}_{[-2^M,0)} + \overline{g}_{-M,0} \mathbb{1}_{[0,2^M)},$$

and, consequently, that

$$\left\|g_{-M}\right\|_{L^2(\mathbb{R})} = 2^{M/2}(|\overline{g}_{-M,-1}| + |\overline{g}_{-M,0}|) \leq 2^{-M/2} \int_{\mathbb{R}} |g(s)|\, ds.$$

This clearly implies that (3.24) is satisfied. $\qquad\square$

We conclude the present section by stating the following important theorem:

Theorem 3.10 (the two variants of the Haar basis of $L^2(\mathbb{R})$).

(a) The sequence of functions
$$\left\{\mathcal{H}_{j,k} : (j,k) \in \mathbb{Z}^2\right\}$$
is an orthonormal basis of $L^2(\mathbb{R})$.

(b) For each fixed $J \in \mathbb{Z}$, the sequence of functions
$$\left\{\mathcal{K}_{J,l} : l \in \mathbb{Z}\right\} \cup \left\{\mathcal{H}_{j,k} : (j,k) \in \mathbb{Z}^2 \text{ and } j \geq J\right\},$$
is an orthonormal basis of $L^2(\mathbb{R})$.

Proof of Theorem 3.10. In view of the fact that the continuous and compactly supported functions form a dense subspace of $L^2(\mathbb{R})$, the theorem can be easily derived from Remark 3.5, Lemma 3.9 and Definition 3.1. $\quad\square$

3.3   Path behavior of Brownian Motion

Starting from the Haar basis $\{\mathcal{H}_{j,k} : (j,k) \in \mathbb{Z}^2\}$ of $L^2(\mathbb{R})$ (see Definition 3.4 and Theorem 3.10), one introduces a first representation of Brownian Motion in Faber-Schauder system. Then, this representation is used in order to study behavior of Brownian paths. Notice that, similarly to Haar basis, there are two variants of Faber-Schauder system (see Definition 3.11). The representation of Brownian Motion to be given in the present section concerns the first one of them. In the next section, one will give in terms of the second variant of Faber-Schauder system another representation of Brownian Motion on which relies the classical midpoint displacement technique for simulating Brownian paths in a fast way.

Definition 3.11 (the two variants of the Faber-Schauder system).
One denotes by τ and θ the two triangle nonnegative functions defined, for all $x \in \mathbb{R}$, as

$$\tau(x) = \sup\left(0, 2^{-1} - |x - 2^{-1}|\right) \quad \text{and} \quad \theta(x) = 2\,\tau\left(2^{-1}(x+1)\right); \quad (3.25)$$

observe that

$$\text{Supp}\,\tau = [0,1]\,,\ \max_{x \in \mathbb{R}} \tau(x) = \tau(2^{-1}) = 2^{-1} \quad (3.26)$$

$$\text{and}\quad \text{Supp}\,\theta = [-1,1]\,,\ \max_{x \in \mathbb{R}} \theta(x) = \theta(0) = 1.$$

(a) The sequence of functions, in the variable $t \in \mathbb{R}$,

$$\left\{\tau(2^j t - k) : (j,k) \in \mathbb{Z}^2\right\}$$

is called the first variant of the Faber-Schauder system.
(b) Assume that $J \in \mathbb{Z}$ is arbitrary and fixed. The sequence of functions, in the variable $t \in \mathbb{R}$,

$$\left\{\theta(2^J t - l) : l \in \mathbb{Z}\right\} \cup \left\{\tau(2^j t - k) : (j,k) \in \mathbb{Z}^2 \text{ and } j \geq J\right\}$$

is called the second variant of the Faber-Schauder system.

One denotes by $\{B(t), t \in \mathbb{R}\}$ a Brownian Motion (BM), with continuous paths, defined on the probability space $(\Omega, \mathcal{A}, \mathbb{P})$. Let $\mathcal{B}_0(\mathbb{R})$ be the δ-ring of the Borel subsets of $\mathbb{R}$ with a finite Lebesgue measure. One denotes by $\mathbb{B}$ the unique Brownian measure on $\mathcal{B}_0(\mathbb{R})$ (see Definition 1.24) which satisfies, for any non-empty interval $[a,b)$,

$$\mathbb{B}\big([a,b)\big) = B(b) - B(a). \quad (3.27)$$

The sequence of the independent real-valued $\mathcal{N}(0,1)$ Gaussian random variables $\{\nu_{j,k} : (j,k) \in \mathbb{Z}^2\}$ is defined as:

$$\nu_{j,k} := \int_{\mathbb{R}} \mathcal{H}_{j,k}(s)\, dB(s) \tag{3.28}$$

$$= -2^{j/2}\Big(B\big(2^{-j}(k+1)\big) - 2B\big(2^{-j}(k+1/2)\big) + B\big(2^{-j}k\big)\Big),$$

where the last equality follows from (3.10) and (3.27).

Remark 3.12. One knows from Lemma A.23 in the Appendix that there exists a random variable C of finite moments of any order, and there is $\Omega_0^* \subseteq \Omega$, an event of probability 1, such that, for each $\omega \in \Omega_0^*$ and for all $(j,k) \in \mathbb{Z}^2$, one has

$$\big|\nu_{j,k}(\omega)\big| \le C(\omega)\sqrt{\log\big(2 + |j| + |k|\big)}. \tag{3.29}$$

Theorem 3.13 (first representation of BM). *The Brownian Motion $\{B(t), t \in \mathbb{R}\}$ can be represented as:*

$$B(t) = \sum_{(j,k)\in\mathbb{Z}^2} 2^{-j/2}\, \nu_{j,k}\tau(2^j t - k), \tag{3.30}$$

where the series is, with probability 1, uniformly convergent in t, on each compact interval of $\mathbb{R}$. More precisely, let $(\mathcal{D}_n)_{n\in\mathbb{Z}_+}$ be any arbitrary nondecreasing sequence of finite subsets of $\mathbb{Z}^2$ satisfying $\mathbb{Z}^2 = \bigcup_{n\in\mathbb{Z}_+} \mathcal{D}_n$. Then, there exists $\widetilde{\Omega}$, an event of probability 1 ($\widetilde{\Omega}$ a priori depends on the choice of $(\mathcal{D}_n)_{n\in\mathbb{Z}_+}$), such that, for all $\omega \in \widetilde{\Omega}$ and positive real number M, one has

$$\sup_{t\in[-M,M]}\Big|B(t,\omega) - \sum_{(j,k)\in\mathcal{D}_n} 2^{-j/2}\, \nu_{j,k}(\omega)\tau(2^j t - k)\Big| \xrightarrow[n\to+\infty]{} 0. \tag{3.31}$$

Proof of Theorem 3.13. One knows from Remark 1.31 that $\{B(t), t \in \mathbb{R}\}$ can be viewed as a primitive of the white noise (Brownien measure) $\mathbb{B}$, namely, for each $t \in \mathbb{R}$, one has

$$B(t) = \int_{\mathbb{R}} \kappa(t,s)\, d\mathbb{B}(s), \tag{3.32}$$

where the deterministic kernel function κ is defined, for all $(s,t) \in \mathbb{R}^2$, as

$$\kappa(t,s) := \mathbb{1}_{[0,t]}(s) \text{ if } t \ge 0 \quad \text{and} \quad \kappa(t,s) := -\mathbb{1}_{[t,0]}(s) \text{ else.} \tag{3.33}$$

Expanding, for each fixed $t \in \mathbb{R}$, the function $\kappa(t,\cdot) : s \mapsto \kappa(t,s)$, on the Haar basis $\{\mathcal{H}_{j,k} : (j,k) \in \mathbb{Z}^2\}$, one gets that

$$\kappa(t,\cdot) = \sum_{(j,k)\in\mathbb{Z}^2} \langle \kappa(t,\cdot), \mathcal{H}_{j,k}\rangle_{L^2(\mathbb{R})}\, \mathcal{H}_{j,k}(\cdot), \tag{3.34}$$

where the series is unconditionally convergent in the sense of the norm $\| \cdot \|_{L^2(\mathbb{R})}$. Moreover, it can easily be derived from (3.4), (3.33), (3.10) and the first equality in (3.25) that

$$\langle \kappa(t, \cdot), \mathcal{H}_{j,k} \rangle_{L^2(\mathbb{R})} = 2^{-j/2} \tau(2^j t - k). \tag{3.35}$$

Next, it follows from (3.32), (3.34), (3.35), (3.28), and the isometry property of Wiener integral (see (1.25)), that $B(t)$ can be expressed as in (3.30) where the series is unconditionally convergent in the sense of the norm $\| \cdot \|_{L^2(\Omega)}$. Finally, the equalities $\tau(-k) = 0$, for every $k \in \mathbb{Z}$, and $\mathbb{E}\big(|B(t_1) - B(t_2)|^2\big) = |t_1 - t_2|$, for all $t_1, t_2 \in \mathbb{R}$, allow one to use Theorem A.22 in the Appendix, in order to derive (3.31).　□

Let us now turn to study of behavior of Brownian paths. We will show that they satisfy the following three results.

Theorem 3.14 (modulus of continuity for BM). *One has, almost surely, for each fixed positive real number M,*

$$\sup_{t_1, t_2 \in [-M, M]} \left\{ \frac{|B(t_1) - B(t_2)|}{|t_1 - t_2|^{1/2} \sqrt{\log\big(2 + |t_1 - t_2|^{-1}\big)}} \right\} < +\infty. \tag{3.36}$$

Theorem 3.15. *For any $u \in \mathbb{R}$ and for all $\rho \in (0, 1]$, one denotes by $\mathrm{osc}_B(u, \rho) := \mathrm{osc}_B\big([u - \rho, u + \rho]\big)$ the oscillation (see Definition 1.43) of an arbitrary path of B on the compact interval $[u - \rho, u + \rho]$. Then, one has almost surely*

$$\limsup_{j \to +\infty} \left\{ 2^{j/2} \, \mathrm{osc}_B(u, 2^{-j}) \right\} \geq 2^{-5/2} \sqrt{\pi}. \tag{3.37}$$

It is worth noticing that (3.37) is valid on an event of probability 1 not depending on u.

Theorem 3.16. *Using the same notations as in Theorem 3.15, one has almost surely, for all $u \in \mathbb{R}$,*

$$\liminf_{\rho \to 0^+} \left\{ \sqrt{\rho^{-1} \log_2(\rho^{-1})} \, \mathrm{osc}_B(u, \rho) \right\} \tag{3.38}$$

$$:= \lim_{\delta \to 0^+} \left\{ \inf_{\rho \in (0, \delta]} \left\{ \sqrt{\rho^{-1} \log_2(\rho^{-1})} \, \mathrm{osc}_B(u, \rho) \right\} \right\} \geq 2^{-9/2}.$$

It is worth noticing that (3.38) is valid on an event of probability 1 not depending on u.

It can easily be derived from Theorem 3.14 and Theorem 3.15 (or Theorem 3.16) the following:

Corollary 3.17. *Let I be an arbitrary fixed compact interval of the real line and $\beta_B(I)$ the global Hölder exponent on I (see Definition 1.34) of an arbitrary path of the Brownian Motion $\{B(t), t \in \mathbb{R}\}$, then, on an event of probability 1 not depending on I, one has $\beta_B(I) = 1/2$. Let τ be an arbitrary fixed real number and $\widetilde{\alpha}_B(\tau)$ and $\alpha_B(\tau)$ the local and pointwise Hölder exponents at τ (see Definitions 1.39 and 1.45 and the inequality (1.59)) of an arbitrary path of $\{B(t), t \in \mathbb{R}\}$, then, on an event of probability 1 not depending on τ, one has $\widetilde{\alpha}_B(\tau) = \alpha_B(\tau) = 1/2$.*

Proof of Theorem 3.14. First, notice that the continuity of paths of B and the fact that the ratio in (3.36) remains unchanged when t_1 and t_2 are interchanged allows one to assume, without any restriction, that they satisfy $0 < t_2 - t_1 < 1$. Let Ω_0^* and $\widetilde{\Omega}$ be the events of probability 1 introduced in Remark 3.12 and Theorem 3.13, one denotes by Ω_1^* the event of probability 1 defined as $\Omega_1^* := \Omega_0^* \cap \widetilde{\Omega}$. For every $j \in \mathbb{Z}$ and $t \in \mathbb{R}$ the integer part of the real number $2^j t$ is denoted by $\check{k}_j(t)$. In view of Theorem 3.13 and of the first equality in (3.26), on the event Ω_1^*, one has

$$\left|B(t_1) - B(t_2)\right| \leq \sum_{j=-\infty}^{+\infty} 2^{-j/2} \left|\nu_{j,\check{k}_j(t_1)} \tau(2^j t_1 - \check{k}_j(t_1)) - \nu_{j,\check{k}_j(t_2)} \tau(2^j t_2 - \check{k}_j(t_2))\right|.$$
(3.39)

In order to prove that (3.36) is satisfied, one splits the sum in (3.39) into three parts: $j \geq j_0 + 1$, $0 \leq j \leq j_0$ and $j \leq -1$, where j_0 is the unique nonnegative integer such that

$$2^{-j_0-1} \leq t_2 - t_1 < 2^{-j_0}.$$
(3.40)

Let us first study the case where $j \geq j_0 + 1$. One can derive from the second equality in (3.26), from (3.29) and from the inequality

$$2^j \geq j + 1, \quad \text{for each } j \in \mathbb{Z}_+,$$
(3.41)

that one has, for all $t \in [-M, M]$,

$$\sum_{j=j_0+1}^{+\infty} 2^{-j/2} |\nu_{j,\check{k}_j(t)}| \tau(2^j t - \check{k}_j(t)) \leq C_1 \sum_{j=j_0+1}^{+\infty} 2^{-j/2} \sqrt{\log\left(2 + j + \check{k}_j(t)\right)}$$

$$\leq C_1 \sum_{j=j_0+1}^{+\infty} 2^{-j/2} \sqrt{\log\left(2 + j + (M+1)2^j\right)} \leq C_2 \sum_{j=j_0+1}^{+\infty} 2^{-j/2} \sqrt{1+j},$$
(3.42)

where C_1 is the random variable C in (3.29) and $C_2 := C_1\sqrt{\log(M+3)}$. Moreover, easy calculations allow to show that

$$\sum_{j=j_0+1}^{+\infty} 2^{-j/2}\sqrt{1+j} = \sum_{n=0}^{+\infty} 2^{-(n+j_0+1)/2}\sqrt{2+j_0+n}$$

$$\leq 2^{-(j_0+1)/2}\sqrt{2+j_0}\sum_{n=0}^{+\infty} 2^{-n/2}\sqrt{1+\frac{n}{2+j_0}} \leq c_3 2^{-(j_0+1)/2}\sqrt{2+j_0}$$

$$\leq c_4(t_2-t_1)^{1/2}\sqrt{\log\left(2+(t_2-t_1)^{-1}\right)}, \tag{3.43}$$

where $c_3 := \sum_{n=0}^{+\infty} 2^{-n/2}\sqrt{1+n} < +\infty$ and $c_4 := c_3\sqrt{3/(\log 2)}$. Notice that the last inequality in (3.43) results from (3.40). Next, setting $C_5 := 2c_4 C_2$ and combining (3.42) and (3.43) with the triangle inequality, one gets that

$$\sum_{j=j_0+1}^{+\infty} 2^{-j/2}\left|\nu_{j,\check{k}_j(t_1)}\tau(2^j t_1 - \check{k}_j(t_1)) - \nu_{j,\check{k}_j(t_2)}\tau(2^j t_2 - \check{k}_j(t_2))\right|$$

$$\leq C_5(t_2-t_1)^{1/2}\sqrt{\log\left(2+(t_2-t_1)^{-1}\right)}. \tag{3.44}$$

Let us now study the case where $j \leq j_0$. Notice that, in view of (3.40), for every integer $j \leq j_0$, one has

$$\check{k}_j(t_2) = \check{k}_j(t_1) \quad \text{or} \quad \check{k}_j(t_2) = \check{k}_j(t_1) + 1. \tag{3.45}$$

Using the triangle inequality, the first equality in (3.25), (3.45) and (3.29) one obtains, for any integer $j \leq j_0$, that

$$\left|\nu_{j,\check{k}_j(t_1)}\tau(2^j t_1 - \check{k}_j(t_1)) - \nu_{j,\check{k}_j(t_2)}\tau(2^j t_2 - \check{k}_j(t_2))\right|$$

$$\leq \left|\nu_{j,\check{k}_j(t_1)}\right|\left|\tau(2^j t_1 - \check{k}_j(t_1)) - \tau(2^j t_2 - \check{k}_j(t_2))\right|$$

$$+\left|\nu_{j,\check{k}_j(t_1)} - \nu_{j,\check{k}_j(t_2)}\right|\tau(2^j t_2 - \check{k}_j(t_2))$$

$$\leq 2C_1\sqrt{\log\left(2+|j|+(M+1)2^j\right)}\left(2^j(t_2-t_1) + \left(2^j t_2 - \max\{2^j t_1, \check{k}_j(t_2)\}\right)\right)$$

$$\leq C_6(t_2-t_1)2^j\sqrt{\log\left(2+|j|+(M+1)2^j\right)}, \tag{3.46}$$

where $C_6 := 4C_1$. Next, setting $C_7 := C_6\sqrt{\log(M+3)}$, $C_8 := (2+\sqrt{2})C_7$, and $C_9 := C_8\sqrt{2/(\log 2)}$, it follows from (3.46), (3.41) and (3.40) that

$$\sum_{j=0}^{j_0} 2^{-j/2}\left|\nu_{j,\check{k}_j(t_1)}\tau(2^j t_1 - \check{k}_j(t_1)) - \nu_{j,\check{k}_j(t_2)}\tau(2^j t_2 - \check{k}_j(t_2))\right|$$

$$\leq C_7(t_2-t_1)\sum_{j=0}^{j_0} 2^{j/2}\sqrt{1+j} \leq C_8(t_2-t_1)2^{j_0/2}\sqrt{1+j_0}$$

$$\leq C_9(t_2-t_1)^{1/2}\sqrt{\log\left(2+(t_2-t_1)^{-1}\right)}. \tag{3.47}$$

Moreover, setting

$$C_{10} := C_6 \sum_{j=-\infty}^{-1} 2^{j/2} \sqrt{\log\left(3 + M + |j|\right)} < +\infty,$$

and using (3.46), one gets that

$$\sum_{j=-\infty}^{-1} 2^{-j/2} \left| \nu_{j,\check{k}_j(t_1)} \tau(2^j t_1 - \check{k}_j(t_1)) - \nu_{j,\check{k}_j(t_2)} \tau(2^j t_2 - \check{k}_j(t_2)) \right|$$

$$\leq C_{10}(t_2 - t_1). \tag{3.48}$$

Finally, putting together (3.39), (3.44), (3.47) and (3.48), one obtains (3.36). □

Proof of Theorem 3.15. Let $\left\{ \nu_{j,k} : (j,k) \in \mathbb{Z}^2 \right\}$ be the sequence of the independent real-valued $\mathcal{N}(0,1)$ Gaussian random variables defined through (3.28). One clearly has, for all $(j,k) \in \mathbb{Z}^2$, that

$$|\nu_{j,k}| \leq 2^{(j+2)/2} \max \left\{ \left| B\left(2^{-j}(k+1)\right) - B\left(2^{-j}(k+1/2)\right) \right|, \right.$$

$$\left. \left| B\left(2^{-j}(k+1/2)\right) - B\left(2^{-j}k\right) \right| \right\}.$$

Thus, using Definition 1.43, for $u \in \mathbb{R}$ and $j \in \mathbb{Z}$, one has

$$2^{(j+2)/2} \operatorname{osc}_B(u, 2^{-j}) \geq |\nu_{j+1,[2^{j+1}u]}|, \tag{3.49}$$

where $[2^{j+1}u]$ denotes the integer part of $2^{j+1}u$. Moreover, one knows from Lemma A.27 in the Appendix that, one has, almost surely,

$$\limsup_{j \to +\infty} \left\{ |\nu_{j+1,[2^{j+1}u]}| \right\} \geq 2^{-3/2}\sqrt{\pi}, \tag{3.50}$$

where the exceptional negligible set does not depend on u. Combining (3.49) and (3.50) one obtains (3.37). □

Proof of Theorem 3.16. First, let us show that for obtaining (3.38) it is enough to derive, almost surely, that

$$\lim_{J \in \mathbb{N},\, J \to +\infty} \left\{ \inf_{j \in \mathbb{N},\, j \geq J} \left\{ 2^{j/2} \sqrt{j}\, \operatorname{osc}_B(u, 2^{-j}) \right\} \right\} \geq 2^{-7/2}, \tag{3.51}$$

where $u \in \mathbb{R}$ is arbitrary, and the exceptional negligible event, on which (3.51) fails to be true, does not depend on u. For all $\delta \in (0, 2^{-1}]$, one sets

$$\tilde{J}(\delta) := [-\log_2(\delta)], \tag{3.52}$$

where $[\,\cdot\,]$ is the integer part function. Notice that $\tilde{J}(\delta)$ is the unique positive integer such that $2^{-\tilde{J}(\delta)-1} < \delta \leq 2^{-\tilde{J}(\delta)}$. The last inequality implies that

$$\inf_{\rho\in(0,\delta]} \left\{ \sqrt{\rho^{-1}\log_2(\rho^{-1})}\,\mathrm{osc}_B(u,\rho) \right\} \tag{3.53}$$

$$\geq \inf_{\rho\in(0,2^{-\tilde{J}(\delta)}]} \left\{ \sqrt{\rho^{-1}\log_2(\rho^{-1})}\,\mathrm{osc}_B(u,\rho) \right\}.$$

Moreover, one has that

$$\inf_{\rho\in(0,2^{-\tilde{J}(\delta)}]} \left\{ \sqrt{\rho^{-1}\log_2(\rho^{-1})}\,\mathrm{osc}_B(u,\rho) \right\} \tag{3.54}$$

$$= \inf_{j\in\mathbb{N},\,j\geq\tilde{J}(\delta)} \left\{ \inf_{\rho\in(2^{-j-1},2^{-j}]} \left\{ \sqrt{\rho^{-1}\log_2(\rho^{-1})}\,\mathrm{osc}_B(u,\rho) \right\} \right\},$$

and using the fact that $\rho \mapsto \mathrm{osc}_B(u,\rho)$ is a nondecreasing function (see Definition 1.43), one gets, for all $j \in \mathbb{N}$, that

$$\inf_{\rho\in(2^{-j-1},2^{-j}]} \left\{ \sqrt{\rho^{-1}\log_2(\rho^{-1})}\,\mathrm{osc}_B(u,\rho) \right\} \tag{3.55}$$

$$\geq 2^{-1}\, 2^{(j+1)/2} \sqrt{j+1}\,\mathrm{osc}_B\big(u,2^{-j-1}\big).$$

Putting together (3.53), (3.54) and (3.55), it follows that

$$\inf_{\rho\in(0,\delta]} \left\{ \sqrt{\rho^{-1}\log_2(\rho^{-1})}\,\mathrm{osc}_B(u,\rho) \right\} \tag{3.56}$$

$$\geq 2^{-1} \inf_{j\in\mathbb{N},\,j\geq\tilde{J}(\delta)+1} \left\{ 2^{j/2}\sqrt{j}\,\mathrm{osc}_B\big(u,2^{-j}\big) \right\}.$$

Moreover, in view of (3.52), one has that

$$\lim_{\delta\to 0^+} \left\{ \inf_{j\in\mathbb{N},\,j\geq\tilde{J}(\delta)+1} \left\{ 2^{j/2}\sqrt{j}\,\mathrm{osc}_B\big(u,2^{-j}\big) \right\} \right\} \tag{3.57}$$

$$= \lim_{J\in\mathbb{N},\,J\to+\infty} \left\{ \inf_{j\in\mathbb{N},\,j\geq J} \left\{ 2^{j/2}\sqrt{j}\,\mathrm{osc}_B(u,2^{-j}) \right\} \right\}.$$

Then, it is clear that (3.38) results from (3.51), (3.56) and (3.57).

From now on, one focuses on the proof of (3.51). One denotes by $\big\{ \varepsilon_{j,k} : (j,k) \in \mathbb{N}\times\mathbb{Z} \big\}$ the sequence of the real-valued $\mathcal{N}(0,1)$ Gaussian random variables defined by

$$\varepsilon_{j,k} := 2^{j/2}\Big(B\big(2^{-j}(k+1)\big) - B\big(2^{-j}k\big)\Big), \quad \text{for all } (j,k)\in\mathbb{N}\times\mathbb{Z}. \tag{3.58}$$

Observe that, for any fixed $j \in \mathbb{N}$, the $\varepsilon_{j,k}$'s, $k \in \mathbb{Z}$, are independent. Let ϕ be the function from $\mathbb{N}$ to $\mathbb{N}$ defined as

$$\phi(j) := j + [\log_2(j)], \quad \text{for each } j \in \mathbb{N}. \tag{3.59}$$

Let $\varepsilon_{\phi(j+3)}^{\max}(u)$ be defined by (A.16) where $N = 1$, $\tau = u$ and j is replaced by $\phi(j+3)$. Using Definition 1.43, (3.58) and (3.59), one gets, for all $j \in \mathbb{N}$, that

$$2^{5/2}\, 2^{j/2}\sqrt{j}\,\mathrm{osc}_B(u,2^{-j}) \geq 2^{\phi(j+3)/2}\,\mathrm{osc}_B(u,2^{-j}) \geq \varepsilon_{\phi(j+3)}^{\max}(u).$$

Thus, (3.51) results from Lemma A.26. $\qquad\qquad\square$

3.4 The theory behind the midpoint displacement technique for simulating Brownian paths

Definition 3.18 (ARIMA type random walk). One says that a sequence of real-valued random variables $\{S_l : l \in \mathbb{Z}\}$ defined on a probability space $(\Omega, \mathcal{A}, \mathbb{P})$ is an ARIMA type random walk when S_0 vanishes almost surely and the increments $S_{l+1} - S_l$, $l \in \mathbb{Z}$, are independent $\mathcal{N}(0,1)$ Gaussian random variables.

Remark 3.19. Let $\{\widetilde{B}(t), t \in \mathbb{R}\}$ be a Brownian Motion (BM), then, for each fixed $J \in \mathbb{Z}_+$, the sequence of the random variables $\{2^{J/2}\widetilde{B}(2^{-J}l) : l \in \mathbb{Z}\}$ is an ARIMA type random walk.

For the sake of clarity, we mention that throughout the present section a BM $\{\widetilde{B}(t), t \in \mathbb{R}\}$ is defined simply by saying that it is a centered Gaussian process with the covariance function satisfying, for all $t', t'' \in \mathbb{R}$,

$$\mathbb{E}\big(\widetilde{B}(t')\widetilde{B}(t'')\big) = 2^{-1}\big(|t'| + |t''| - |t' - t''|\big). \tag{3.60}$$

The midpoint displacement technique provides a simple iterative way for constructing and simulating such a process starting from $\{S_{0,l} : l \in \mathbb{Z}\}$ an arbitrary initial real-valued random sequence of the type ARIMA random walk and from $\{\delta_{j,k} : (j,k) \in \mathbb{Z}_+ \times \mathbb{Z}\}$ an arbitrary sequence of real-valued independent $\mathcal{N}(0,1)$ Gaussian random variables. Notice that the sequences $\{S_{0,l} : l \in \mathbb{Z}\}$ and $\{\delta_{j,k} : (j,k) \in \mathbb{Z}_+ \times \mathbb{Z}\}$ are defined on the same probability space $(\Omega, \mathcal{A}, \mathbb{P})$ and they are assumed to be independent.

More precisely, one denotes by $\{\widetilde{B}_0(t), t \in \mathbb{R}\}$ the centered Gaussian stochastic process with continuous piecewise linear paths defined as the linear interpolation between the random points $(l, S_{0,l})$, $l \in \mathbb{Z}$. It can easily be derived from (3.25) and the two last equalities in (3.26) that

$$\widetilde{B}_0(t) := \sum_{l \in \mathbb{Z}} S_{0,l}\,\theta(t - l), \quad \text{for every } t \in \mathbb{R}. \tag{3.61}$$

For all $J \in \mathbb{Z}_+$, one denotes by $\{\widetilde{B}_{J+1}(t), t \in \mathbb{R}\}$ the centered Gaussian stochastic process with continuous piecewise linear paths defined through the recurrence relation

$$\widetilde{B}_{J+1}(t) := \widetilde{B}_J(t) + \sum_{k \in \mathbb{Z}} 2^{-J/2}\delta_{J,k}\tau(2^J t - k), \quad \text{for each } t \in \mathbb{R}. \tag{3.62}$$

In other words, $\{\widetilde{B}_{J+1}(t), t \in \mathbb{R}\}$ is defined as

$$\widetilde{B}_{J+1}(t) := \widetilde{B}_0(t) + \sum_{j=0}^{J}\sum_{k \in \mathbb{Z}} 2^{-j/2}\delta_{j,k}\tau(2^j t - k), \quad \text{for each } t \in \mathbb{R}. \tag{3.63}$$

It can be shown by induction that paths of $\{\widetilde{B}_{J+1}(t), t \in \mathbb{R}\}$ are linear functions on any dyadic interval of order $J+1$, that is on any interval of the form $[2^{-J-1}l, 2^{-J-1}(l+1)]$, where $l \in \mathbb{Z}$ is arbitrary. Therefore, one can derive from (3.25) and the two last equalities in (3.26) that

$$\widetilde{B}_{J+1}(t) = \sum_{l \in \mathbb{Z}} S_{J+1,l}\, \theta(2^{J+1}t - l), \quad \text{for all } t \in \mathbb{R}. \tag{3.64}$$

Moreover, (3.62), (3.64), (3.25) and (3.26) imply that the random coefficients $S_{J+1,l}$, $l \in \mathbb{Z}$, satisfy the recurrence relation

$$S_{J+1,l} = \begin{cases} S_{J,l/2}, & \text{when } l \text{ is even,} \\[2mm] 2^{-1}\big(S_{J,[l/2]} + S_{J,[l/2]+1}\big) + 2^{-J/2-1}\,\delta_{J,l}, & \text{else,} \end{cases} \tag{3.65}$$

where $[l/2]$ is the integer part of $l/2$. The midpoint displacement technique for simulating Brownian paths is based on the crucial equalities (3.64) and (3.65) and on the following fundamental theorem.

Theorem 3.20. *When J goes to $+\infty$, for almost all fixed $\omega \in \Omega$ (that is with probability 1), the continuous piecewise linear function $\widetilde{B}_J(\cdot, \omega)$: $t \mapsto \widetilde{B}_J(t, \omega)$ is uniformly convergent, on each compact interval I of $\mathbb{R}$, to a continuous function $\widetilde{B}(\cdot, \omega)$: $t \mapsto \widetilde{B}(t, \omega)$. The rate of convergence is $2^{-J/2}\sqrt{J}$, that is one has*

$$\sup_{J \in \mathbb{Z}_+} \left\{ \left(2^{-J/2}\sqrt{J+1}\right)^{-1} \left\|\widetilde{B}_J(\cdot, \omega) - \widetilde{B}(\cdot, \omega)\right\|_{I,\infty} \right\} < +\infty, \tag{3.66}$$

where $\|\cdot\|_{I,\infty}$ denotes the uniform (semi)-norm [1] on the interval I. Moreover, the stochastic process $\{\widetilde{B}(t), t \in \mathbb{R}\}$ is a Brownian Motion.

The proof of the theorem relies on the following two results.

Lemma 3.21. *For almost all fixed $\omega \in \Omega$ and for any fixed compact interval I of $\mathbb{R}$, one has*

$$\sup_{j \in \mathbb{Z}_+} \left\{ \left(\sqrt{j+1}\right)^{-1} \left\|\sum_{k \in \mathbb{Z}} |\delta_{j,k}(\omega)| \tau(2^j \cdot - k)\right\|_{I,\infty} \right\} < +\infty. \tag{3.67}$$

---

[1] Recall that, for any real-valued bounded function f on I, one has

$$\|f\|_{I,\infty} := \sup_{t \in I} |f(t)|.$$

Proof of Lemma 3.21. First notice that, one knows from Lemma A.23 in the Appendix that one has almost surely

$$\sup_{(j,k)\in\mathbb{Z}_+\times\mathbb{Z}} \left\{ \left(\sqrt{\log(2+j+|k|)} \right)^{-1} |\delta_{j,k}| \right\} < +\infty. \tag{3.68}$$

Let a fixed integer $K_0 \geq 2$ be such that $I \subseteq [-K_0, K_0]$. Then, it follows from the definition of the semi-norm $\|\cdot\|_{I,\infty}$ and from the first two equalities in (3.26) that

$$\left\| \sum_{k\in\mathbb{Z}} |\delta_{j,k}(\omega)| \tau(2^j \cdot -k) \right\|_{I,\infty} \leq 2^{-1} \sup \left\{ |\delta_{j,k}| : k \in \mathbb{Z} \text{ and } |k| \leq 2^j K_0 \right\}. \tag{3.69}$$

Thus, combining (3.68) and (3.69) one gets (3.67). ☐

Proposition 3.22. *Almost surely, for all $J \in \mathbb{Z}$ and $t \in \mathbb{R}$, one has that*

$$\sum_{l\in\mathbb{Z}} S_{J,l}\, \theta(2^J t - l) = \sum_{l\in\mathbb{Z}} 2^{J/2}(S_{J,l+1} - S_{J,l})\langle \kappa(t,\cdot), \mathcal{K}_{J,l}\rangle_{L^2(\mathbb{R})}, \tag{3.70}$$

where the functions κ and $\mathcal{K}_{J,l}$ are defined through (3.33) and (3.9).

Proof of Proposition 3.22. First observe that in view of (3.25) and (3.7), for all $(J,l) \in \mathbb{Z}_+ \times \mathbb{Z}$ and $t \in \mathbb{R}$, one has

$$\theta(2^J t - l) = \int_{-\infty}^{2^J t - l} \left(\mathcal{K}(y+1) - \mathcal{K}(y) \right) dy.$$

Therefore, setting $s = 2^{-J}(y+l)$ and using (3.9), one gets that

$$\theta(2^J t - l) = 2^{J/2} \int_{-\infty}^{t} \left(\mathcal{K}_{J,l-1}(s) - \mathcal{K}_{J,l}(s) \right) ds. \tag{3.71}$$

From now on, one assumes that $t < 0$, the other case $t \geq 0$ can be treated in a rather similar way. Using the fact that $S_{J,0} = S_{0,0} \overset{\text{a.s.}}{=} 0$ and the equalities $\text{Supp}\, \theta(2^J \cdot -l) = \left[2^{-J}(l-1), 2^{-J}(l+1) \right]$ and $\text{Supp}\, \mathcal{K}_{J,l} = \left[2^{-J}l, 2^{-J}(l+1) \right]$, for every $l \in \mathbb{Z}$, it turns out that for showing (3.70) it is enough to prove that

$$\sum_{l=-\infty}^{-1} S_{J,l}\, \theta(2^J t - l) \overset{\text{a.s.}}{=} \sum_{l=-\infty}^{-1} 2^{J/2}(S_{J,l+1} - S_{J,l})\langle \kappa(t,\cdot), \mathcal{K}_{J,l}\rangle_{L^2(\mathbb{R})}. \tag{3.72}$$

Next, observe that, when $l < 0$, one can derive from (3.9) that

$$\int_{-\infty}^{0} \left(\mathcal{K}_{J,l-1}(s) - \mathcal{K}_{J,l}(s) \right) ds = 0.$$

Thus, using (3.71), (3.33) and (3.4) one obtains that

$$\theta(2^J t - l) = 2^{J/2}\left(\langle \kappa(t,\cdot), \mathcal{K}_{J,l-1}\rangle_{L^2(\mathbb{R})} - \langle \kappa(t,\cdot), \mathcal{K}_{J,l}\rangle_{L^2(\mathbb{R})} \right). \tag{3.73}$$

Finally, (3.72) follows from (3.73), the equality $S_{J,0} \overset{\text{a.s.}}{=} 0$ and standard calculations. ☐

Proof of Theorem 3.20. Let I be an arbitrary compact interval of $\mathbb{R}$. It can easily be derived from Lemma 3.21 that, almost surely, one has

$$
\sup_{J \in \mathbb{Z}_+} \left\{ \left(2^{-J/2} \sqrt{J+1} \right)^{-1} \sum_{j=J+1}^{+\infty} 2^{-j/2} \left\| \sum_{k \in \mathbb{Z}} |\delta_{j,k}(\omega)| \tau(2^j \cdot -k) \right\|_{I,\infty} \right\} < +\infty.
$$
(3.74)

This implies that the series of continuous functions of general term $\sum_{k \in \mathbb{Z}} 2^{-j/2} \delta_{j,k}(\omega) \tau(2^j \cdot -k)$, $j \in \mathbb{Z}_+$, is almost surely normally convergent with respect to the (semi)-norm $\| \cdot \|_{I,\infty}$. Thus, in view of (3.63), it turns out that, when J goes to $+\infty$, for almost all fixed $\omega \in \Omega$ the continuous function $\widetilde{B}_J(\cdot, \omega)$ is uniformly convergent on I to a continuous function $\widetilde{B}(\cdot, \omega)$. Also, observe that (3.66) easily follows from (3.74) and (3.63).

It remains to us to show that the stochastic process $\{\widetilde{B}(t), t \in \mathbb{R}\}$ is a Brownian Motion (BM). It results from (3.61) and (3.63) and from the hypotheses on the sequences of random variables $\{S_{0,l} : l \in \mathbb{Z}\}$ and $\{\delta_{j,k} : (j,k) \in \mathbb{Z}_+ \times \mathbb{Z}\}$ that $\{\widetilde{B}_J(t), (J,t) \in \mathbb{Z}_+ \times \mathbb{R}\}$ is a centered Gaussian process. Then, the fact that $\widetilde{B}_J(t)$ converges almost surely to $\widetilde{B}(t)$ and Lemma A.4 in the Appendix imply that $\{\widetilde{B}(t), t \in \mathbb{R}\}$ is a centered Gaussian process and that convergence of $\widetilde{B}_J(t)$ to $\widetilde{B}(t)$ also holds in quadratic mean. Thus using (3.63), (3.35), the fact that $\{\delta_{j,k} : (j,k) \in \mathbb{Z}_+ \times \mathbb{Z}\}$ is a sequence of independent $\mathcal{N}(0,1)$ Gaussian random variables, and the fact that this sequence is independent of the process $\{\widetilde{B}_0(t), t \in \mathbb{R}\}$, one gets, for all $t', t'' \in \mathbb{R}$,

$$
\mathbb{E}\big(\widetilde{B}(t')\widetilde{B}(t'')\big)
$$
(3.75)

$$
= \mathbb{E}\big(\widetilde{B}_0(t')\widetilde{B}_0(t'')\big) + \sum_{(j,k) \in \mathbb{Z}_+ \times \mathbb{Z}} \langle \kappa(t', \cdot), \mathcal{H}_{j,k} \rangle_{L^2(\mathbb{R})} \langle \kappa(t'', \cdot), \mathcal{H}_{j,k} \rangle_{L^2(\mathbb{R})}.
$$

Moreover, (3.61), Proposition 3.22 and the fact $\{S_{0,l+1} - S_{0,l} : l \in \mathbb{Z}\}$ is a sequence of independent $\mathcal{N}(0,1)$ Gaussian random variables imply that

$$
\mathbb{E}\big(\widetilde{B}_0(t')\widetilde{B}_0(t'')\big) = \sum_{l \in \mathbb{Z}} \langle \kappa(t', \cdot), \mathcal{K}_{0,l} \rangle_{L^2(\mathbb{R})} \langle \kappa(t'', \cdot), \mathcal{K}_{0,l} \rangle_{L^2(\mathbb{R})}.
$$
(3.76)

Recall that one knows from Theorem 3.10 that

$$
\{\mathcal{K}_{0,l} : l \in \mathbb{Z}\} \cup \{\mathcal{H}_{j,k} : (j,k) \in \mathbb{Z}_+ \times \mathbb{Z}\}
$$

is and orthonormal basis of $L^2(\mathbb{R})$. Therefore, one can derive from (3.75) and (3.76) that

$$
\mathbb{E}\big(\widetilde{B}(t')\widetilde{B}(t'')\big) = \langle \kappa(t', \cdot), \kappa(t'', \cdot) \rangle_{L^2(\mathbb{R})}.
$$

Finally, using (3.4), (3.33) and standard computations, one can show that

$$
\langle \kappa(t', \cdot), \kappa(t'', \cdot) \rangle_{L^2(\mathbb{R})} = 2^{-1}\big(|t'| + |t''| - |t' - t''|\big),
$$

which proves that $\{\widetilde{B}(t), t \in \mathbb{R}\}$ is a BM. $\qquad \square$

Chapter 4

Orthonormal wavelet bases

4.1 Introduction

The integer $N \geq 1$ is arbitrary and fixed.

Definition 4.1 (Orthonormal wavelet basis). An orthonormal wavelet basis of the Hilbert space $L^2(\mathbb{R}^N)$ is an orthonormal basis of this space (see Definition 3.1) obtained by dilations and translations of $2^N - 1$ functions usually denoted by $\psi_1, \ldots, \psi_{2^N-1}$ and called the mother wavelets. More precisely such a basis is of the form:

$$\left\{ 2^{jN/2}\, \psi_p(2^j x - k) : p \in \{1, \ldots, 2^N - 1\}, \; j \in \mathbb{Z} \text{ and } k \in \mathbb{Z}^N \right\}.$$

Usually, one denotes by $\psi_{p,j,k}$ the wavelet function on $\mathbb{R}^N$, $x \mapsto 2^{jN/2}\, \psi_p(2^j x - k)$. Observe that in the univariate case $N = 1$, an orthonormal wavelet basis is generated by only one mother wavelet denoted ψ; then $\psi_{j,k}$ denotes the function on $\mathbb{R}$, $x \mapsto 2^{j/2}\, \psi(2^j x - k)$.

It is clear that orthonormal wavelet bases are generalizations of Haar basis (see Theorem 3.10 and Definition 3.4). It is worth noticing that they can offer at least the following two major advantages with respect to Haar basis:

(a) in contrast with discontinuous Haar functions $\mathcal{H}_{j,k}$, the wavelet functions $\psi_{l,j,k}$ can be very smooth and even infinitely differentiable;

(b) in contrast with Haar functions $\mathcal{H}_{j,k}$ which have only one vanishing moment, the functions $\psi_{p,j,k}$ can have a great number of vanishing moments and even infinitely many of them, that is one has

$$\int_{\mathbb{R}^N} x_1^{q_1} x_2^{q_2} \ldots x_N^{q_N} \psi_{p,j,k}(x_1, x_2, \ldots, x_N) \, dx_1 dx_2 \ldots d_N = 0,$$

for all $(q_1, q_2, \ldots, q_N) \in \mathbb{Z}_+^N$ such that $q_1 + q_2 + \ldots + q_N \leq Q$ where the threshold Q can be very large and even equal to $+\infty$.

The first methods which have been proposed by [Meyer (1985-1986); Stromberg (1982)] for constructing orthonormal wavelet bases rely on many tricky computations and therefore seem to be a bit miraculous. However when the notion of Multiresolution Analysis (MRA) was introduced by Mallat and Meyer in the fall of 1986 the construction of such bases became quite natural.

The notion of a Riesz basis of a Hilbert space extends that of orthonormal basis (see Definition 3.1). Section 4.2 of the present chapter deals with such bases and more particularly with those of the type $\{h(x - k) : k \in \mathbb{Z}^N\}$, where $h \in L^2(\mathbb{R}^N)$, which play a fundamental role in the setting of MRA's. Section 4.3 presents the important notion of MRA and provides classical examples of MRA's. Finally, Section 4.4 explains how this notion plays a fundamental role in the construction of orthonormal wavelet bases. A particular attention is given to orthonormal wavelet bases of Lemarié-Meyer type [Lemarié and Meyer (1986); Meyer (1990, 1992)].

Before ending this introduction, we mention that detailed presentations of wavelet theory, its applications and its connections to other areas can be found in many books. We refer for instance to [Daubechies (1992); Hernández and Weiss (1996); Jaffard *et al.* (2001); Mallat (1998); Meyer (1990, 1992); Wojtaszczyk (1997)] to list but a few. Also we mention that in the rest of this chapter we use some notions and notations which are introduced in Section A.5 of the Appendix.

4.2   Riesz bases

Definition 4.2 (Riesz basis). Let $\mathbb{H}$ be a Hilbert space over $\mathbb{C}$ (the field of complex numbers) equipped with the norm $\| \cdot \|$. Let Λ be an arbitrary countable set. One says that a sequence $\{e_\lambda : \lambda \in \Lambda\}$ of elements of $\mathbb{H}$ is a Riesz basis of this space if and only if $\{e_\lambda : \lambda \in \Lambda\}$ satisfies the following two properties.

(i) The (finite) linear combinations of the e_λ's are dense in $\mathbb{H}$, that is one has:

$$\mathbb{H} = \overline{\mathrm{span}\{e_\lambda : \lambda \in \Lambda\}}.$$

(ii) $\{e_\lambda : \lambda \in \Lambda\}$ is a Riesz sequence which means that there are two constants $0 < c < c'$ such that for each complex-valued sequence $(a_\lambda)_{\lambda \in \Lambda}$

with a finite number of non vanishing terms, one has

$$c \sum_{\lambda \in \Lambda} |a_\lambda|^2 \leq \left\| \sum_{\lambda \in \Lambda} a_\lambda e_\lambda \right\|^2 \leq c' \sum_{\lambda \in \Lambda} |a_\lambda|^2. \tag{4.1}$$

The following proposition is a natural result which easily follows from Definition 4.2.

Proposition 4.3. *Assume that $\{e_\lambda : \lambda \in \Lambda\}$ is a Riesz basis of a Hilbert space $\mathbb{H}$. One denotes by $l^2(\Lambda)$ the usual Hilbert space of the complex-valued sequences $(b_\lambda)_{\lambda \in \Lambda}$ such that $\sum_{\lambda \in \Lambda} |b_\lambda|^2 < +\infty$. Then, for any given $(b_\lambda)_{\lambda \in \Lambda}$ in $l^2(\Lambda)$, the series $\sum_{\lambda \in \Lambda} b_\lambda e_\lambda$ is unconditionally convergent in $\mathbb{H}$. More precisely, there exists a unique $z \in \mathbb{H}$ such that one has*

$$\lim_{n \to +\infty} \left\| z - \sum_{\lambda \in \Lambda_n} b_\lambda e_\lambda \right\| = 0,$$

where $(\Lambda_n)_{n \in \mathbb{Z}_+}$ denotes any arbitrary nondecreasing sequence of nonempty finite subsets of Λ which satisfies $\Lambda = \bigcup_{n \in \mathbb{Z}_+} \Lambda_n$.

On the other hand, letting $\tilde{z}$ be an arbitrary element of $\mathbb{H}$, there exists a unique sequence $(\tilde{b}_\lambda)_{\lambda \in \Lambda}$ of $l^2(\Lambda)$ such that $\tilde{z} = \sum_{\lambda \in \Lambda} \tilde{b}_\lambda e_\lambda$.

Remark 4.4. In view of Definition 4.2 and Proposition 4.3, it can easily be shown that any Riesz basis of $\mathbb{H}$ is the image of an orthonormal basis of $\mathbb{H}$ by an isomorphism of $\mathbb{H}$.

In the wavelet frame, it is important to know when a sequence in $L^2(\mathbb{R}^N)$ generated by integer translations of some function h, that is a sequence of the type $\{h(x - k) : k \in \mathbb{Z}^N\}$, is a Riesz sequence. Theorem 4.6 below provides a necessary and sufficient condition in terms of $\widehat{h}$ the Fourier transform of h. We recall that throughout this book the Fourier transform is defined by using the convention that when $f \in L^1(\mathbb{R}^N)$ its Fourier transform $\widehat{f}$ is such that

$$\widehat{f}(\xi) := \frac{1}{(2\pi)^{N/2}} \int_{\mathbb{R}^N} e^{-ix \cdot \xi} f(x) \, dx, \quad \text{for all } \xi \in \mathbb{R}^N,$$

where $x \cdot \xi$ denotes the usual inner product of the vectors x and ξ of $\mathbb{R}^N$.

The following proposition recalls some useful properties of the Fourier transform.

Proposition 4.5 (some useful properties of Fourier transform).

(i) *The Fourier transform of a function belonging to $L^1(\mathbb{R}^N)$ is a continuous function on $\mathbb{R}^N$ which vanishes at infinity.*

(ii) *The Fourier transform map induces a bijective isometry from $L^2(\mathbb{R}^N)$ into itself. Thus, the $L^2(\mathbb{R}^N)$-norm is preserved under Fourier transform, that is one has:*

$$\int_{\mathbb{R}^N} |g(x)|^2 \, dx = \int_{\mathbb{R}^N} |\widehat{g}(\xi)|^2 \, d\xi, \quad \text{for every } g \in L^2(\mathbb{R}^N). \quad (4.2)$$

Notice that this is equivalent to say that the inner product associated with the $L^2(\mathbb{R}^N)$-norm is preserved under Fourier transform, that is one has:

$$\int_{\mathbb{R}^N} f(x)\overline{g(x)} \, dx = \int_{\mathbb{R}^N} \widehat{f}(\xi)\overline{\widehat{g}(\xi)} \, d\xi, \quad \text{for all } f, g \in L^2(\mathbb{R}^N). \quad (4.3)$$

We mention that the equality (4.3) is usually called the Plancherel formula.

(iii) *Let g be an arbitrary function belonging to $L^1(\mathbb{R}^N) \cup L^2(\mathbb{R}^N)$ and let $x_0 \in \mathbb{R}^N$ be arbitrary and fixed. The Fourier transform of the translated version of g, $x \mapsto g(x - x_0)$, is given by $e^{-ix_0 \cdot \xi} \widehat{g}(\xi)$, for almost all $\xi \in \mathbb{R}^N$.*

(iv) *Let g be an arbitrary function belonging to $L^1(\mathbb{R}^N) \cup L^2(\mathbb{R}^N)$ and let δ_0 be an arbitrary and fixed positive real number. The Fourier transform of the dilated version of g, $x \mapsto g(\delta_0 x)$, is given by $\delta_0^{-N} \widehat{g}(\delta_0^{-1}\xi)$, for almost all $\xi \in \mathbb{R}^N$.*

(v) *Let g_1 and g_2 be two arbitrary functions belonging to $L^1(\mathbb{R}^N)$. Then the Fourier transform of the convolution product $g_1 * g_2$ is the product (in the usual sense) $\widehat{g_1}\widehat{g_2}$.*

Theorem 4.6. *Let $h \in L^2(\mathbb{R}^N)$ and let $0 < c \le c'$ be two constants, then the following two assertions are equivalent.*

(i) *For each complex-valued sequence $(a_k)_{k \in \mathbb{Z}^N}$ with a finite number of non vanishing terms, one has*

$$(2\pi)^N c \sum_{k \in \mathbb{Z}^N} |a_k|^2 \le \int_{\mathbb{R}^N} \left| \sum_{k \in \mathbb{Z}^N} a_k h(x - k) \right|^2 dx \le (2\pi)^N c' \sum_{k \in \mathbb{Z}^N} |a_k|^2. \quad (4.4)$$

(ii) *For almost all $\xi \in \mathbb{R}^N$, one has:*

$$c \le \sum_{k \in \mathbb{Z}^N} \left| \widehat{h}(\xi + 2\pi k) \right|^2 \le c'. \quad (4.5)$$

Observe that (i) means that $\{ h(x - k) \, : \, k \in \mathbb{Z}^N \}$ is a Riesz sequence in $L^2(\mathbb{R}^N)$.

The following result easily follows from Theorem 4.6.

Corollary 4.7. *Let* $h \in L^2(\mathbb{R}^N)$. *The following two assertions are equivalent.*

(i) $\{h(x-k) : k \in \mathbb{Z}^N\}$ *is an orthonormal sequence in* $L^2(\mathbb{R}^N)$.
(ii) *For almost all* $\xi \in \mathbb{R}^N$, *one has:*

$$\sum_{k \in \mathbb{Z}^N} \left| \widehat{h}(\xi + 2\pi k) \right|^2 = (2\pi)^{-N}. \tag{4.6}$$

Proof of Corollary 4.7. Observe that to say that $\{h(x-k) : k \in \mathbb{Z}^N\}$ is an orthonormal sequence in $L^2(\mathbb{R}^N)$ is equivalent to say that: for all sequence $(d_k)_{k \in \mathbb{Z}^N}$ with a finite number of non vanishing terms, one has:

$$\int_{\mathbb{R}^N} \left| \sum_{k \in \mathbb{Z}^N} d_k h(x-k) \right|^2 dx = \sum_{k \in \mathbb{Z}^N} |d_k|^2.$$

Thus, Corollary 4.7 is nothing else than Theorem 4.6 in the particular case where the constants c and c' in it equal to $(2\pi)^{-N}$. $\qquad\square$

In order to show that Theorem 4.6 holds one needs the following lemma.

Lemma 4.8. *Let* $h \in L^2(\mathbb{R}^N)$ *be arbitrary. Then, for each complex-valued sequence* $(a_k)_{k \in \mathbb{Z}^N}$ *with a finite number of non vanishing terms, one has*

$$\int_{\mathbb{R}^N} \left| \sum_{k \in \mathbb{Z}^N} a_k h(x-k) \right|^2 dx = \int_{\mathbb{T}^N} |\lambda_a(\xi)|^2 \Big(\sum_{l \in \mathbb{Z}^N} \left| \widehat{h}(\xi + 2\pi l) \right|^2 \Big) d\xi, \tag{4.7}$$

where $\mathbb{T}^N$ *denotes the* N-*torus, and* λ_a *the trigonometric polynomial defined as:*

$$\lambda_a(\xi) = \sum_{k \in \mathbb{Z}^N} a_k e^{-ik \cdot \xi}, \quad \text{for all } \xi \in \mathbb{R}^N. \tag{4.8}$$

Proof of Lemma 4.8. The lemma easily follows from parts (ii) and (iii) of Proposition 4.5, and from the $(2\pi\mathbb{Z})^N$-periodicity of the trigonometric polynomial λ_a. $\qquad\square$

Proof of Theorem 4.6. First observe that using (4.8) and elementary properties of the exponential functions $e^{-ik \cdot \xi}$, $k \in \mathbb{Z}^N$, one gets that

$$\int_{\mathbb{T}^N} |\lambda_a(\xi)|^2 \, d\xi = (2\pi)^N \sum_{k \in \mathbb{Z}^N} |a_k|^2. \tag{4.9}$$

One clearly has that $(ii) \Rightarrow (i)$, since (4.4) easily follows from (4.5), (4.7) and (4.9).

Let us now show that $(i) \Rightarrow (ii)$. To this end, one denotes by B the function of $L^1(\mathbb{T}^N)$ defined, for almost all $\xi \in \mathbb{R}^N$, as

$$B(\xi) := \sum_{l \in \mathbb{Z}^N} \left| \widehat{h}(\xi + 2\pi l) \right|^2. \tag{4.10}$$

Next, we suppose that $\xi_0 \in \mathbb{R}^N$ is arbitrary and fixed, and that the sequence $(a_k)_{k \in \mathbb{Z}^N}$ has been chosen in such a way that the associated trigonometric polynomial λ_a (see (4.8)) satisfies, for all $\xi \in \mathbb{R}^N$,

$$|\lambda_a(\xi)|^2 = \mathbb{K}_m(\xi_0 - \xi), \tag{4.11}$$

where $\mathbb{K}_m$ is the same trigonometric polynomial as in Proposition A.37, and the integer $m \geq 1$ is arbitrary and fixed. For this choice of the sequence $(a_k)_{k \in \mathbb{Z}^N}$, in view of (4.7), (4.10), (4.11) and of the Definition A.32 of the convolution product in $L^1(\mathbb{T}^N)$, it turns out that

$$\int_{\mathbb{R}^N} \left| \sum_{k \in \mathbb{Z}^N} a_k h(x-k) \right|^2 dx = (\mathbb{K}_m * B)(\xi_0). \tag{4.12}$$

Next using (4.4), (4.9), (4.11), (A.47) and (4.12), one gets that, for all integer $m \geq 1$ and $\xi_0 \in \mathbb{R}^N$,

$$c \leq (\mathbb{K}_m * B)(\xi_0) \leq c', \tag{4.13}$$

where c and c' are the same constants as in (4.4). Moreover, one knows from Theorem A.34 that the sequence $(\mathbb{K}_m * B)_{m \geq 1}$ converges to B in $L^1(\mathbb{T}^N)$. As a consequence, there exists a subsequence $q \mapsto m_q$

$$(\mathbb{K}_{m_q} * B)(\xi_0) \xrightarrow[q \to +\infty]{} B(\xi_0), \quad \text{for almost all } \xi_0 \in \mathbb{R}^N. \tag{4.14}$$

Finally, combining (4.10) with (4.13) and (4.14), one obtains (4.5).   □

The following proposition provides in the Fourier domain a characterization of an arbitrary closed subspace of $L^2(\mathbb{R}^N)$ spanned by an arbitrary Riesz sequence of the type $\{ h(x-k) : k \in \mathbb{Z}^N \}$.

Proposition 4.9. *Assume that S is an arbitrary closed subspace of $L^2(\mathbb{R}^N)$ and that $\{ h(x-k) : k \in \mathbb{Z}^N \}$ is a Riesz basis of S. Then the following two results hold.*

(i) Let f be any arbitrary function belonging to S, then $\widehat{f}$ the Fourier transform of f can be uniquely expressed as:

$$\widehat{f}(\xi) = \lambda_f(\xi)\widehat{h}(\xi), \quad \text{for almost all } \xi \in \mathbb{R}^N, \tag{4.15}$$

where $\lambda_f \in L^2(\mathbb{T}^N)$ (see Definition A.29 and part (ii) of Remark A.31).

(ii) Let $\mu \in L^2(\mathbb{T}^N)$ be arbitrary and let Φ be the Borel function defined as:

$$\Phi(\xi) := \mu(\xi)\widehat{h}(\xi), \quad \text{for almost all } \xi \in \mathbb{R}^N. \tag{4.16}$$

Then Φ belongs to $L^2(\mathbb{R}^N)$, moreover denoting by F its inverse Fourier transform, i.e. $\widehat{F} = \Phi$, one has

$$F(x) = \sum_{k \in \mathbb{Z}^N} c_{-k}(\mu)h(x - k), \tag{4.17}$$

where the series is unconditionally convergent in $L^2(\mathbb{R}^N)$, and $\big\{c_k(\mu) : k \in \mathbb{Z}^N\big\}$ is the sequence of the Fourier coefficient of μ (see part (ii) of Remark A.31). Observe that (4.17) implies that $F \in S$.

Remark 4.10. We use the same notations as in Proposition 4.9. One can easily derive from it and from part *(iii)* of Proposition 4.5 what follows: when a function f belongs to S, then all its integer translations, that is all of the functions $x \mapsto f(x - l)$, $l \in \mathbb{Z}^N$, belong to S as well.

Proof of Proposition 4.9. Let us first derive part *(i)* of the proposition. In view of the fact that $\{h(x - k) : k \in \mathbb{Z}^N\}$ is a Riesz basis of S, one knows from Proposition 4.3 that, any function $f \in S$, can be written in a unique way as:

$$f(x) = \sum_{k \in \mathbb{Z}^N} b_k h(x - k), \tag{4.18}$$

where the series converges unconditionally in $L^2(\mathbb{R}^N)$, and where the sequence $(b_k)_{k \in \mathbb{Z}^N}$ belongs to $l^2(\mathbb{Z}^N)$. Then it follows from (4.18) and from parts *(ii)* and *(iii)* of Proposition 4.5 that

$$\widehat{f}(\xi) = \sum_{k \in \mathbb{Z}^N} b_k e^{-ik \cdot \xi} \widehat{h}(\xi), \tag{4.19}$$

where the series converges unconditionally in $L^2(\mathbb{R}^N)$. Let us now show that

$$\sum_{k \in \mathbb{Z}^N} b_k e^{-ik \cdot \xi} \widehat{h}(\xi) = \lambda_f(\xi)\widehat{h}(\xi), \quad \text{for almost all } \xi, \tag{4.20}$$

where the function λ_f is defined as:

$$\lambda_f(\xi) := \sum_{k \in \mathbb{Z}^N} b_k e^{-ik \cdot \xi}.$$

Observe that, one knows from part *(ii)* of Remark A.31 that the latter series is unconditionally convergent in $L^2(\mathbb{T}^N)$, and consequently that λ_f belongs to this space. In order to derive (4.20), it is sufficient to prove that

$$\lim_{n \to +\infty} \int_{\mathbb{R}^N} \Big| \lambda_f(\xi) - \sum_{|k| \le n} b_k e^{-ik \cdot \xi} \Big|^2 |\widehat{h}(\xi)|^2 \, d\xi = 0, \tag{4.21}$$

where $|k|$ denotes the Euclidian norm of k. Using the $(2\pi\mathbb{Z})^N$-periodicity of the function

$$\xi \mapsto \left| \lambda_f(\xi) - \sum_{|k| \leq n} b_k e^{-ik\cdot\xi} \right|^2$$

and the second inequality in (4.5), one obtains that

$$\int_{\mathbb{R}^N} \left| \lambda_f(\xi) - \sum_{|kl \leq n} b_k e^{-ik\cdot\xi} \right|^2 |\widehat{h}(\xi)|^2 \, d\xi$$

$$= \int_{\mathbb{T}^N} \left| \lambda_f(\xi) - \sum_{|k| \leq n} b_k e^{-ik\cdot\xi} \right|^2 \Big(\sum_{k \in \mathbb{Z}^N} |\widehat{h}(\xi + 2\pi k)|^2 \Big) \, d\xi$$

$$\leq c' \int_{\mathbb{T}^N} \left| \lambda_f(\xi) - \sum_{|k| \leq n} b_k e^{-ik\cdot\xi} \right|^2 \, d\xi$$

and the latter inequality clearly implies that (4.21) holds.

Let us now derive part (ii) of the proposition. Using (4.16), the $(2\pi\mathbb{Z})^N$-periodicity of the function μ, the second inequality in (4.5), and the fact that μ belongs to $L^2(\mathbb{T}^N)$, one gets that

$$\int_{\mathbb{R}^N} |\Phi(\xi)|^2 \, d\xi = \int_{\mathbb{R}^N} |\mu(\xi)|^2 |\widehat{h}(\xi)|^2 \, d\xi = \int_{\mathbb{T}^N} |\mu(\xi)|^2 \Big(\sum_{k \in \mathbb{Z}^N} |\widehat{h}(\xi + 2\pi k)|^2 \Big) \, d\xi$$

$$\leq c' \int_{\mathbb{T}^N} |\mu(\xi)|^2 \, d\xi < +\infty,$$

which shows that Φ belongs to $L^2(\mathbb{R}^N)$. In order to prove (4.17) it is enough to prove that

$$\lim_{n \to +\infty} \int_{\mathbb{R}^N} \left| F(x) - \sum_{|k| \leq n} c_{-k}(\mu) h(x - k) \right|^2 \, d\xi = 0. \qquad (4.22)$$

Using the equality $\Phi = \widehat{F}$, (4.16), and parts (ii) and (iii) of Proposition 4.5, it turns out that (4.22) is equivalent to

$$\lim_{n \to +\infty} \int_{\mathbb{R}^N} \left| \mu(\xi) - \sum_{|k| \leq n} c_{-k}(\mu) e^{-ik\cdot\xi} \right|^2 |\widehat{h}(\xi)|^2 \, d\xi = 0. \qquad (4.23)$$

Finally, (4.23) can be obtained in the same way as (4.21).   □

The following proposition provides a very important way for orthonormalising a Riesz basis of a closed subspace of $L^2(\mathbb{R}^N)$ of the type $\{h(x-k) : k \in \mathbb{Z}^N\}$.

Proposition 4.11. *Assume that S is an arbitrary closed subspace of $L^2(\mathbb{R}^N)$ and that $\{h(x - k) : k \in \mathbb{Z}^N\}$ is a Riesz basis of S. Let φ be* *the function of $L^2(\mathbb{R}^N)$ defined through its Fourier transform $\widehat{\varphi}$ which is given by:*

$$\widehat{\varphi}(\xi) := \left((2\pi)^N \sum_{k \in \mathbb{Z}^N} \left| \widehat{h}(\xi + 2\pi k) \right|^2 \right)^{-1/2} \times \widehat{h}(\xi), \quad \textit{for almost all } \xi \in \mathbb{R}^N.$$

(4.24)

Then, the sequence $\{ \varphi(x - l) : l \in \mathbb{Z}^N \}$ is an orthonormal basis of S.

Proof of Proposition 4.11. First observe that one knows from (4.5) that the $(2\pi\mathbb{Z})^N$-periodic functions

$$\xi \mapsto \left((2\pi)^N \sum_{k \in \mathbb{Z}^N} \left| \widehat{h}(\xi + 2\pi k) \right|^2 \right)^{-1/2} \quad \text{and} \quad \xi \mapsto \left((2\pi)^N \sum_{k \in \mathbb{Z}^N} \left| \widehat{h}(\xi + 2\pi k) \right|^2 \right)^{1/2}$$

are well-defined and belong to $L^2(\mathbb{T}^N)$. Then it follows from (4.24), part (ii) of Proposition 4.9 and Remark 4.10 that all of the functions $x \mapsto \varphi(x - l)$, $l \in \mathbb{Z}^N$, belong to S. Thus, setting

$$S' := \overline{\text{span}\{ \varphi(x - l) : l \in \mathbb{Z}^N \}},$$

one has that $S' \subseteq S$. Moreover, it easily results from (4.24) that

$$\sum_{q \in \mathbb{Z}^N} \left| \widehat{\varphi}(\xi + 2\pi q) \right|^2 = (2\pi)^{-N}, \quad \text{for almost all } \xi \in \mathbb{R}^N.$$

Thus, Corollary 4.7 implies that the sequence $\{ \varphi(x - l) : l \in \mathbb{Z}^N \}$ is an orthonormal basis of S'. It remains to us to show the inclusion:

$$S := \overline{\text{span}\{ h(x - k) : k \in \mathbb{Z}^N \}} \subseteq S'. \tag{4.25}$$

Noticing that (4.24) entails that

$$\widehat{h}(\xi) = \left(\sum_{k \in \mathbb{Z}^N} \left| \widehat{h}(\xi + 2\pi k) \right|^2 \right)^{1/2} \times \widehat{\varphi}(\xi), \quad \text{for almost all } \xi \in \mathbb{R}^N,$$

and using again part (ii) of Proposition 4.9 and Remark 4.10, one can prove that all of the functions $x \mapsto h(x - k)$, $k \in \mathbb{Z}^N$, belong to S'. Therefore, the inclusion in (4.25) holds. $\qquad\square$

4.3 Multiresolution Analyses (MRA's)

Definition 4.12 (Multiresolution Analysis (MRA)).
A MRA of $L^2(\mathbb{R}^N)$ is a sequence $(V_j)_{j \in \mathbb{Z}}$ of closed subspaces of $L^2(\mathbb{R}^N)$ satisfying the following five properties:

(a) The sequence $(V_j)_{j \in \mathbb{Z}}$ is increasing in the sense of the inclusion, that is, for every $j \in \mathbb{Z}$, one has $V_j \subset V_{j+1}$.

(b) The intersection $\bigcap_{j\in\mathbb{Z}} V_j$ reduces to $\{0\}$.
(c) The union $\bigcup_{j\in\mathbb{Z}} V_j$ forms a dense subspace in $L^2(\mathbb{R}^N)$, in other words its closure satisfies $\overline{\bigcup_{j\in\mathbb{Z}} V_j} = L^2(\mathbb{R}^N)$.
(d) For every $j \in \mathbb{Z}$, the function $x \mapsto f(x)$ belongs to V_j if and only if its dilated version $x \mapsto f(2x)$ belongs to V_{j+1}.
(e) There exists a function $h \in V_0$ such that the sequence of its integer translations $\{h(x - k) : k \in \mathbb{Z}^N\}$ is a Riesz basis of V_0. Notice that, thanks to Proposition 4.11, one can derive from this Riesz basis an orthonormal basis $\{\varphi(x - l) : l \in \mathbb{Z}^N\}$ of V_0. The function φ is called the scaling function since it satisfies the two scales equation (4.27).

Remark 4.13.

(i) There are many sequences of subspaces of $L^2(\mathbb{R}^N)$ which satisfies properties (a), (b) and (c); whereas properties (d) and (e) are specific to the the concept of MRA.
(ii) Property (d) implies that, for any arbitrary fixed $j \in \mathbb{Z}$, the space V_j is the dilated version of the space V_0 given by:
$$V_j = \{f(2^j x) : f(x) \in V_0\}. \tag{4.26}$$
Therefore, the space V_0 is called the reference space.
(iii) Property (e) entails that, for each arbitrary fixed $j \in \mathbb{Z}$, the sequences $\{h(2^j x - k) : k \in \mathbb{Z}^N\}$ and $\{2^{jN/2}\varphi(2^j x - l) : l \in \mathbb{Z}^N\}$ are respectively a Riesz basis and an orthonormal basis of the space V_j. Thus, using the inclusion $V_0 \subset V_1$, one gets that
$$\varphi(x) = \sum_{l\in\mathbb{Z}^N} d_l 2^N \varphi(2x - l), \tag{4.27}$$
where $(d_l)_{l\in\mathbb{Z}^N}$ is the sequence belonging to $l^2(\mathbb{Z}^N)$ defined as:
$$d_l := \int_{\mathbb{R}^N} \varphi(x)\overline{\varphi(2x - l)}\,dx, \quad \text{for all } l \in \mathbb{Z}^N. \tag{4.28}$$
Observe that the series in (4.27) is unconditionally convergent in $L^2(\mathbb{R}^N)$.
(iv) The equality (4.27) can be expressed in the Fourier domain as:
$$\widehat{\varphi}(2\xi) = m_0(\xi)\widehat{\varphi}(\xi), \quad \text{for almost all } \xi \in \mathbb{R}^N, \tag{4.29}$$
where m_0 is the $(2\pi\mathbb{Z})^N$-periodic function of $L^2(\mathbb{T}^N)$ defined as:
$$m_0(\xi) := \sum_{l\in\mathbb{Z}^N} d_l e^{-il\cdot\xi}. \tag{4.30}$$
Of course, the convergence of the series in (4.30) has to be understood in the sense of the $L^2(\mathbb{T}^N)$ norm.

Let us now present the classical tensor product method which allows to construct, for every integer $N \geq 2$, a MRA $(V_j^N)_{j\in\mathbb{Z}}$ of $L^2(\mathbb{R}^N)$ starting from $(V_j^1)_{j\in\mathbb{Z}}$ a MRA of $L^2(\mathbb{R})$.

Definition 4.14. Assume that the integers $P \geq 1$ and $Q \geq 1$ are arbitrary. Let G be an arbitrary closed subspace of $L^2(\mathbb{R}^P)$ and let T be an arbitrary closed subspace of $L^2(\mathbb{R}^Q)$. The tensor product of G and T is the closed subspace of $L^2(\mathbb{R}^P \times \mathbb{R}^Q) = L^2(\mathbb{R}^{P+Q})$ spanned by all the functions $(x,y) \mapsto f(x)g(y)$, where $f \in L^2(\mathbb{R}^P)$ and $g \in L^2(\mathbb{R}^Q)$. It is denoted by $G \otimes T$.

The following lemma easily results from Definitions 3.1 and 4.14.

Lemma 4.15. *The closed subspaces $G \subseteq L^2(\mathbb{R}^P)$ and $T \subseteq L^2(\mathbb{R}^Q)$ are the same as in Definition 4.14. Let Λ and Γ be two arbitrary countable sets. Assume that the sequence of functions $\{g_\lambda : \lambda \in \Lambda\}$ is an orthonormal basis of G and that the sequence of functions $\{t_\gamma : \gamma \in \Gamma\}$ is an orthonormal basis of T. Then the sequence of functions $\{(x,y) \mapsto g_\lambda(x)t_\gamma(y) : (\lambda,\gamma) \in \Lambda \times \Gamma\}$ forms an orthonormal basis of the tensor product $G \otimes T$.*

Proposition 4.16. *We assume that $(V_j^1)_{j\in\mathbb{Z}}$ is a MRA of $L^2(\mathbb{R})$, and that the reference space V_0^1 is equipped with a Riesz basis (resp. an orthonormal basis) generated by the integer translations of a function h^1 (resp. φ^1). For each fixed $j \in \mathbb{Z}$ and integer $N \geq 2$, the closed subspace V_j^N of $L^2(\mathbb{R}^N)$ is defined through the induction relation:*

$$V_j^N := V_j^{N-1} \otimes V_j^1. \tag{4.31}$$

Then $(V_j^N)_{j\in\mathbb{Z}}$ is a MRA of $L^2(\mathbb{R}^N)$, the integer translations of the function $(x_1,\ldots,x_N) \mapsto h^N(x_1,\ldots,x_N) := \prod_{n=1}^N h^1(x_n)$ (resp. of the function $(x_1,\ldots,x_N) \mapsto \varphi^N(x_1,\ldots,x_N) := \prod_{n=1}^N \varphi^1(x_n))$ form a Riesz basis (resp. an orthonormal basis) of V_0^N.

The following lemma will play an important role in the proof of Proposition 4.16.

Lemma 4.17. *The integers $N \geq 2$ and $j \in \mathbb{Z}$ are arbitrary. Assume that a function, from $\mathbb{R}^{N-1} \times \mathbb{R}$ into $\mathbb{C}$, $(x,y) \mapsto F(x,y)$ belongs to V_j^N defined through (4.31). Then, for almost all fixed $y \in \mathbb{R}$, the function $x \mapsto F(x,y)$ belongs to V_j^{N-1}, and, for almost all fixed $x \in \mathbb{R}^{N-1}$, the function $y \mapsto F(x,y)$ belongs to V_j^1.*

Proof of Lemma 4.17. One knows from (4.31) and from Definition 4.14 that there exists $(n_q)_{q \in \mathbb{N}}$ a sequence of positive integers and there are functions $f_{q,l} \in V_j^{N-1}$ and $g_{q,l} \in V_j^1$, $q \in \mathbb{N}$ and $l \in \{1, \ldots, n_q\}$, such that

$$\lim_{q \to +\infty} \int_{\mathbb{R}} \left(\int_{\mathbb{R}^{N-1}} \left| F(x,y) - \sum_{l=1}^{n_q} f_{q,l}(x) g_{q,l}(y) \right|^2 dx \right) dy = 0.$$

Then, it follows from Fatou Lemma that

$$\int_{\mathbb{R}} \left(\liminf_{q \to +\infty} \int_{\mathbb{R}^{N-1}} \left| F(x,y) - \sum_{l=1}^{n_q} f_{q,l}(x) g_{q,l}(y) \right|^2 dx \right) dy = 0,$$

and consequently that, for almost all fixed $y \in \mathbb{R}$,

$$\liminf_{q \to +\infty} \int_{\mathbb{R}^{N-1}} \left| F(x,y) - \sum_{l=1}^{n_q} f_{q,l}(x) g_{q,l}(y) \right|^2 dx = 0. \tag{4.32}$$

Finally, using the fact that the function

$$x \mapsto \sum_{l=1}^{n_q} f_{q,l}(x) g_{q,l}(y)$$

belongs to V_j^{N-1}, it results from (4.32) that the function $x \mapsto F(x,y)$ belongs to V_j^{N-1}. Using similar arguments, one can show that, for almost all fixed $x \in \mathbb{R}^{N-1}$, the function $y \mapsto F(x,y)$ belongs to V_j^1. □

Proof of Proposition 4.16. It is clear that $(V_j^N)_{j \in \mathbb{Z}}$ satisfies the properties (a) and (d) and that

$$V_0^N = \overline{\text{span}\{h^N(x-k) : k \in \mathbb{Z}^N\}}. \tag{4.33}$$

The fact that $\{h^N(x-k) : k \in \mathbb{Z}^N\}$ is a Riesz sequence in $L^2(\mathbb{R}^N)$, easily follows from Theorem 4.6 and from the fact that h^{N-1} and h^1 satisfy (4.5). Combining this result with (4.33), it turns out that $(V_j^N)_{j \in \mathbb{Z}}$ possesses the property (e). We mention in passing that, in view of (4.31) and of Lemma 4.15, the fact that $\{\varphi^N(x-k) : k \in \mathbb{Z}^N\}$ is an orthonormal basis of V_0^N can easily be derived from the fact that $\{\varphi^1(x-k) : k \in \mathbb{Z}\}$ is an orthonormal basis of V_0^1. On another hand, by making use of Lemma 4.17, it can easily be shown that $(V_j^N)_{j \in \mathbb{Z}}$ has the property (b). Finally, $(V_j^N)_{j \in \mathbb{Z}}$ satisfies the property (c) since $L^2(\mathbb{R}^N) = L^2(\mathbb{R}^{N-1}) \otimes L^2(\mathbb{R})$. □

Let us now state three propositions which provide three important examples of MRA's of $L^2(\mathbb{R})$. Their proofs will be given later.

Proposition 4.18 (the Haar MRA). *Assume that, for every* $j \in \mathbb{Z}$,

$$V_j^1 := \left\{ f \in L^2(\mathbb{R}) \; : \; \forall k \in \mathbb{Z}, \; f_{/_{[\frac{k}{2^j}, \frac{k+1}{2^j})}} = constant \right\}, \qquad (4.34)$$

where $f_{/_{[\frac{k}{2^j}, \frac{k+1}{2^j})}}$ *denotes the restriction of* f *to the dyadic interval* $[\frac{k}{2^j}, \frac{k+1}{2^j})$, *also assume that*

$$h^1 := \mathbb{1}_{[0,1)}. \qquad (4.35)$$

Then $(V_j^1)_{j \in \mathbb{Z}}$ *forms a MRA of* $L^2(\mathbb{R})$ *usually called the Haar MRA. Observe that in this case not only* $\left\{ h^1(x-k) \; : \; k \in \mathbb{Z} \right\}$ *is a Riesz basis of* V_0^1 *but it is also an orthonormal basis of this space, in fact one has* $h^1 = \varphi^1$.

The main advantage of the Haar MRA is its simplicity. However, in this case the function h^1 which generates V_0^1 is discontinuous and this might be a drawback. So, let us now introduce "a regularised version" of the Haar MRA.

Proposition 4.19 ("a regularised version" of the Haar MRA). *Assume that the integer* $n \geq 1$ *is arbitrary and fixed. For every* $j \in \mathbb{Z}$, *one sets*

$$V_j^{1,n} := \left\{ f \in L^2(\mathbb{R}) \cap \mathcal{C}^{n-1}(\mathbb{R}) \; : \qquad (4.36) \right.$$

$$\left. \forall k \in \mathbb{Z}, \; f_{/_{[\frac{k}{2^j}, \frac{k+1}{2^j})}} = a \; polynomial \; of \; degree \leq n \right\},$$

where $\mathcal{C}^{n-1}(\mathbb{R})$ *denotes the space of the* $n-1$ *times continuously differentiable functions from* $\mathbb{R}$ *into* $\mathbb{C}$. *Also, assume that the function* $h^{1,n}$ *is the B-spline of order* n, *which means that it is defined through the convolution product:*

$$h^{1,n} := \mathbb{1}_{[0,1)} * \mathbb{1}_{[0,1)} * \cdots * \mathbb{1}_{[0,1)}, \quad (there \; are \; n+1 \; functions \; \mathbb{1}_{[0,1)}). \quad (4.37)$$

Then $(V_j^{1,n})_{j \in \mathbb{Z}}$ *forms a MRA of* $L^2(\mathbb{R})$. *Observe that* $h^{1,n}$ *is compactly supported and such that*

$$\text{Supp} \, h^{1,n} = [0, n+1]. \qquad (4.38)$$

Also, observe that $\left\{ h^{1,n}(x-k) \; : \; k \in \mathbb{Z} \right\}$ *is not an orthonormal basis of* $V_0^{1,n}$ *but only a Riesz basis of this space.*

Remark 4.20 (on the "regularised" Haar MRA when $n = 1$**).**

(i) The function $h^{1,1} := \mathbb{1}_{[0,1)} * \mathbb{1}_{[0,1)}$ has a triangular shape. More precisely, it satisfies:

$$h^{1,1}(x) = \sup \left(0, 1 - |x-1| \right), \quad \text{for all } x \in \mathbb{R}. \qquad (4.39)$$

(ii) The function $x \mapsto h^{1,1}(x + 1)$ is nothing else than the function θ defined in (3.25) and used in the setting of the midpoint displacement technique for simulating Brownian paths (see (3.61), (3.64) and Theorem 3.20).

(iii) Denoting by f an arbitrary continuous piecewise linear function belonging to the space $V_0^{1,1}$, then the coefficients of f in the Riesz basis $\{h^{1,1}(x - k) : k \in \mathbb{Z}\}$ are simply the values of f at the integers. More precisely, the sequence $\big(f(k)\big)_{k \in \mathbb{Z}}$ belongs to $l^2(\mathbb{Z})$ and it can easily be derived from (4.39) that one has

$$f(x) = \sum_{k \in \mathbb{Z}} f(k + 1) h^{1,1}(x - k), \qquad (4.40)$$

where the series is unconditionally convergent in $L^2(\mathbb{R})$. Observe that on each compact interval of $\mathbb{R}$ this series reduces to a finite sum.

Before giving the third important example of MRA one needs to make a few recalls on the Schwartz class $\mathcal{S}(\mathbb{R}^N)$, where the integer $N \geq 1$ is arbitrary.

Definition 4.21 (the Schwartz class $\mathcal{S}(\mathbb{R}^N)$). The class $\mathcal{S}(\mathbb{R}^N)$ is the space of the infinitely differentiable functions f from $\mathbb{R}^N$ into $\mathbb{C}$ which have a rapid decrease at infinity as well as all their partial derivatives of any order. That is, for each fixed integer $n \geq 0$ and multi-index $q = (q_1, \ldots, q_N) \in \mathbb{Z}_+^N$, one has

$$\sup_{x \in \mathbb{R}^N} \left\{ \left(1 + |x|\right)^n \left|\partial^q f(x)\right| \right\} < +\infty, \qquad (4.41)$$

where the partial derivative operator ∂^q is defined as:

$$\partial^q := \frac{\partial^{q_1 + q_2 + \ldots + q_N}}{(\partial x_1)^{q_1} (\partial x_2)^{q_2} \cdots (\partial x_N)^{q_N}}, \qquad (4.42)$$

$x_1, x_2, \ldots, x_N$ being the coordinates of x. It is clear $\mathcal{S}(\mathbb{R}^N) \subset L^p(\mathbb{R}^N)$, for all $p \in [1, \infty]$.

The proof of the following proposition can be found in e.g. [Schwartz (1966); Stein and Weiss (1971)].

Proposition 4.22 (stability of $\mathcal{S}(\mathbb{R}^N)$ by Fourier transform). *An arbitrary function f belongs to $\mathcal{S}(\mathbb{R}^N)$ if and only if its Fourier transform $\widehat{f}$ belongs to $\mathcal{S}(\mathbb{R}^N)$.*

We are now in position to introduce the third important example of MRA.

Proposition 4.23 (the Meyer MRA). *Let ν be an infinitely differentiable function from $\mathbb{R}$ into $[0,1]$ which satisfies*

$$\nu(\eta) = \begin{cases} 0, & \text{for all } \eta \in]-\infty,0], \\[2mm] 1, & \text{for all } \eta \in [1,+\infty], \end{cases} \tag{4.43}$$

and

$$\nu(\eta) + \nu(1-\eta) = 1, \quad \text{for every } \eta \in \mathbb{R}. \tag{4.44}$$

One denotes by φ^1 the function of the Schwartz class $\mathcal{S}(\mathbb{R})$ whose Fourier transform $\widehat{\varphi^1}$ is given, for all $\xi \in \mathbb{R}$, by:

$$\widehat{\varphi^1}(\xi) = \begin{cases} (2\pi)^{-1/2}, & \text{if } |\xi| \le \frac{2\pi}{3}, \\[3mm] (2\pi)^{-1/2}\cos\left[\frac{\pi}{2}\nu\left(\frac{3}{2\pi}|\xi|-1\right)\right], & \text{if } \frac{2\pi}{3} \le |\xi| \le \frac{4\pi}{3}, \\[3mm] 0, & \text{else.} \end{cases} \tag{4.45}$$

Then $\left\{\varphi^1(x-k) : k \in \mathbb{Z}\right\}$ is an orthonormal sequence in $L^2(\mathbb{R})$. Moreover, assuming that $(V_j^1)_{j\in\mathbb{Z}}$ is the sequence of the closed subspaces of $L^2(\mathbb{R})$ defined as:

$$V_0^1 := \overline{\text{span}\left\{\varphi^1(x-l) : l \in \mathbb{Z}\right\}} \tag{4.46}$$

and

$$V_j^1 := \left\{f(2^j x) : f(x) \in V_0^1\right\}, \quad \text{for all } j \in \mathbb{Z}\setminus\{0\}. \tag{4.47}$$

Then $(V_j^1)_{j\in\mathbb{Z}}$ forms a MRA of $L^2(\mathbb{R})$ for which one has $h^1 = \varphi^1$.

Remark 4.24. Let c be the positive constant defined as:

$$c := \left(\int_0^1 \exp\left(-s^{-1}(1-s)^{-1}\right)ds\right)^{-1}.$$

Elementary computations allow one to show that the function ν from $\mathbb{R}$ into $[0,1]$, defined, for all $x \in \mathbb{R}$, as: $\nu(x) := 0$ when $x \le 0$, $\nu(x) := 1$ when $x \ge 1$, and

$$\nu(x) := c\int_0^x \exp\left(-s^{-1}(1-s)^{-1}\right)ds, \quad \text{when } x \in (0,1),$$

is infinitely differentiable on $\mathbb{R}$ and satisfies (4.43) and (4.44).

Remark 4.25. Assume that $\widehat{\varphi^1}$ is the compactly supported and infinitely differentiable function from $\mathbb{R}$ into $\left[0, (2\pi)^{-1/2}\right]$ defined through (4.45). One denotes by m_0^1 the 2π-periodic function [1] defined, for each $\xi \in \mathbb{R}$, as:

$$m_0^1(\xi) := (2\pi)^{1/2} \sum_{l \in \mathbb{Z}} \widehat{\varphi^1}\big(2(\xi + 2\pi l)\big). \tag{4.48}$$

Then, one has

$$\widehat{\varphi^1}(\xi) = m_0^1(2^{-1}\xi)\widehat{\varphi^1}(2^{-1}\xi), \quad \text{for all } \xi \in \mathbb{R}. \tag{4.49}$$

Observe that when $\xi \in [-\rho, \rho]$, the fixed positive real number ρ being arbitrary, then the right-hand side of (4.48) reduces to a finite sum; more precisely all the terms such that $|l| \geq \rho + 1$ vanishes. Therefore, the 2π-periodic function m_0^1 is infinitely differentiable on $\mathbb{R}$ and thus it belongs to the space $L^2(\mathbb{T})$.

Proof of Remark 4.25. Observe that (4.48) entails that

$$m_0^1(2^{-1}\xi)\widehat{\varphi^1}(2^{-1}\xi) = (2\pi)^{1/2} \sum_{l \in \mathbb{Z}} \widehat{\varphi^1}(\xi + 4\pi l)\widehat{\varphi^1}(2^{-1}\xi). \tag{4.50}$$

Also, observe that one knows from (4.45) that the left-hand side and the right-hand side of (4.49) reduce to zero when $|\xi| \geq 8\pi/3$. Thus, from now on, one assumes that $|\xi| < 8\pi/3$. Then, for all $l \in \mathbb{Z} \setminus \{0\}$, one has

$$|\xi + 4\pi l| \geq 4\pi|l| - |\xi| > 4\pi - \frac{8\pi}{3} = \frac{4\pi}{3}.$$

As a consequence, (4.50) and (4.45) imply that

$$m_0^1(2^{-1}\xi)\widehat{\varphi^1}(2^{-1}\xi) = (2\pi)^{1/2} \widehat{\varphi^1}(\xi)\widehat{\varphi^1}(2^{-1}\xi).$$

Then, using again (4.45), one gets (4.49). $\qquad\square$

Remark 4.26. For each $j \in \mathbb{Z}$, let V_j^1 be the subspace of $L^2(\mathbb{R})$ defined through (4.47) and (4.46). One denotes by $\widehat{V_j^1}$ the subspace of $L^2(\mathbb{R})$ obtained as the image of V_j^1 under the Fourier transform isometry from $L^2(\mathbb{R})$ into itself; in other words, a function G in $L^2(\mathbb{R})$ belongs to $\widehat{V_j^1}$ if and only if there exists $F \in V_j^1$ such that $G = \widehat{F}$.

The following two results hold.

---

[1] We recall that a function D from $\mathbb{R}$ into $\mathbb{C}$ is said to be 2π-periodic if one has:

$$D(\xi + 2\pi) = D(\xi), \quad \text{for almost all } \xi \in \mathbb{R}. \tag{$*$}$$

Notice that the equality $(*)$ even holds for all $\xi \in \mathbb{R}$, in the case of the function m_0 defined in (4.48).

(i) Assume that G is an arbitrary function of $L^2(\mathbb{R})$ which is supported by the interval $\left[-2^j(2\pi/3), 2^j(2\pi/3)\right]$, in other words, one has $G(\xi) = 0$, for almost all $\xi \in \mathbb{R} \setminus \left[-2^j(2\pi/3), 2^j(2\pi/3)\right]$, then G belongs to $\widehat{V_j^1}$.

(ii) Let $\widetilde{G}$ be an arbitrary function belonging to $\widehat{V_j^1}$, then $\widetilde{G}$ is supported by the interval $\left[-2^j(4\pi/3), 2^j(4\pi/3)\right]$.

Proof of Remark 4.26. First observe that it follows from (4.47) that

$$\widehat{V_j^1} = \left\{ F(2^{-j}\xi) \,:\, F(\xi) \in \widehat{V_0^1} \right\}.$$

Thus, it is enough to show that the remark holds in the case where $j = 0$.

Using (4.46), the fact that $\left\{\varphi^1(x - k) \,:\, k \in \mathbb{Z}\right\}$ is an orthonormal sequence in $L^2(\mathbb{R})$, and Proposition 4.9, one has that

$$\widehat{V_0^1} = \left\{ \theta(\xi)\widehat{\varphi^1}(\xi) \,:\, \theta \in L^2(\mathbb{T}) \right\}. \tag{4.51}$$

Then, it turns out that part (ii) of the remark is a straightforward consequence of (4.51) and of the fact that $\widehat{\varphi^1}$ is supported by the interval $[-4\pi/3, 4\pi/3]$ (see (4.45)).

Now, let G be an arbitrary function of $L^2(\mathbb{R})$ which is supported by the interval $[-2\pi/3, 2\pi/3]$, that is one has.

$$G(\xi) = 0, \quad \text{for almost all } \xi \in \mathbb{R} \setminus [-2\pi/3, 2\pi/3]. \tag{4.52}$$

In order to prove that part (i) of the remark is satisfied one has to show that G belongs to $\widehat{V_0^1}$. One denotes by θ_G the 2π-periodic function defined as:

$$\theta_G(\xi) := \sum_{p \in \mathbb{Z}} G(\xi + 2\pi p), \quad \text{for almost all } \xi \in \mathbb{R}.$$

It easily follows from (4.52) that θ_G is well-defined, and one has:

$$\theta_G(\xi) = G(\xi), \quad \text{for almost all } \xi \in [-4\pi/3, 4\pi/3], \tag{4.53}$$

which, among other things, implies that $\theta_G \in L^2(\mathbb{T})$. Finally, it results from (4.45), (4.53) and (4.52) that one has: $G(\xi) = \theta_G(\xi)\widehat{\varphi^1}(\xi)$, for almost all $\xi \in \mathbb{R}$. Then, one knows from (4.51) that G belongs to $\widehat{V_0^1}$. $\qquad\square$

Let us now turn to the proofs of the three propositions we have stated. We skip the proof of Proposition 4.18 since it is rather similar to that of Proposition 4.19.

Proof of Proposition 4.19. We assume that $n \geq 1$ is arbitrary and fixed.

Observe that, for each $j \in \mathbb{Z}$, any dyadic interval of order $j + 1$, that is of the form $[\frac{k}{2^{j+1}}, \frac{k+1}{2^{j+1}})$ where $k \in \mathbb{Z}$, is included in a unique dyadic interval of order j. Therefore $(V_j^{1,n})_{j \in \mathbb{Z}}$ has the property (a).

Let us show that $(V_j^{1,n})_{j \in \mathbb{Z}}$ satisfies the property (b). Assume that $f \in V_j^{1,n}$, for all $j \in \mathbb{Z}$. Then, there are polynomials P_j^+ and P_j^- of degree less than or equal to n such that

$$f_{/_{[0,2^{-j})}} = P_j^+ \quad \text{and} \quad f_{/_{[-2^{-j},0)}} = P_j^-, \quad \text{for all } j \in \mathbb{Z}. \tag{4.54}$$

Next, noticing that (4.54) implies that all the polynomials P_j^+, $j \in \mathbb{Z}$, (resp. P_j^-, $j \in \mathbb{Z}$) are equal to the same polynomial P^+ (resp. P^-), it turns out that

$$f_{/_{[0,+\infty)}} = P^+ \quad \text{and} \quad f_{/_{(-\infty,0)}} = P^-.$$

Thus, using the fact that f belongs to $L^2(\mathbb{R})$ and the fact that the only polynomial which belongs to $L^2([0,+\infty))$ (resp. to $L^2((-\infty,0))$) is the vanishing polynomial, one gets $f = 0$.

It easily follows from (4.36) that $(V_j^{1,n})_{j \in \mathbb{Z}}$ satisfies the property (d).

From now on, for the sake of simplicity, one restricts to the case $n = 1$, the general case where n is an arbitrary positive integer can be treated by using rather similar ideas, yet it requires much more technicality.

In order to prove that $(V_j^{1,1})_{j \in \mathbb{Z}}$ possesses the property (c), it is enough to show that for each $\varepsilon > 0$ and for all continuous and compactly supported function g from $\mathbb{R}$ into $\mathbb{C}$ with a support included in $[-p,p]$, the positive integer p being arbitrary and fixed, there exist $j_0 = j_0(\varepsilon) \in \mathbb{Z}_+$ and a compactly supported function f_{ε,j_0} belonging to $V_{j_0}^{1,1}$ and having a support included in $[-p-1, p+1]$ which satisfies

$$\sup_{x \in \mathbb{R}} |g(x) - f_{\varepsilon,j_0}(x)| \leq \varepsilon(2p+2)^{-1/2}. \tag{4.55}$$

We know from the uniform continuity of g on the real line that there exists $j_0 \in \mathbb{Z}_+$, such that, for all $(x,y) \in \mathbb{R}^2$, one has $|g(x) - g(y)| \leq \varepsilon(2p+2)^{-1/2}$ as soon as $|x - y| \leq 2^{-j_0}$. Thus, assuming that f_{ε,j_0} is the piecewise linear, continuous, and compactly supported function from $\mathbb{R}$ into $\mathbb{C}$, with a support included in $[-p-1, p+1]$, obtained by linear interpolations of the points whose coordinates are $(k2^{-j_0}, g(k2^{-j_0}))$, $k \in \mathbb{Z}$, one gets (4.55).

Finally, let us show that $(V_j^{1,1})_{j \in \mathbb{Z}}$ satisfies the property (e). One can easily derive from (4.36) and (4.39) that any function $f \in V_0^{1,1}$ can be expressed, for every $x \in \mathbb{R}$, as the series in (4.40) and that this series reduces to a finite sum on each compact interval of $\mathbb{R}$. Then, using the fact that $f \in L^2(\mathbb{R})$ and standard computations it can be shown that the (finite) linear combinations of the functions $h^{1,1}(x - k)$, $k \in \mathbb{Z}$, are dense in $V_0^{1,1}$ in the sense of the $L^2(\mathbb{R})$-norm. It remains to us to prove that $\{h^{1,1}(x - k) : k \in \mathbb{Z}\}$ is a Riesz sequence which is equivalent to prove that $|\widehat{h^{1,1}}|^2$ satisfies (4.5). In view of (4.37) and of part (v) of Propositon 4.5, $|\widehat{h^{1,1}}|^2$ can be expressed as:

$$\left|\widehat{h^{1,1}}(\xi)\right|^2 = \left(\frac{1}{2\pi}\right)^2 \frac{\sin^4(2^{-1}\xi)}{(2^{-1}\xi)^4}, \quad \text{for all } \xi \in \mathbb{R}, \tag{4.56}$$

with the classical convention that

$$\frac{\sin(0)}{0} := 1. \tag{4.57}$$

Next, observe that the function

$$\xi \mapsto D(\xi) := \sum_{k \in \mathbb{Z}} \left|\widehat{h^{1,1}}(\xi + 2\pi k)\right|^2 \tag{4.58}$$

is 2π-periodic. Therefore, one can restrict to $\xi \in [-\pi, \pi]$. It can easily be derived from (4.56) and (4.57) that the series in (4.58) is uniformly convergent on this interval, which in turn implies that the 2π-periodic function D is continuous. Therefore, it is bounded from above by a finite positive constant. It remains to us to show that D is bounded from below by a positive constant. Suppose ad absurdum that $\inf_{\xi \in [-\pi, \pi]} D(\xi) = 0$. Using the continuity of D and the fact that $[-\pi, \pi]$ is a compact interval, the last equality implies that there is $\xi_0 \in [-\pi, \pi]$ such that $D(\xi_0) = 0$. Then, using (4.58) and (4.56), one gets that

$$\frac{\sin^4(2^{-1}\xi_0)}{(2^{-1}\xi_0 + \pi k)^4} = 0, \quad \text{for all } k \in \mathbb{Z},$$

which is clearly not true (see (4.57)). □

Proof of Proposition 4.23. Let us first show that $\{\varphi^1(x - k) : k \in \mathbb{Z}\}$ is an orthonornal sequence in $L^2(\mathbb{R})$. In view of Corollary 4.7, it is enough to prove that

$$F(\xi) := \sum_{k \in \mathbb{Z}} \left|\widehat{\varphi^1}(\xi + 2\pi k)\right|^2 = (2\pi)^{-1}, \quad \text{for almost all } \xi \in \mathbb{R}. \tag{4.59}$$

Thanks to the 2π-periodicity of the function F, one can restrict to $\xi \in [0, 2\pi)$. Then, it results from (4.45) that

$$F(\xi) = \left|\widehat{\varphi^1}(\xi - 2\pi)\right|^2 + \left|\widehat{\varphi^1}(\xi)\right|^2. \tag{4.60}$$

Let us consider the following three cases:

(I) $\xi \in [0, 2\pi/3]$ i.e. $\xi - 2\pi \in [-2\pi, -4\pi/3]$,

(II) $\xi \in (2\pi/3, 4\pi/3]$ i.e. $\xi - 2\pi \in (-4\pi/3, -2\pi/3]$,

(III) $\xi \in (4\pi/3, 2\pi)$ i.e. $\xi - 2\pi \in (-2\pi/3, 0)$.

In the case (I), using (4.45) one has $\widehat{\varphi^1}(\xi) = (2\pi)^{-1/2}$ and $\widehat{\varphi^1}(\xi - 2\pi) = 0$; thus it follows from (4.60) that (4.59) holds. The case (III) is similar to the case (I) except that ξ and $\xi - 2\pi$ are interchanged. In the case (II) (4.59) holds as well, since one can derive from (4.60), (4.45) and (4.44) that

$$
\begin{aligned}
(2\pi)F(\xi) &= \cos^2\left[\frac{\pi}{2}\nu\left(\frac{3}{2\pi}(2\pi - \xi) - 1\right)\right] + \cos^2\left[\frac{\pi}{2}\nu\left(\frac{3}{2\pi}\xi - 1\right)\right] \\
&= \cos^2\left[\frac{\pi}{2}\nu\left(1 - (\frac{3}{2\pi}\xi - 1)\right)\right] + \cos^2\left[\frac{\pi}{2}\nu\left(\frac{3}{2\pi}\xi - 1\right)\right] \\
&= \cos^2\left[\frac{\pi}{2}\left(1 - \nu(\frac{3}{2\pi}\xi - 1)\right)\right] + \cos^2\left[\frac{\pi}{2}\nu\left(\frac{3}{2\pi}\xi - 1\right)\right] \\
&= \sin^2\left[\frac{\pi}{2}\nu\left(\frac{3}{2\pi}\xi - 1\right)\right] + \cos^2\left[\frac{\pi}{2}\nu\left(\frac{3}{2\pi}\xi - 1\right)\right] = 1.
\end{aligned}
$$

Next, observe that in view of (4.46) and (4.47), it is clear that $(V_j^1)_{j\in\mathbb{Z}}$ has the property (e) with $h^1 = \varphi^1$ and the property (d).

Let us now show that $(V_j^1)_{j\in\mathbb{Z}}$ satisfies the property (a). In view of (4.47), it is enough to prove that

$$V_{-1}^1 \subset V_0^1. \tag{4.61}$$

Observe that (4.47) and the fact that $\{\varphi^1(x-l) : l \in \mathbb{Z}\}$ is an orthonormal basis of V_0^1 imply that $\{2^{j/2}\,\varphi^1(2^j x - l) : l \in \mathbb{Z}\}$ is an orthonormal basis of V_j^1, for every $j \in \mathbb{Z}$. Thus, for showing (4.61), it is sufficient to prove that:

$$\varphi^1(2^{-1}x - k) \in V_0^1, \quad \text{for all } k \in \mathbb{Z}. \tag{4.62}$$

Next, observe that one knows from the equality:

$$\varphi^1(2^{-1}x - k) = \varphi^1\big(2^{-1}(x - 2k)\big), \quad \text{for all } k \in \mathbb{Z},$$

from Remark 4.10, and from the fact that $\{\varphi^1(x - l) : l \in \mathbb{Z}\}$ is an orthonormal basis of V_0^1, that (4.62) is satisfied as soon as one has:

$$\varphi^1(2^{-1}x) \in V_0^1. \tag{4.63}$$

Next noticing that the Fourier transform of the function $\varphi^1(2^{-1}x)$ is the function $2\,\widehat{\varphi^1}(2\xi)$ (see part (iv) of Proposition 4.5), and using Remark 4.25, Proposition 4.9, and the fact that $\{\varphi^1(x - l) : l \in \mathbb{Z}\}$ is an orthonormal basis of V_0^1, one gets (4.63).

Finally, a straightforward consequence of part (ii) of Proposition 4.5 and of Remark 4.26 is that $(V_j^1)_{j\in\mathbb{Z}}$ satisfies the properties (b) and (c). $\square$

4.4 Construction of wavelet bases through MRA's

In order to state the two main results of the present section, first one needs to fix several notations. The integer $N \geq 1$ is arbitrary and fixed.

- When $N \geq 2$, $(V_j^N)_{j \in \mathbb{Z}}$ denotes a MRA of $L^2(\mathbb{R}^N)$ obtained from an arbitrary MRA $(V_j^1)_{j \in \mathbb{Z}}$ of $L^2(\mathbb{R})$ by making use of the tensor product method presented in Proposition 4.16. The reference space V_0^N is equipped with the orthonormal basis $\{\varphi^N(x - l) : l \in \mathbb{Z}^N\}$.
- $(W_j^N)_{j \in \mathbb{Z}}$ is the sequence of the closed and pairwise orthogonal subspaces [2] of $L^2(\mathbb{R}^N)$ such that, for each $j \in \mathbb{Z}$, W_j^N is the orthogonal complement of V_j^N in V_{j+1}^N. In other words, one has

$$V_{j+1}^N = V_j^N \overset{\perp}{\oplus} W_j^N, \quad \text{for every } j \in \mathbb{Z}. \tag{4.64}$$

Notice that one can easily derive from (4.26) and (4.64) that

$$W_j^N = \{f(2^j x) : f \in W_0^N\}, \quad \text{for all } j \in \mathbb{Z}. \tag{4.65}$$

Also notice that, when $N \geq 2$, using (4.31), (4.64) and the bilinearity property of tensor product, one has that

$$V_{j+1}^N = \left(V_j^{N-1} \overset{\perp}{\oplus} W_j^{N-1}\right) \otimes \left(V_j^1 \overset{\perp}{\oplus} W_j^1\right)$$

$$= V_j^N \overset{\perp}{\oplus} \left(\left(V_j^{N-1} \otimes W_j^1\right) \overset{\perp}{\oplus} \left(W_j^{N-1} \otimes V_j^1\right) \overset{\perp}{\oplus} \left(W_j^{N-1} \otimes W_j^1\right)\right),$$

which implies that

$$W_j^N = \left(V_j^{N-1} \otimes W_j^1\right) \overset{\perp}{\oplus} \left(W_j^{N-1} \otimes V_j^1\right) \overset{\perp}{\oplus} \left(W_j^{N-1} \otimes W_j^1\right). \tag{4.66}$$

Thus, letting $\mathbb{L}_N$ be the finite set of cardinality $2^N - 1$ defined as:

$$\mathbb{L}_N := \{0, 1\}^N \setminus \{(0, \ldots, 0)\}, \tag{4.67}$$

it can be shown by induction on $N \geq 2$ and by using (4.66) that

$$W_j^N = \overset{\perp}{\underset{\varepsilon \in \mathbb{L}_N}{\bigoplus}} W_j^{(\varepsilon)}, \tag{4.68}$$

---

[2] Let $\mathbb{H}$ be an arbitrary Hilbert space equipped with the inner product $\langle \cdot, \cdot \rangle$. One says that two subspaces W' and W'' of $\mathbb{H}$ are orthogonal if one has:

$$\langle w', w'' \rangle = 0, \quad \text{for all } w' \in W' \text{ and } w'' \in W''.$$

Moreover, one says that W'' is the orthogonal complement of W' in a subspace W of $\mathbb{H}$ (or equivalently that W' is the orthogonal complement of W'' in W), when each $w \in W$ can be expressed in a unique way as $w = w' + w''$. The notation

$$W = W' \overset{\perp}{\oplus} W''$$

means that W'' is the orthogonal complement of W' in W.

where, for any $\varepsilon = (\varepsilon_1, \ldots, \varepsilon_N) \in \mathbb{L}_N$,

$$W_j^{(\varepsilon)} := \bigotimes_{n=1}^{N} W_j^{\varepsilon_n}, \tag{4.69}$$

with the convention that

$$W_j^0 := V_j^1. \tag{4.70}$$

Also notice that, in view of Proposition 4.33 (given below), it follows from (4.64) and from the properties (a), (b) and (c) of the MRA $(V_j^N)_{j \in \mathbb{Z}}$ that the space $L^2(\mathbb{R}^N)$ can be expressed as the direct sums:

$$L^2(\mathbb{R}^N) = \overset{\perp}{\bigoplus_{j \in \mathbb{Z}}} W_j^N \tag{4.71}$$

and

$$L^2(\mathbb{R}^N) = V_J^N \overset{\perp}{\oplus} \left(\overset{\perp}{\bigoplus_{j \geq J}} W_j^N \right), \quad \text{for each fixed } J \in \mathbb{Z}. \tag{4.72}$$

Let us mention that some recalls on the notion of infinite direct sum of closed and pairwise orthogonal subspaces of a Hilbert space are made in Proposition 4.32 (given below).

• According to part (iv) of Remark 4.13, there exists a unique 2π-periodic function belonging to $L^2(\mathbb{T})$, denoted by m_0^1 in the present section, such that $\widehat{\varphi^1}$, the Fourier transform of the scaling function φ^1, satisfies:

$$\widehat{\varphi^1}(2\xi) = m_0^1(\xi)\widehat{\varphi^1}(\xi), \quad \text{for almost all } \xi \in \mathbb{R}. \tag{4.73}$$

It is worth mentioning that the orthonormality property of the functions $\varphi^1(x - l)$, $l \in \mathbb{Z}$, implies that:

$$\left| m_0^1(\xi) \right|^2 + \left| m_0^1(\xi + \pi) \right|^2 = 1, \quad \text{for almost all } \xi \in \mathbb{R}. \tag{4.74}$$

Indeed, using Corollary 4.7 and (4.73), one has that

$$(2\pi)^{-1} = \sum_{k \in \mathbb{Z}} \left| \widehat{\varphi^1}(2\xi + 2\pi k) \right|^2$$

$$= \left| m_0^1(\xi) \right|^2 \sum_{p \in \mathbb{Z}} \left| \widehat{\varphi^1}(\xi + 2\pi p) \right|^2 + \left| m_0^1(\xi + \pi) \right|^2 \sum_{p \in \mathbb{Z}} \left| \widehat{\varphi^1}(\xi + \pi + 2\pi p) \right|^2$$

$$= \left(\left| m_0^1(\xi) \right|^2 + \left| m_0^1(\xi + \pi) \right|^2 \right) \times (2\pi)^{-1}.$$

- One denotes by m_1^1 the 2π-periodic function belonging to $L^2(\mathbb{T})$, defined as:

$$m_1^1(\xi) = e^{i\xi}\,\overline{m_0^1(\xi + \pi)}, \quad \text{for almost all } \xi \in \mathbb{R}. \qquad (4.75)$$

Thus, one clearly has that

$$m_0^1(\xi)\overline{m_1^1(\xi)} + m_0^1(\xi + \pi)\overline{m_1^1(\xi + \pi)} = 0, \quad \text{for almost all } \xi \in \mathbb{R}. \quad (4.76)$$

Moreover, it easily follows from (4.74) and (4.75) that

$$\left|m_1^1(\xi)\right|^2 + \left|m_1^1(\xi + \pi)\right|^2 = 1, \quad \text{for almost all } \xi \in \mathbb{R}. \qquad (4.77)$$

- One knows from part (ii) of Proposition 4.9 and from the fact that $\left\{\varphi^1(x - l) : l \in \mathbb{Z}\right\}$ is an orthonormal basis of V_0^1 that the function $\xi \mapsto m_1^1(\xi)\widehat{\varphi^1}(\xi)$ belongs to $L^2(\mathbb{R})$ and, more precisely, that it is the Fourier transform of some function belonging to V_1^1. Thus, there exists a function, denoted by ψ^1, belonging to V_1^1 whose Fourier transform satisfies:

$$\widehat{\psi^1}(2\xi) = m_1^1(\xi)\widehat{\varphi^1}(\xi), \quad \text{for almost all } \xi \in \mathbb{R}. \qquad (4.78)$$

In view of Theorem 4.27, stated below, the function ψ^1 is called a (univariate) mother wavelet.

- For each $\varepsilon = (\varepsilon_1, \ldots, \varepsilon_N) \in \mathbb{L}_N$ (see (4.67)), one denotes by $\psi^{(\varepsilon)}$ the function of $L^2(\mathbb{R}^N)$ defined as:

$$\psi^{(\varepsilon)}(x_1, \ldots, x_N) := \prod_{n=1}^{N} \psi^{\varepsilon_n}(x_n), \quad \text{for all } (x_1, \ldots, x_N) \in \mathbb{R}^N, \quad (4.79)$$

with the convention that

$$\psi^0 := \varphi^1. \qquad (4.80)$$

In view of Corollary 4.28, stated below, the function $\psi^{(\varepsilon)}$ is called a (multivariate) mother wavelet.

We are now in position to state the two main results of this section. The first one is the following theorem which can probably be considered as the keystone of the wavelet theory.

Theorem 4.27 (Mallat and Meyer, 1986). *We recall that the function* φ^1, *called a (univariate) scaling function, is such that* $\left\{\varphi^1(x - l) : l \in \mathbb{Z}\right\}$ *is an orthonormal basis of the reference space* V_0^1 *of a MRA* $(V_j^1)_{j \in \mathbb{Z}}$ *of* $L^2(\mathbb{R})$. *Also, we recall that the function* ψ^1, *called a (univariate) mother wavelet, is defined through (4.78). The following three results hold.*

(i) *For each fixed $j \in \mathbb{Z}$, the sequence*

$$\left\{ 2^{j/2} \psi^1 (2^j x - k) \; : \; k \in \mathbb{Z} \right\}$$

is an orthonormal basis of the space W_j^1 defined through (4.64) with $N = 1$.

(ii) *The sequence*

$$\left\{ 2^{j/2} \psi^1 (2^j x - k) \; : \; j \in \mathbb{Z} \text{ and } k \in \mathbb{Z} \right\}$$

is an orthonormal basis of $L^2(\mathbb{R})$.

(iii) *For every fixed $J \in \mathbb{Z}$, the sequence*

$$\left\{ 2^{J/2} \varphi^1 (2^J x - l) \; : \; l \in \mathbb{Z} \right\} \cup \left\{ 2^{j/2} \psi^1 (2^j x - k) \; : \; j \in \mathbb{Z}, \; j \geq J \text{ and } k \in \mathbb{Z} \right\}$$

is an orthonormal basis of $L^2(\mathbb{R})$.

The proof of Theorem 4.27 will be given in the sequel. The second main result of this section is the following one.

Corollary 4.28. *The integer $N \geq 2$ is arbitrary. Let $\mathbb{L}_N$ be the finite set of cardinality $2^N - 1$ defined in (4.67). We recall that, for each $\varepsilon \in \mathbb{L}_N$, the function $\psi^{(\varepsilon)}$, called a (multivariate) mother wavelet, is defined in (4.79) and by using the convention (4.80). Also, we recall that the (multivariate) scaling function φ^N, which already has been introduced in Proposition 4.16, is defined as*

$$\varphi^N(x_1, \ldots, x_N) := \prod_{n=1}^{N} \varphi^1(x_n) = \prod_{n=1}^{N} \psi^0(x_n), \quad \text{for all } (x_1, \ldots, x_N) \in \mathbb{R}^N,$$

where the last equality results from the convention (4.80). The following three results hold.

(i) *For each fixed $\varepsilon \in \mathbb{L}_N$ and $j \in \mathbb{Z}$, the sequence*

$$\left\{ 2^{jN/2} \psi^{(\varepsilon)} (2^j x - k) \; : \; k \in \mathbb{Z}^N \right\}$$

is an orthonormal basis of the space $W_j^{(\varepsilon)}$ defined in (4.69).

(ii) *The sequence*

$$\left\{ 2^{jN/2} \psi^{(\varepsilon)} (2^j x - k) \; : \; \varepsilon \in \mathbb{L}_N, \; j \in \mathbb{Z} \text{ and } k \in \mathbb{Z}^N \right\}$$

is an orthonormal basis of $L^2(\mathbb{R}^N)$.

(iii) *For every fixed $J \in \mathbb{Z}$, the sequence*

$$\left\{ 2^{JN/2} \varphi^N (2^J x - l) \; : \; l \in \mathbb{Z}^N \right\}$$
$$\cup \left\{ 2^{jN/2} \psi^{(\varepsilon)} (2^j x - k) \; : \; \varepsilon \in \mathbb{L}_N, \; j \in \mathbb{Z}, \; j \geq J \text{ and } k \in \mathbb{Z} \right\}$$

is an orthonormal basis of $L^2(\mathbb{R}^N)$.

At last, notice that $\mathbb{L}_N$ *is frequently identified with* $\{1,\ldots,2^N-1\}$, *therefore the* 2^N-1 *mother wavelets* $\psi^{(\varepsilon)}$, $\varepsilon \in \mathbb{L}_N$, *are frequently denoted by* ψ_p, $p \in \{1,\ldots,2^N-1\}$.

Proof of Corollary 4.28. Part (i) of the corollary can easily be obtained by using part (iii) of Remark 4.13 with $N = 1$, part (i) of Theorem 4.27, (4.69), and Lemma 4.15. Part (ii) of the corollary is a straightforward consequence of its part (i), and also of (4.68) and (4.71). At last, one knows from Proposition 4.16 and from part (iii) of Remark 4.13, that $\{2^{JN/2}\varphi^N(2^J x - l) : l \in \mathbb{Z}^N\}$ is an orthonormal basis of V_J^N, for each fixed $J \in \mathbb{Z}$. Combining the latter fact with part (i) of the corollary, and also with (4.68), and with (4.72), one gets part (iii) of the corollary. □

Before proving Theorem 4.27, let us give the following two results which basically show that the Fourier transforms of the univariate and multivariate Meyer mother wavelets satisfy a very useful property: they are compactly supported infinitely differentiable functions which vanish in a neighborhood of 0 (the zero element of $\mathbb{R}^N$). As, a consequence, in view of Proposition 4.22, it turns out that the Meyer wavelets belong to the Schwartz class $\mathcal{S}(\mathbb{R}^N)$ (see Definition 4.21).

Proposition 4.29 (univariate Meyer mother wavelet). *Assume that* φ^1 *is the univariate scaling function* [3] *associated with the Meyer MRA introduced in Proposition 4.23. Let* ψ^1 *be the corresponding mother wavelet which is defined through (4.78), (4.75), and (4.48). Then,* $\widehat{\psi^1}$ *(the Fourier transform of* ψ^1*) is the infinitely differentiable compactly supported function vanishing in a neighborhood of* 0, *and satisfying, for all* $\xi \in \mathbb{R}$,

$$\widehat{\psi^1}(\xi) = \begin{cases} (2\pi)^{-1/2}\,e^{i\xi/2}\sin\left[\frac{\pi}{2}\nu\left(\frac{3}{2\pi}|\xi|-1\right)\right], & \text{if } \frac{2\pi}{3} \leq |\xi| \leq \frac{4\pi}{3}, \\ (2\pi)^{-1/2}\,e^{i\xi/2}\cos\left[\frac{\pi}{2}\nu\left(\frac{3}{4\pi}|\xi|-1\right)\right], & \text{if } \frac{4\pi}{3} \leq |\xi| \leq \frac{8\pi}{3}, \\ 0, & \text{else,} \end{cases}$$

(4.81)

where ν *is the same infinitely differentiable function from* $\mathbb{R}$ *into* $[0,1]$ *as in Proposition 4.23. As a consequence, the Meyer mother wavelet* ψ^1 *belongs to the Schwartz class* $\mathcal{S}(\mathbb{R})$ *(see Definition 4.21).*

---

[3]Such a scaling function is called a univariate Meyer scaling function.

Definition 4.30 (multivariate Meyer mother wavelets). The integer $N \geq 2$ is arbitrary. The multivariate Meyer mother wavelets, which are
also called the Lemarié-Meyer mother wavelets [Lemarié and Meyer (1986)],
are the $2^N - 1$ functions $\psi^{(\varepsilon)}$, $\varepsilon \in \mathbb{L}_N$ (see (4.67)), defined on $\mathbb{R}^N$ by (4.79),
where $\psi^0 = \varphi^1$ is assumed to be a univariate Meyer scaling function and ψ^1
is the corresponding univariate Meyer mother wavelet. Observe that the
set $\mathbb{L}_N$ is frequently identified with $\{1, \ldots, 2^N - 1\}$, and thus the Lemarié-
Meyer mother wavelets are frequently denoted by ψ_p, $p \in \{1, \ldots, 2^N - 1\}$,
rather than $\psi^{(\varepsilon)}$, $\varepsilon \in \mathbb{L}_N$.

It easily results from Proposition 4.29 that:

Corollary 4.31. *The Fourier transforms* $\widehat{\psi}_1, \ldots, \widehat{\psi}_{2^N - 1}$ *of the Lemarié-
Meyer mother wavelets are infinitely differentiable compactly supported
functions vanishing in a neighborhood of* 0 *(the zero element of* $\mathbb{R}^N$ *). More
precisely, one has that*

$$\mathrm{Supp}\, \widehat{\psi}_p \subseteq \left[-\frac{8\pi}{3}, \frac{8\pi}{3} \right]^N \setminus \left(-\frac{2\pi}{3}, \frac{2\pi}{3} \right)^N, \quad \text{for all } p \in \{1, \ldots, 2^N - 1\}.$$
$$(4.82)$$

Thus, the Lemarié-Meyer mother wavelets $\psi_1, \ldots, \psi_{2^N - 1}$ *belong to the
Schwartz class* $\mathcal{S}(\mathbb{R}^N)$ *(see Definition 4.21).*

Proof of Proposition 4.29. First, observe that one knows from Remark 4.25 and from (4.75) that the 2π-periodic function m_1^1 is infinitely
differentiable on the real line. Moreover, one knows from Proposition 4.23
that the function $\widehat{\varphi^1}$ is infinitely differentiable on the real line. Thus, the
function $\widehat{\psi^1}$, which is defined as

$$\widehat{\psi^1}(\xi) := m_1^1(2^{-1}\xi)\widehat{\varphi^1}(2^{-1}\xi), \quad \text{for all } \xi \in \mathbb{R},$$
$$(4.83)$$

is infinitely differentiable on the real line.

Let us now prove that (4.81) holds. It easily follows from (4.45) and
(4.83) that

$$\widehat{\psi^1}(\xi) = 0, \quad \text{when } |\xi| \geq 8\pi/3.$$
$$(4.84)$$

So, from now on, one assumes that $|\xi| < 8\pi/3$. Thus, for all $l \in \mathbb{Z} \setminus \{-1, 0\}$,
one has

$$\left| \xi + 2\pi(2l + 1) \right| \geq 2\pi|2l + 1| - |\xi| > 6\pi - \frac{8\pi}{3} = \frac{10\pi}{3} > \frac{4\pi}{3}.$$

Then, using (4.75), (4.48), (4.45), and the fact that $\widehat{\varphi^1}$ is an even real-valued function, one gets that

$$m_1^1(2^{-1}\xi) = (2\pi)^{1/2}e^{i\xi/2}\sum_{l\in\mathbb{Z}}\widehat{\varphi^1}\big(\xi + 2\pi(2l+1)\big)$$

$$= (2\pi)^{1/2}e^{i\xi/2}\Big(\widehat{\varphi^1}\big(|\xi| - 2\pi\big) + \widehat{\varphi^1}\big(|\xi| + 2\pi\big)\Big). \quad (4.85)$$

Next, let us consider the following three cases:

(I) $|\xi| \in [0, 2\pi/3]$ i.e. $|\xi| - 2\pi \in [-2\pi, -4\pi/3]$ i.e. $|\xi| + 2\pi \in [2\pi, 8\pi/3]$,
(II) $|\xi| \in (2\pi/3, 4\pi/3]$ i.e. $|\xi| - 2\pi \in (-4\pi/3, -2\pi/3]$
  i.e. $|\xi| + 2\pi \in (8\pi/3, 10\pi/3]$,
(III) $|\xi| \in (4\pi/3, 8\pi/3)$ i.e. $|\xi| - 2\pi \in (-2\pi/3, 2\pi/3)$
  i.e. $|\xi| + 2\pi \in (10\pi/3, 14\pi/3)$.

In the case (I), one can easily derive from (4.85), (4.45) and (4.83), that

$$\widehat{\psi^1}(\xi) = 0. \quad (4.86)$$

In the case (II), using (4.85), (4.45), (4.83) and (4.44), one has that

$$\widehat{\psi^1}(\xi) = e^{i\xi/2}\,\widehat{\varphi^1}\big(|\xi| - 2\pi\big)$$

$$= (2\pi)^{-1/2}e^{i\xi/2}\,\cos\left[\frac{\pi}{2}\nu\Big(\frac{3}{2\pi}(2\pi - |\xi|) - 1\Big)\right]$$

$$= (2\pi)^{-1/2}e^{i\xi/2}\,\cos\left[\frac{\pi}{2}\nu\Big(2 - \frac{3}{2\pi}|\xi|\Big)\right]$$

$$= (2\pi)^{-1/2}e^{i\xi/2}\,\cos\left[\frac{\pi}{2}\nu\Big(1 - (\frac{3}{2\pi}|\xi| - 1)\Big)\right]$$

$$= (2\pi)^{-1/2}e^{i\xi/2}\,\cos\left[\frac{\pi}{2}\Big(1 - \nu\big(\frac{3}{2\pi}|\xi| - 1\big)\Big)\right]$$

$$= (2\pi)^{-1/2}e^{i\xi/2}\,\sin\left[\frac{\pi}{2}\nu\Big(\frac{3}{2\pi}|\xi| - 1\Big)\right]. \quad (4.87)$$

In the case (III), it follows from (4.85), (4.45) and (4.83), that

$$\widehat{\psi^1}(\xi) = e^{i\xi/2}\,\widehat{\varphi^1}\big(2^{-1}\xi\big) = (2\pi)^{-1/2}e^{i\xi/2}\,\cos\left[\frac{\pi}{2}\nu\Big(\frac{3}{4\pi}|\xi| - 1\Big)\right]. \,(4.88)$$

Finally, putting together (4.84), (4.86), (4.87) and (4.88), one obtains (4.81). $\qquad\square$

In order to be able to prove Theorem 4.27, one needs some preliminary results.

Proposition 4.32. *Let $\mathbb{H}$ be a Hilbert space over $\mathbb{C}$ (the field of complex numbers) equipped with the norm $\|\cdot\|$. Let Λ be an arbitrary countable set. Let $(S_\lambda)_{\lambda \in \Lambda}$ be a sequence of closed and pairwise orthogonal subspaces of $\mathbb{H}$. The direct sum of the S_λ's, defined as:*

$$y \in \overset{\perp}{\underset{j \in \mathbb{Z}}{\bigoplus}} S_\lambda \Longleftrightarrow y = \sum_{\lambda \in \Lambda} x_\lambda, \text{ where } x_\lambda \in S_\lambda, \text{ for all } \lambda \in \Lambda, \qquad (4.89)$$

is a closed subspace of $\mathbb{H}$. Notice that the x_λ's depend on y, and that the series in (4.89) is assumed to be (unconditionally) convergent in the sense of the norm $\|\cdot\|$. Also, notice that

$$\|y\|^2 = \sum_{\lambda \in \Lambda} \|x_\lambda\|^2. \qquad (4.90)$$

Proof of Proposition 4.32. Let $(y^n)_{n \in \mathbb{N}}$ be an arbitrary sequence of elements of the direct sum of the S_λ's which converges to some element z of $\mathbb{H}$. One has to show that z belongs to the direct sum of the S_λ's. Observe that each y^n can be expressed as

$$y^n = \sum_{\lambda \in \Lambda} x_\lambda^n, \text{ where } x_\lambda^n \in S_\lambda, \text{ for all } \lambda \in \Lambda. \qquad (4.91)$$

Using (4.90) and the fact that $(y^n)_{n \in \mathbb{N}}$ is a Cauchy sequence, it can be shown that, for every fixed $\lambda \in \Lambda$, $(x_\lambda^n)_{n \in \mathbb{N}}$ is a Cauchy sequence as well. Therefore, there exists $x_\lambda \in S_\lambda$ such that

$$\lim_{n \to +\infty} \|x_\lambda^n - x_\lambda\|^2 = 0. \qquad (4.92)$$

Next, observe that one has

$$\sum_{\lambda \in \Lambda} \|x_\lambda\|^2 < +\infty. \qquad (4.93)$$

Indeed, using Fatou Lemma, (4.91) and (4.90), one gets that

$$\sum_{\lambda \in \Lambda} \|x_\lambda\|^2 = \sum_{\lambda \in \Lambda} \left(\lim_{n \to +\infty} \|x_\lambda^n\|^2 \right) \leq \lim_{n \to +\infty} \sum_{\lambda \in \Lambda} \|x_\lambda^n\|^2$$
$$= \lim_{n \to +\infty} \|y^n\|^2 = \|z\|^2 < +\infty.$$

In view of (4.93),

$$z' := \sum_{\lambda \in \Lambda} x_\lambda,$$

is a well-defined element of the direct sum of the S_λ's. It remains to us to show that $z' = z$, in other words that the sequence $(y^n)_{n\in\mathbb{N}}$ converges to z', that is

$$\lim_{n\to+\infty} \sum_{\lambda\in\Lambda} \|x^n_\lambda - x_\lambda\|^2 = 0.$$

To this end, in view of the fact that $(y^n)_{n\in\mathbb{N}}$ converges to z, it is enough to show that

$$\sum_{\lambda\in\Lambda} \|x^n_\lambda - x_\lambda\|^2 \le \|y^n - z\|^2, \quad \text{for all } n \in \mathbb{N}. \tag{4.94}$$

It follows from (4.91) and (4.90) that for every $n, n' \in \mathbb{N}$, one has

$$\sum_{\lambda\in\Lambda} \|x^n_\lambda - x^{n'}_\lambda\|^2 = \|y^n - y^{n'}\|^2.$$

Thus, using again Fatou Lemma, one obtains that

$$\sum_{\lambda\in\Lambda} \|x^n_\lambda - x_\lambda\|^2 = \sum_{\lambda\in\Lambda} \left(\lim_{n'\to+\infty} \|x^n_\lambda - x^{n'}_\lambda\|^2 \right)$$
$$\le \lim_{n'\to+\infty} \sum_{\lambda\in\Lambda} \|x^n_\lambda - x^{n'}_\lambda\|^2 = \lim_{n'\to+\infty} \|y^n - y^{n'}\|^2 = \|y^n - z\|^2,$$

which shows that (4.94) is satisfied $\qquad\square$

Proposition 4.33. *One denotes by $\mathbb{H}$ a Hilbert space over $\mathbb{C}$ equipped with the inner product $\langle\cdot,\cdot\rangle$. Let $(V_j)_{j\in\mathbb{Z}}$ be a sequence of closed subspaces of $\mathbb{H}$ which satisfies the properties (a), (b) and (c) of a MRA (see Definition 4.12); notice that, for the property (c), $L^2(\mathbb{R}^N)$ has to be replaced by $\mathbb{H}$. Let $(W_j)_{j\in\mathbb{Z}}$ be the sequence of the closed and pairwise orthogonal subspaces of $\mathbb{H}$ such that, for each $j \in \mathbb{Z}$, W_j is the orthogonal complement of V_j in V_{j+1}. In other words, one has*

$$V_{j+1} = V_j \overset{\perp}{\oplus} W_j, \quad \text{for every } j \in \mathbb{Z}. \tag{4.95}$$

Then the space $\mathbb{H}$ can be decomposed as

$$\mathbb{H} = V_J \overset{\perp}{\oplus} \left(\overset{\perp}{\underset{j\ge J}{\bigoplus}} W_j \right), \quad \text{for each fixed } J \in \mathbb{Z}, \tag{4.96}$$

as well as

$$\mathbb{H} = \overset{\perp}{\underset{j\in\mathbb{Z}}{\bigoplus}} W_j. \tag{4.97}$$

Proof of Proposition 4.33. For each fixed $J \in \mathbb{Z}$, one sets

$$Z_J := V_J \overset{\perp}{\oplus} \left(\overset{\perp}{\underset{j \geq J}{\bigoplus}} W_j \right). \tag{4.98}$$

One knows from Proposition 4.32 that Z_J is a closed subspace of $\mathbb{H}$. Thus, in order to show that (4.96) is satisfied, it is enough to prove that the orthogonal complement of Z_J in $\mathbb{H}$ reduces to $\{0\}$. That is:

$$\forall h \in \mathbb{H}, \ \big(\forall z \in Z_J, \ \langle h, z \rangle = 0 \big) \Longrightarrow h = 0. \tag{4.99}$$

It can easily be shown, by induction on Q and by using (4.98) and (4.95), that $Z_J = Z_{J+Q}$, for every $Q \in \mathbb{N}$. Then, it follows from (4.98) and from the property (a) of $(V_j)_{j \in \mathbb{Z}}$, that one has $\bigcup_{j \in \mathbb{Z}} V_j \subset Z_J$. Thus, using its property (c), one obtains (4.99).

Now, one sets

$$Z' := \overset{\perp}{\underset{j \in \mathbb{Z}}{\bigoplus}} W_j \quad \text{and} \quad Z'_J := \overset{\perp}{\underset{j \geq J}{\bigoplus}} W_j, \quad \text{for each fixed } J \in \mathbb{Z}. \tag{4.100}$$

Observe that Z'_J is orthogonal to V_J and that $Z'_J \subset Z'$. Using the same arguments as in the case of Z_J, it turns out that in order to show that (4.97) is satisfied, it is enough to prove that

$$\forall h' \in \mathbb{H}, \ \big(\forall z' \in Z', \ \langle h', z' \rangle = 0 \big) \Longrightarrow h' = 0. \tag{4.101}$$

So, let h' be an arbitrary fixed element of $\mathbb{H}$ such that

$$\forall z' \in Z', \ \langle h', z' \rangle = 0. \tag{4.102}$$

One knows from (4.96) and from the second equality in (4.100) that, for every fixed $J \in \mathbb{Z}$, one has $h' = v_J + d_J$, where $v_J \in V_J$, $d_J \in Z'_J$ and $v_J \perp d_J$. Thus, taking in (4.102) $z' = d_J$, it follows that $d_J = 0$ and consequently that $h' = v_J \in V_J$. Then, the fact that J has been chosen in an arbitrary way implies that $h' \in \bigcap_{j \in \mathbb{Z}} V_j = \{0\}$, where the equality comes from the property (b) of $(V_j)_{j \in \mathbb{Z}}$. Thus, one gets that $h' = 0$ which shows that (4.101) is satisfied. $\qquad \square$

Lemma 4.34. *Let $F, G \in L^2(\mathbb{R}^N)$, the following two assertions are equivalent.*

(i) One has

$$\int_{\mathbb{R}^N} F(x) \overline{G(x - l)} \, dx = 0, \quad \text{for all } l \in \mathbb{Z}^N.$$

(ii) One has

$$\sum_{k\in\mathbb{Z}^N} \widehat{F}(\xi + 2\pi k)\overline{\widehat{G}(\xi + 2\pi k)} = 0, \quad \text{for almost all } \xi \in \mathbb{R}^N.$$

Proof of Lemma 4.34. One assumes that $l \in \mathbb{Z}^N$ is arbitrary and fixed. Using the Plancherel formula given in (4.3), part *(ii)* of Proposition 4.5, and the $(2\pi\mathbb{Z})^N$-periodicity of the function $\xi \mapsto e^{il\cdot\xi}$, one obtains that

$$\int_{\mathbb{R}^N} F(x)\overline{G(x - l)}\, dx = \int_{\mathbb{R}^N} e^{il\cdot\xi}\widehat{F}(\xi)\overline{\widehat{G}(\xi)}\, d\xi$$

$$= \int_{\mathbb{T}^N} e^{il\cdot\xi}\Big(\sum_{k\in\mathbb{Z}^N} \widehat{F}(\xi + 2\pi k)\overline{\widehat{G}(\xi + 2\pi k)}\Big)\, d\xi.$$

Thus, in view of (A.43), it turns out that the quantity

$$(2\pi)^{-N} \int_{\mathbb{R}^N} F(x)\overline{G(x - l)}\, dx$$

is equal to the Fourier coefficient of order $-l$ of the $(2\pi\mathbb{Z})^N$-periodic function

$$\xi \mapsto \sum_{k\in\mathbb{Z}^N} \widehat{F}(\xi + 2\pi k)\overline{\widehat{G}(\xi + 2\pi k)}.$$

Then, in view of Corollary A.39, it is clear that the two assertions *(i)* and *(ii)* are equivalent. $\qquad\qquad\square$

We are now in position to prove Theorem 4.27.

Proof of Theorem 4.27. First, we recall that, for every fixed $k \in \mathbb{Z}$, the function $\psi^1(x - k)$ belongs to V_1^1, since (4.78), part *(ii)* of Proposition 4.9 and the fact that $\{\varphi^1(x - l) : l \in \mathbb{Z}\}$ is an orthonormal basis of V_0^1 imply that the function $\psi^1(2^{-1}x - k)$ belongs to V_0^1.

Let us now prove that the function $\psi^1(x - k)$ belongs to W_0^1, for all $k \in \mathbb{Z}$. In view of (4.64), with $N = 1$, It is sufficient to show that

$$\int_{\mathbb{R}} \psi^1(x - k)\overline{\varphi^1(x - l)}\, dx = 0, \quad \text{for all } l \in \mathbb{Z},$$

which is clearly equivalent to

$$\int_{\mathbb{R}} \psi^1(x)\overline{\varphi^1(x - l)}\, dx = 0, \quad \text{for all } l \in \mathbb{Z}.$$

In view of Lemma 4.34 (with $N = 1$), it is enough to show that

$$\sum_{k\in\mathbb{Z}} \widehat{\varphi^1}(\xi + 2\pi k)\overline{\widehat{\psi^1}(\xi + 2\pi k)} = 0, \quad \text{for almost all } \xi \in \mathbb{R}. \qquad (4.103)$$

Using (4.73), (4.78), the fact that $m_0^1(\eta)$ and $m_1^1(\eta)$ are 2π-periodic functions, the equality [4]

$$\sum_{q\in\mathbb{Z}}\left|\widehat{\varphi^1}(\eta + 2\pi q)\right|^2 = (2\pi)^{-1}, \quad \text{for almost all } \eta \in \mathbb{R}, \qquad (4.104)$$

and (4.76), one obtains, for almost all $\xi \in \mathbb{R}$, that

$$\sum_{k\in\mathbb{Z}}\widehat{\varphi^1}(\xi + 2\pi k)\overline{\widehat{\psi^1}(\xi + 2\pi k)}$$

$$= m_0^1(2^{-1}\xi)\overline{m_1^1(2^{-1}\xi)}\sum_{q\in\mathbb{Z}}\left|\widehat{\varphi^1}(2^{-1}\xi + 2\pi q)\right|^2$$

$$+m_0^1(2^{-1}\xi + \pi)\overline{m_1^1(2^{-1}\xi + \pi)}\sum_{q\in\mathbb{Z}}\left|\widehat{\varphi^1}(2^{-1}\xi + \pi + 2\pi q)\right|^2$$

$$= (2\pi)^{-1}\left(m_0^1(2^{-1}\xi)\overline{m_1^1(2^{-1}\xi)} + m_0^1(2^{-1}\xi + \pi)\overline{m_1^1(2^{-1}\xi + \pi)}\right) = 0.$$

Thus one gets (4.103).

Let us now show that $\{\psi^1(x - k) : k \in \mathbb{Z}\}$ is an orthonormal sequence in $L^2(\mathbb{R})$. In view of Corollary 4.7, it is sufficient to prove that

$$\sum_{k\in\mathbb{Z}}\left|\widehat{\psi^1}(\xi + 2\pi k)\right|^2 = (2\pi)^{-1}, \quad \text{for almost all } \xi \in \mathbb{R}.$$

This is can be done similarly to (4.103), by using (4.78), the fact $m_1^1(\eta)$ is a 2π-periodic function, (4.104) and (4.77).

Let us now show that

$$W_0^1 = \overline{\text{span}\{\psi^1(x - k) : k \in \mathbb{Z}\}}. \qquad (4.105)$$

One already knows that

$$\overline{\text{span}\{\psi^1(x - k) : k \in \mathbb{Z}\}} \subseteq W_0^1.$$

Thus, in view of the equalities

$$V_0^1 = \overline{\text{span}\{\varphi^1(x - k) : k \in \mathbb{Z}\}} \quad \text{and} \quad V_1^1 = V_0^1 \overset{\perp}{\oplus} W_0^1,$$

and of the fact that $\{\varphi^1(x - k), \psi^1(x - k) : k \in \mathbb{Z}\}$ is an orthonormal sequence in $L^2(\mathbb{R})$, in order to obtain (4.105), it is enough to show that

$$V_1^1 = \overline{\text{span}\{\varphi^1(x - k), \psi^1(x - k) : k \in \mathbb{Z}\}}.$$

This can be done by showing that the orthogonal complement of

$$\overline{\text{span}\{\varphi^1(x - k), \psi^1(x - k) : k \in \mathbb{Z}\}}$$

[4]Observe that this equality results from Corollary 4.7, with $N = 1$, since $\{\varphi^1(x - l) : l \in \mathbb{Z}\}$ is an orthonormal sequence in $L^2(\mathbb{R})$.

in V_1^1 reduces to $\{0\}$. That is, for any $f \in V_1^1$ satisfying

$$\int_\mathbb{R} \varphi^1(x-k)\overline{f(x)}\,dx = \int_\mathbb{R} \psi^1(x-k)\overline{f(x)}\,dx = 0, \quad \text{for all } k \in \mathbb{Z}, \quad (4.106)$$

one has

$$f(x) = 0, \quad \text{for almost all } x \in \mathbb{R}. \quad (4.107)$$

Observe that the fact that the function $f(2^{-1}x)$ belongs to V_0^1, and part (i) of Proposition 4.9 imply that there is $\lambda_f \in L^2(\mathbb{T})$ such that one has

$$\widehat{f}(\xi) = \lambda_f(2^{-1}\xi)\widehat{\varphi^1}(2^{-1}\xi), \quad \text{for almost all } \xi \in \mathbb{R}. \quad (4.108)$$

Thus, using (4.106), the Plancherel formula given in (4.3), (4.108), (4.73) and (4.78), one gets that

$$\int_\mathbb{R} e^{-i2k\xi}\,m_0^1(\xi)\lambda_f(\xi)|\widehat{\varphi^1}(\xi)|^2\,d\xi = 0, \quad \text{for all } k \in \mathbb{Z}, \quad (4.109)$$

and

$$\int_\mathbb{R} e^{-i2k\xi}\,m_1^1(\xi)\lambda_f(\xi)|\widehat{\varphi^1}(\xi)|^2\,d\xi = 0, \quad \text{for all } k \in \mathbb{Z}. \quad (4.110)$$

Then, (4.109), (4.110), (4.104) and the 2π-periodicity of the functions $e^{-i2k\xi}\,m_0^1(\xi)\lambda_f(\xi)$ and $e^{-i2k\xi}\,m_1^1(\xi)\lambda_f(\xi)$ imply that

$$\int_\mathbb{T} e^{-i2k\xi}\,m_0^1(\xi)\lambda_f(\xi)\,d\xi = \int_\mathbb{T} e^{-i2k\xi}\,m_1^1(\xi)\lambda_f(\xi)\,d\xi = 0, \quad \text{for all } k \in \mathbb{Z}.$$

Thus, one obtains that

$$\int_\mathbb{T} e^{-ip\xi}\Big(m_0^1(\xi)\lambda_f(\xi) + m_0^1(\xi+\pi)\lambda_f(\xi+\pi)\Big)\,d\xi = 0, \quad \text{for all } p \in \mathbb{Z},$$

and

$$\int_\mathbb{T} e^{-ip\xi}\Big(m_1^1(\xi)\lambda_f(\xi) + m_1^1(\xi+\pi)\lambda_f(\xi+\pi)\Big)\,d\xi = 0, \quad \text{for all } p \in \mathbb{Z}.$$

This means that all the Fourier coefficients of the two π-periodic functions of $L^1(\mathbb{T})$,

$$\xi \mapsto m_0^1(\xi)\lambda_f(\xi) + m_0^1(\xi+\pi)\lambda_f(\xi+\pi)$$

and

$$\xi \mapsto m_1^1(\xi)\lambda_f(\xi) + m_1^1(\xi+\pi)\lambda_f(\xi+\pi),$$

vanish. Therefore using Corollary A.39 one gets, for almost all $\xi \in \mathbb{R}$, that

$$\begin{cases} m_0^1(\xi)\lambda_f(\xi) + m_0^1(\xi+\pi)\lambda_f(\xi+\pi) = 0 \\ m_1^1(\xi)\lambda_f(\xi) + m_1^1(\xi+\pi)\lambda_f(\xi+\pi) = 0. \end{cases} \quad (4.111)$$

Moreover, it follows (4.75) and (4.74) that, for almost all $\xi \in \mathbb{R}$,

$$\det \begin{pmatrix} m_0^1(\xi) & m_0^1(\xi + \pi) \\ m_1^1(\xi) & m_1^1(\xi + \pi) \end{pmatrix} = -e^{i\xi} \neq 0.$$

Thus (4.111) entails that $\lambda_f(\xi) = 0$, for almost all $\xi \in \mathbb{R}$. Then (4.108) implies that $\widehat{f}(\xi) = 0$, for almost all $\xi \in \mathbb{R}$, and consequently that (4.107) holds.

We have proved that $\{\psi^1(x-k) : k \in \mathbb{Z}\}$ is an orthonormal basis of W_0^1. Then (4.65) implies that, for all fixed $j \in \mathbb{Z}$, $\{2^{j/2}\psi^1(2^j x - k) : k \in \mathbb{Z}\}$ is an orthonormal basis of W_j^1, which shows that part (i) of the theorem holds. Combining this result with (4.71), where $N = 1$, we obtain part (ii) of the theorem. Finally, we know from part (iii) of Remark 4.13, where $N = 1$, that, for all fixed $J \in \mathbb{Z}$, $\{2^{J/2}\varphi^1(2^J x - l) : l \in \mathbb{Z}\}$ is an orthonormal basis of V_J^1. Combining this result with part (i) of the theorem and with (4.72), where $N = 1$, it follows that part (iii) of the theorem is satisfied. $\square$

Chapter 5

Wavelet series representation of Generator of Multifractional Brownian Field

5.1 Introduction

The integer $N \geq 1$ is arbitrary and fixed. $(0,1)$ denotes the open interval of the real line whose endpoints are 0 and 1. Recall that the so-called Generator of Multifractional Brownian Field (MuBF), which already has been introduced in Definition 1.78, is the real-valued centered Gaussian field, denoted by $X := \{X(u,v), (u,v) \in \mathbb{R}^N \times (0,1)\}$, and defined as:

$$X(u,v) := \int_{\mathbb{R}^N} \mathbb{K}(u,v,\xi)\, d\widehat{\mathbb{W}}(\xi), \quad \text{for all couple } (u,v) \in \mathbb{R}^N \times (0,1),$$
(5.1)

where $\widehat{\mathbb{W}}$ denotes the orthogonally scattered random measure introduced in (1.28), and $\mathbb{K}$ is the deterministic kernel function defined as:

$$\mathbb{K}(u,v,\xi) := \frac{1}{(2\pi)^{N/2}} \frac{e^{iu\cdot\xi} - 1}{|\xi|^{v+N/2}}, \quad \text{for all } (u,v,\xi) \in \mathbb{R}^N \times (0,1) \times \mathbb{R}^N \setminus \{0\}.$$
(5.2)

Notice that the normalizing constant $(2\pi)^{-N/2}$ is not important at all, yet it will allow us to simplify our notations.

This field X is very important at least for the following two reasons.

(1) In view of Definition 1.82, any Multifractional Brownian Field (MuBF) $Y := \{Y(t), t \in \mathbb{R}^N\}$, of an arbitrary deterministic Hurst functional parameter $H(\cdot)$ with values in a compact interval, denoted by $[\underline{H}, \overline{H}]$ and included in $(0,1)$, is obtained starting from X by setting:

$$Y(t) := X(t, H(t)) = \frac{1}{(2\pi)^{N/2}} \int_{\mathbb{R}^N} \frac{e^{it\cdot\xi} - 1}{|\xi|^{H(t)+N/2}}\, d\widehat{\mathbb{W}}(\xi), \quad \text{for all } t \in \mathbb{R}^N.$$
(5.3)

(2) For each fixed $v \in (0,1)$, $X(\cdot,v) := \{X(u,v), u \in \mathbb{R}^N\}$ is the classical

Fractional Brownian Field (FBF) of Hurst parameter v (see Definition 1.17 and Proposition 1.28).

As we have already mentioned in Remark 1.81, the field X is always identified with its modification with continuous paths provided by Proposition 1.80.

The rest of this chapter is organized in the following way. In Section 5.2, starting from a Lemarié-Meyer orthonormal wavelet basis of $L^2(\mathbb{R}^N)$, one constructs a wavelet series representation of X, which can be more or less viewed as an extension of the representation (3.30) of Brownian Motion in the Faber-Schauder system. Also, one shows that the series is almost surely uniformly convergent in (u, v) on each compact subset of $\mathbb{R} \times (0, 1)$, and one studies some important properties of the wavelet functions appearing in it (which should not be confused with Lemarié-Meyer wavelets). In Section 5.3, one proves that the rate of convergence of this series in uniform norm is optimal in the sense of [Kühn and Linde (2002)].

5.2   Construction of the wavelet series representation

First one needs to introduce some notations.

Definition 5.1. The sequence of functions
$$\left\{\psi_{p,j,k} : p \in \{1, \ldots, 2^N - 1\}, \ j \in \mathbb{Z} \text{ and } k \in \mathbb{Z}^N\right\},$$
where, for all $(p, j, k, x) \in \{1, \ldots, 2^N - 1\} \times \mathbb{Z} \times \mathbb{Z}^N \times \mathbb{R}^N$,
$$\psi_{p,j,k}(x) := 2^{jN/2}\, \psi_p(2^j x - k), \tag{5.4}$$
denotes a Lemarié-Meyer orthonormal wavelet basis of $L^2(\mathbb{R}^N)$ (see Definition 4.30 and Corollary 4.31). Recall that the $2^N - 1$ functions $\psi_1, \ldots, \psi_{2^N-1}$ generating the basis are called the Lemarié-Meyer mother wavelets, and that they belong to the Schwartz class $\mathcal{S}(\mathbb{R}^N)$ (see Definition 4.21), and that their Fourier transforms are infinitely differentiable compactly supported functions satisfying (4.82). Observe that, thanks to the equality (5.4) and to its equivalent version [1] in the frequency domain: for all $(p, j, k, \xi) \in \{1, \ldots, 2^N - 1\} \times \mathbb{Z} \times \mathbb{Z}^N \times \mathbb{R}^N$,
$$\widehat{\psi}_{p,j,k}(\xi) = 2^{-jN/2}\, e^{-2^{-j}k\cdot\xi}\, \widehat{\psi}_p(2^{-j}\xi), \tag{5.5}$$
these nice properties $(\psi_p, \widehat{\psi}_p \in \mathcal{S}(\mathbb{R}^N)$ and $\widehat{\psi}_p$ compactly supported) are also satisfied by all the wavelets $\psi_{p,j,k}$; yet, (4.82) has to be replaced by
$$\text{Supp}\,\widehat{\psi}_{p,j,k} \subseteq \Delta_j, \quad \text{for each } (p, j, k) \in \{1, \ldots, 2^N - 1\} \times \mathbb{Z} \times \mathbb{Z}^N, \tag{5.6}$$

---

[1]The fact that the equalities (5.5) and (5.4) are equivalent easily follows from parts *(iii)* and *(iv)* of Proposition 4.5.

where

$$\Delta_j := \left[-\frac{2^{j+3}\pi}{3}, \frac{2^{j+3}\pi}{3} \right]^N \setminus \left(-\frac{2^{j+1}\pi}{3}, \frac{2^{j+1}\pi}{3} \right)^N. \quad (5.7)$$

Remark 5.2. Notice that an important consequence of (5.6) and (5.7) is that the set $\mathrm{Supp}\,\widehat{\psi}_{p',j',k'} \cap \mathrm{Supp}\,\widehat{\psi}_{p'',j'',k''}$ is Lebesgue negligible, for all $(p',j',k') \in \{1,\ldots,2^N-1\} \times \mathbb{Z} \times \mathbb{Z}^N$ and $(p'',j'',k'') \in \{1,\ldots,2^N-1\} \times \mathbb{Z} \times \mathbb{Z}^N$ such that $j'' \notin \{j'-1, j', j'+1\}$.

Definition 5.3. For each fixed $p \in \{1,\ldots,2^N-1\}$, one denotes by Ψ_p the deterministic function defined on $\mathbb{R}^N \times \mathbb{R}$ as:

$$\Psi_p(y,v) = \frac{1}{(2\pi)^{N/2}} \int_{\mathbb{R}^N} e^{iy\cdot\eta} \frac{\widehat{\psi}_p(\eta)}{|\eta|^{v+N/2}}\, d\eta, \quad \text{for all } (y,v) \in \mathbb{R}^N \times \mathbb{R}. \quad (5.8)$$

It easily follows from Corollary 4.31, Definition 4.30, Proposition 4.29 and the dominated convergence theorem that Ψ_p is a well-defined real-valued continuous function. Soon (see Propositions 5.10, 5.11 and 5.13), we will show that Ψ_p has much better properties.

Definition 5.4. For all $(p,j,k) \in \{1,\ldots,2^N-1\} \times \mathbb{Z} \times \mathbb{Z}^N$, one sets

$$\varepsilon_{p,j,k} := \int_{\mathbb{R}^N} \overline{\widehat{\psi}_{p,j,k}(\xi)}\, d\widehat{\mathbb{W}}(\xi) = \int_{\mathbb{R}^N} \psi_{p,j,k}(s)\, d\mathbb{W}(s), \quad (5.9)$$

where the last equality results from (1.31) and from the fact that the function $\psi_{p,j,k}$ is real-valued. Notice that it easily follows from Definition 1.24 and from the orthonormality property of the functions $\psi_{p,j,k}$, that

$$\left\{ \varepsilon_{p,j,k} : p \in \{1,\ldots,2^N-1\}, \ j \in \mathbb{Z} \text{ and } k \in \mathbb{Z}^N \right\},$$

is a sequence of independent real-valued $\mathcal{N}(0,1)$ Gaussian random variables.

Now, we are in position to state the main result of this section

Theorem 5.5 (wavelet representation of Generator of MuBF).
The Generator of Multifractional Brownian Field $X := \{X(u,v), (u,v) \in \mathbb{R}^N \times (0,1)\}$ (see (5.1) and (5.2)) can be represented as:

$$X(u,v) = \sum_{p=1}^{2^N-1} \sum_{(j,k)\in\mathbb{Z}\times\mathbb{Z}^N} 2^{-jv} \varepsilon_{p,j,k} \big(\Psi_p(2^j u - k, v) - \Psi_p(-k, v) \big), \quad (5.10)$$

where the series is, with probability 1, uniformly convergent in (u,v), on each compact interval of $\mathbb{R}^N \times (0,1)$. More precisely, let $(\mathcal{D}_n)_{n\in\mathbb{Z}_+}$ be any arbitrary nondecreasing sequence of finite subsets of $\mathbb{Z} \times \mathbb{Z}^N$, satisfying $\mathbb{Z} \times \mathbb{Z}^N = \bigcup_{n\in\mathbb{Z}_+} \mathcal{D}_n$; then, there exists $\widetilde{\Omega}$, an event of probability 1 ($\widetilde{\Omega}$ a priori *depends on the choice of* $(\mathcal{D}_n)_{n\in\mathbb{Z}_+}$), *such that for all* $\omega \in \widetilde{\Omega}$, *for every compact interval* J *of* $\mathbb{R}^N$, *and for each real numbers* M_1 *and* M_2 *satisfying* $0 < M_1 < M_2 < 1$, *one has*

$$
\sup_{(u,v)\in J\times[M_1,M_2]} \Big| X(u,v,\omega) \tag{5.11}
$$

$$
- \sum_{p=1}^{2^N-1} \sum_{(j,k)\in\mathcal{D}_n} 2^{-jv} \varepsilon_{p,j,k}(\omega)\big(\Psi_p(2^j u - k, v) - \Psi_p(-k, v)\big) \Big| \xrightarrow[n\to+\infty]{} 0.
$$

Remark 5.6. Wavelet series representations of Fractional Brownian Motion have been introduced, since the early 90's, by several different authors in more or less rigorous ways, an excellent reference on them is the seminal article [Meyer *et al.* (1999)] (see also [Doukhan *et al.* (2003)]) which, among many other things, provides a completely rigorous framework. In the case of Fractional Brownian Field, as far as we know, a wavelet series representation was introduced for the first time in [Benassi *et al.* (1997)].

Proof of Theorem 5.5. Using the fact that the sequence of functions

$$
\big\{\psi_{p,j,k} : p \in \{1,\dots,2^N-1\},\ j \in \mathbb{Z} \text{ and } k \in \mathbb{Z}^N\big\}
$$

is an orthonormal basis of $L^2(\mathbb{R}^N)$ (see Definition 5.1), and the isometry property of the Fourier transform map from $L^2(\mathbb{R}^N)$ into itself (see part (ii) of Proposition 4.5), it follows that the sequence of functions

$$
\Big\{\overline{\widehat{\psi}_{p,j,k}} : p \in \{1,\dots,2^N-1\},\ j \in \mathbb{Z} \text{ and } k \in \mathbb{Z}^N\Big\},
$$

is also an orthonormal basis of $L^2(\mathbb{R}^N)$. Expanding, for each fixed couple $(u,v) \in \mathbb{R}^N \times (0,1)$, the function $\mathbb{K}(u,v,\cdot) : \xi \mapsto \mathbb{K}(u,v,\xi)$ (see (5.2)), on the latter basis, one gets that

$$
\mathbb{K}(u,v,\cdot) = \sum_{p=1}^{2^N-1} \sum_{(j,k)\in\mathbb{Z}\times\mathbb{Z}^N} \langle \mathbb{K}(u,v,\cdot), \overline{\widehat{\psi}_{p,j,k}}\rangle_{L^2(\mathbb{R}^N)} \overline{\widehat{\psi}_{p,j,k}}(\cdot), \tag{5.12}
$$

where the series is unconditionally convergent in the sense of the norm $\|\cdot\|_{L^2(\mathbb{R}^N)}$. Moreover, using the definition of the inner product $\langle\cdot,\cdot\rangle_{L^2(\mathbb{R}^N)}$, (5.2), (5.5), the change of variable $\eta = 2^{-j}\xi$, and (5.8), one obtains that

$$
\langle \mathbb{K}(u,v,\cdot), \overline{\widehat{\psi}_{p,j,k}}\rangle_{L^2(\mathbb{R}^N)} = \frac{2^{-jN/2}}{(2\pi)^{N/2}} \int_{\mathbb{R}^N} \Big(\frac{e^{iu\cdot\xi} - 1}{|\xi|^{v+N/2}}\Big) e^{-2^{-j}k\cdot\xi} \,\widehat{\psi}_p(2^{-j}\xi)\,d\xi
$$

$$
= 2^{-jv}\big(\Psi_p(2^j u - k, v) - \Psi_p(-k, v)\big). \tag{5.13}
$$

Next, it follows from (5.1), (5.12), (5.9), (5.13) and the isometry property of the integral $\int_{\mathbb{R}^N} (\,\cdot\,)d\widehat{W}$ (see (1.30)), that $X(u,v)$ can be expressed as in (5.10) where the series is unconditionally convergent in the sense of the norm $\|\cdot\|_{L^2(\Omega)}$. Finally, the inequality (1.116) allows one to use Theorem A.22 in the Appendix, in order to derive (5.11). $\square$

Thanks to Theorem 5.5, one can define the low and high frequency parts of X, the Generator of MuBF in the following ways.

Definition 5.7 (low frequency part of Generator of MuBF). The low frequency part of X is the centered Gaussian field denoted by $X^{\mathrm{lf}} := \{X^{\mathrm{lf}}(u,v),\,(u,v)\in\mathbb{R}^N\times(0,1)\}$ and defined, for all $(u,v)\in\mathbb{R}^N\times(0,1)$, as:

$$X^{\mathrm{lf}}(u,v) = \sum_{p=1}^{2^N-1}\sum_{j=-\infty}^{-1}\sum_{k\in\mathbb{Z}^N} 2^{-jv}\,\varepsilon_{p,j,k}\big(\Psi_p(2^ju-k,v)-\Psi_p(-k,v)\big). \quad (5.14)$$

Definition 5.8 (high frequency part of Generator of MuBF). The high frequency part of X is the centered Gaussian field denoted by $X^{\mathrm{hf}} := \{X^{\mathrm{hf}}(u,v),\,(u,v)\in\mathbb{R}^N\times(0,1)\}$ and defined, for all $(u,v)\in\mathbb{R}^N\times(0,1)$, as:

$$X^{\mathrm{hf}}(u,v) = \sum_{p=1}^{2^N-1}\sum_{j=0}^{+\infty}\sum_{k\in\mathbb{Z}^N} 2^{-jv}\,\varepsilon_{p,j,k}\big(\Psi_p(2^ju-k,v)-\Psi_p(-k,v)\big). \quad (5.15)$$

Remark 5.9. Using the same method as in the proof of Theorem 5.5, it can be shown that the series in (5.14) and (5.15) are, with probability 1, uniformly convergent in (u,v), on each compact interval of $\mathbb{R}^N\times(0,1)$. Therefore, the continuity property of the functions Ψ_p implies that the two fields $X^{\mathrm{lf}} := \{X^{\mathrm{lf}}(u,v),\,(u,v)\in\mathbb{R}^N\times(0,1)\}$ and $X^{\mathrm{hf}} := \{X^{\mathrm{hf}}(u,v),\,(u,v)\in\mathbb{R}^N\times(0,1)\}$ have almost surely continuous paths. Also, notice that one clearly has, almost surely, that

$$X(u,v) = X^{\mathrm{lf}}(u,v) + X^{\mathrm{hf}}(u,v), \quad \text{for every } (u,v)\in\mathbb{R}^N\times(0,1). \quad (5.16)$$

The following three propositions provide some useful properties of the functions Ψ_p, $p\in\{1,\dots,2^N-1\}$, which were introduced in Definition 5.3.

Proposition 5.10. *The functions* Ψ_p, $p\in\{1,\dots,2^N-1\}$, *are infinitely differentiable on* $\mathbb{R}^N\times\mathbb{R}$. *Moreover, for each* $\gamma=(\gamma_1,\dots,\gamma_N)\in\mathbb{Z}_+^N$ *and* $m\in\mathbb{Z}_+$, *denoting by* $\partial_y^\gamma\partial_v^m$ *the partial derivative operator defined as:*

$$\partial_y^\gamma\partial_v^m := \frac{\partial^{\gamma_1+\dots+\gamma_N+m}}{(\partial y_1)^{\gamma_1}\dots(\partial y_N)^{\gamma_N}(\partial v)^m}, \quad (5.17)$$

then, the partial derivative $\partial_y^\gamma \partial_v^m \Psi_p$ is given by: for all $(y, v) \in \mathbb{R}^N \times \mathbb{R}$,

$$(\partial_y^\gamma \partial_v^m \Psi_p)(y, v) = \frac{1}{(2\pi)^{N/2}} \int_{\mathbb{R}^N} e^{iy \cdot \eta} \frac{(i\eta)^\gamma (-\log|\eta|)^m \widehat{\psi_p}(\eta)}{|\eta|^{v+N/2}} \, d\eta, \qquad (5.18)$$

with the convention that

$$(i\eta)^\gamma := \prod_{n=1}^{N} (i\eta_n)^{\gamma_n}, \quad \text{for every } \eta = (\eta_1, \dots, \eta_N) \in \mathbb{R}^N. \qquad (5.19)$$

Proof of Proposition 5.10. First observe that for each $\eta \in \mathbb{R}^N \setminus \{0\}$, $\gamma \in \mathbb{Z}^N$ and $m \in \mathbb{Z}_+$ arbitrary and fixed, applying the partial derivative operator $\partial_y^\gamma \partial_v^m$ to the integrand in (5.8), one obtains the integrand in (5.18). Also, observe that the latter integrand is clearly a continuous function on $\mathbb{R}^N \times \mathbb{R}$ with respect to (y, v). Thus, in view of Proposition A.20, in order to obtain Proposition 5.10, it is enough to show that, under the condition $|y| + |v| \le B$, the positive constant B being arbitrary, there exists a function $g_{\gamma, m, B}$, not depending on (y, v), belonging to $L^1(\mathbb{R}^N)$, and such that one has

$$\left| e^{iy \cdot \eta} \frac{(i\eta)^\gamma (-\log|\eta|)^m \widehat{\psi_p}(\eta)}{|\eta|^{v+N/2}} \right| \le g_{\gamma, m, B}(\eta), \quad \text{for almost all } \eta \in \mathbb{R}^N.$$

It is clear that such a function $g_{\gamma, m, B}$ exists; an natural choice of it is the continuous compactly supported function defined as $g_{\gamma, m, B}(0) := 0$ and, for all $\eta \in \mathbb{R}^N \setminus \{0\}$,

$$g_{\gamma, m, B}(\eta) := |\eta|^{\gamma_1 + \dots + \gamma_N - N/2} \left| \log|\eta| \right|^m \left| \widehat{\psi_p}(\eta) \right| \left(|\eta|^B + |\eta|^{-B} \right).$$

$\square$

Proposition 5.11. *For all $p \in \{1, \dots, 2^N - 1\}$, $\gamma \in \mathbb{Z}_+^N$ and $m \in \mathbb{Z}_+$, the function $\partial_y^\gamma \partial_v^m \Psi_p$ is well-localized in y uniformly in v restricted to any arbitrary compact interval of $\mathbb{R}$. More precisely, for every fixed nonnegative integer L, and positive real number M, one has that*

$$\sup \left\{ \left(\prod_{n=1}^{N} (3+|y_n|)^L \right) |(\partial_y^\gamma \partial_v^m \Psi_p)(y, v)| \ : \ (y, v) \in \mathbb{R}^N \times [-M, M] \right\} < +\infty. \qquad (5.20)$$

Proof of Proposition 5.11. One assumes that $y = (y_1, \dots, y_N) \in \mathbb{R}^N$ and $v \in [-M, M]$ are arbitrary and fixed. Let $\delta = (\delta_1, \dots, \delta_N) \in \{-1, 1\}^N$ be such that, for all $n \in \{1, \dots, N\}$, one has $\delta_n = 1$ if $y_n \ge 0$ and $\delta_n = -1$ else. Then, for each $\eta = (\eta_1, \dots, \eta_N) \in \mathbb{R}^N$, the inner product $y \cdot \eta :=$

$\sum_{n=1}^{N} y_n \eta_n$ can be expressed as: $y \cdot \eta = \sum_{n=1}^{N} \delta_n(3 + |y_n|)\eta_n - 3\delta \cdot \eta$. Thus, in view of (5.18), one has

$$(\partial_y^\gamma \partial_v^m \Psi_p)(y, v) = \int_{\mathbb{R}^N} \Big(\prod_{n=1}^{N} e^{i\delta_n(3+|y_n|)\eta_n} \Big) F_{\gamma,m,\delta,p}(v, \eta)\, d\eta, \qquad (5.21)$$

where, $F_{\gamma,m,\delta,p}(v, 0) := 0$ and, for all $\eta \in \mathbb{R}^N \setminus \{0\}$,

$$F_{\gamma,m,\delta,p}(v, \eta) := e^{-i3\delta \cdot \eta} \frac{(i\eta)^\gamma (-\log|\eta|)^m \widehat{\psi}_p(\eta)}{|\eta|^{v+N/2}}. \qquad (5.22)$$

Observe that the function $\eta \mapsto F_{\gamma,m,\delta,p}(v, \eta)$ is infinitely differentiable on $\mathbb{R}^N$ and that (5.6) and (5.5) imply that it is with compact support satisfying

$$\operatorname{Supp} F_{\gamma,m,\delta,p}(v, \cdot) \subseteq \Delta_0, \qquad (5.23)$$

where Δ_0 is the compact subset of $\mathbb{R}^N \setminus \{0\}$ defined in (5.7) with $j = 0$. Next, one denotes by $\langle L \rangle$ the multi-index of $\mathbb{Z}^N$ whose components are all equal to the same nonnegative integer L. Using (5.21), integrations by parts, and (5.23), one gets that

$$\big|(\partial_y^\gamma \partial_v^m \Psi_p)(y, v)\big| \leq \Big(\prod_{n=1}^{N} (3 + |y_n|)^{-L} \Big) \int_{\Delta_0} \big|(\partial_\eta^{\langle L \rangle} F_{\gamma,m,\delta,p})(v, \eta)\big|\, d\eta.$$
$$(5.24)$$

Moreover, using (5.22), the general Leibniz rule and classical computations, for all $\eta \in \mathbb{R}^N \setminus \{0\}$, one obtains

$$\big|(\partial_\eta^{\langle L \rangle} F_{\gamma,m,\delta,p})(v, \eta)\big| \leq \frac{G_{\gamma,m,\delta,p}^{N,L}(\eta)}{|\eta|^v},$$

where $G_{\gamma,m,\delta,p}^{N,L}$ is a continuous function on $\mathbb{R}^N$, not depending on v, and with a compact support included in Δ_0. Thus, denoting by c the finite constant defined as

$$c := \sup \Big\{ \big|G_{\gamma,m,\delta,p}^{N,L}(\eta)\big| \; : \; (\delta, \eta) \in \{-1, 1\}^N \times \Delta_0 \Big\},$$

it follows that

$$\int_{\Delta_0} \big|(\partial_\eta^{\langle L \rangle} F_{\gamma,m,\delta,p})(v, \eta)\big|\, d\eta \leq c \int_{\Delta_0} \big(|\eta|^{-M} + |\eta|^M \big)\, d\eta < +\infty. \qquad (5.25)$$

Finally, combining (5.24) and (5.25), one gets (5.20). $\qquad\square$

Remark 5.12 (vanishing moments property for $(\partial_y^\gamma \partial_v^m \Psi_p)(\cdot, v)$).
Using Propositions 5.11 and 5.10, classical properties of Fourier transform, and (4.82), for all $\lambda \in \mathbb{Z}_+^N$, $\gamma \in \mathbb{Z}_+^N$, $m \in \mathbb{Z}_+$ and $v \in \mathbb{R}$, one has

$$\int_{\mathbb{R}^N} y^\lambda \, (\partial_y^\gamma \partial_v^m \Psi_p)(y, v)\, dy = 0\,.$$

Proposition 5.13. *Let Ψ_p be the same real-valued function on $\mathbb{R}^N \times \mathbb{R}$ as in Definition 5.3, where $p \in \{1, \dots, 2^N - 1\}$ is arbitrary. For each fixed $v \in \mathbb{R}$, one denotes by $\Psi_p^{(v)}$ the function, from $\mathbb{R}^N$ into $\mathbb{R}$,*

$$y \mapsto \Psi_p^{(v)}(y) := \Psi_p(y, v). \tag{5.26}$$

Observe that one knows from Propositions 5.10 and 5.11 that $\Psi_p^{(v)}$ belongs to the Schwartz class $\mathcal{S}(\mathbb{R}^N)$. More generally, for every $(j, k) \in \mathbb{Z} \times \mathbb{Z}^N$, one denotes by $\Psi_{p,j,k}^{(v)}$ the function of $\mathcal{S}(\mathbb{R}^N)$ defined as:

$$\Psi_{p,j,k}^{(v)}(y) := 2^{jN/2}\, \Psi_p^{(v)}(2^j y - k) = 2^{jN/2}\, \Psi_p(2^j y - k, v). \tag{5.27}$$

Then, for any fixed $v \in \mathbb{R}$, the two sequences of functions

$$\left\{ \Psi_{p,j,k}^{(v)} : p \in \{1, \dots, 2^N - 1\},\ j \in \mathbb{Z} \text{ and } k \in \mathbb{Z}^N \right\}$$

and

$$\left\{ \Psi_{p,j,k}^{(-v-N)} : p \in \{1, \dots, 2^N - 1\},\ j \in \mathbb{Z} \text{ and } k \in \mathbb{Z}^N \right\}$$

are two biorthogonal wavelet bases of $L^2(\mathbb{R}^N)$. That is they satisfy the following two properties.

(i) The biorthogonality property: for all $(p', j', k') \in \{1, \dots, 2^N - 1\} \times \mathbb{Z} \times \mathbb{Z}^N$ and $(p'', j'', k'') \in \{1, \dots, 2^N - 1\} \times \mathbb{Z} \times \mathbb{Z}^N$, one has:

$$\langle \Psi_{p',j',k'}^{(v)}, \Psi_{p'',j'',k''}^{(-v-N)} \rangle_{L^2(\mathbb{R}^N)} = \begin{cases} 1, & \text{if } (p', j', k') = (p'', j'', k''), \\ 0, & \text{else.} \end{cases} \tag{5.28}$$

(ii) The Riesz basis property: for each $f \in L^2(\mathbb{R}^N)$, there exist two unique sequences, belonging to the space $l^2\big(\{1, \dots, 2^N - 1\} \times \mathbb{Z} \times \mathbb{Z}^N\big)$ and denoted by $(b_{p,j,k}^{(v)})_{p,j,k}$ and $(b_{p,j,k}^{(-v-N)})_{p,j,k}$, such that one has

$$f = \sum_{p=1}^{2^N-1} \sum_{(j,k)\in\mathbb{Z}\times\mathbb{Z}^N} b_{p,j,k}^{(-v-N)} \Psi_{p,j,k}^{(v)} = \sum_{p=1}^{2^N-1} \sum_{(j,k)\in\mathbb{Z}\times\mathbb{Z}^N} b_{p,j,k}^{(v)}, \Psi_{p,j,k}^{(-v-N)}, \tag{5.29}$$

where the series are unconditionally convergent in the sense of the $L^2(\mathbb{R}^N)$ norm.

Notice that, there exist two constants $0 < c' \leq c'' < +\infty$, not depending on the sequences $(b_{p,j,k}^{(v)})_{p,j,k}$ and $(b_{p,j,k}^{(-v-N)})_{p,j,k}$, such that one has

$$c' \sum_{p=1}^{2^N-1} \sum_{(j,k)\in\mathbb{Z}\times\mathbb{Z}^N} |b_{p,j,k}^{(-v-N)}|^2 \leq \|f\|_{L^2(\mathbb{R}^N)}^2 \leq c'' \sum_{p=1}^{2^N-1} \sum_{(j,k)\in\mathbb{Z}\times\mathbb{Z}^N} |b_{p,j,k}^{(-v-N)}|^2$$

and

$$c' \sum_{p=1}^{2^N-1} \sum_{(j,k)\in\mathbb{Z}\times\mathbb{Z}^N} |b^{(v)}_{p,j,k}|^2 \leq \|f\|^2_{L^2(\mathbb{R}^N)} \leq c'' \sum_{p=1}^{2^N-1} \sum_{(j,k)\in\mathbb{Z}\times\mathbb{Z}^N} |b^{(v)}_{p,j,k}|^2.$$

Also notice that, it easily follows from (5.29) and (5.28) that, for all $(p,j,k) \in \{1,\ldots,2^N-1\} \times \mathbb{Z} \times \mathbb{Z}^N$, *one has*

$$b^{(-v-N)}_{p,j,k} = \langle f, \Psi^{(-v-N)}_{p,j,k} \rangle_{L^2(\mathbb{R}^N)} \quad and \quad b^{(v)}_{p,j,k} = \langle f, \Psi^{(v)}_{p,j,k} \rangle_{L^2(\mathbb{R}^N)}. \quad (5.30)$$

Proof of Proposition 5.13. First observe that, in view of the isometry property of the Fourier transform map from $L^2(\mathbb{R}^N)$ into itself (see part (ii) of Proposition 4.5), in order to prove Proposition 5.13, it is enough to show that its parts (i) and (ii) hold when the functions $\Psi^{(v)}_{p',j',k'}$, $\Psi^{(-v-N)}_{p'',j'',k''}$, $\Psi^{(v)}_{p,j,k}$ and $\Psi^{(-v-N)}_{p,j,k}$ are replaced by their Fourier transforms. Also, observe that one can easily derive from (5.26), (5.27), (5.8) and (5.5) that, for all $v \in \mathbb{R}$, $(p,j,k) \in \{1,\ldots,2^N-1\} \times \mathbb{Z} \times \mathbb{Z}^N$, and $\eta \in \mathbb{R}^N$, one has

$$\widehat{\Psi}^{(v)}_{p,j,k}(\eta) = 2^{-jN/2} e^{-i2^{-j}k\cdot\eta} \widehat{\Psi}^{(v)}_p(2^{-j}\eta) \quad (5.31)$$

$$= 2^{-jN/2} e^{-i2^{-j}k\cdot\eta} \frac{\widehat{\psi}_p(2^{-j}\eta)}{|2^{-j}\eta|^{v+N/2}} = \frac{\widehat{\psi}_{p,j,k}(\eta)}{|2^{-j}\eta|^{v+N/2}}.$$

Then, using the definition of the inner product $\langle\cdot,\cdot\rangle_{L^2(\mathbb{R}^N)}$, and (5.31), it follows that, for all $(p',j',k') \in \{1,\ldots,2^N-1\}\times\mathbb{Z}\times\mathbb{Z}^N$ and $(p'',j'',k'') \in \{1,\ldots,2^N-1\}\times\mathbb{Z}\times\mathbb{Z}^N$, one has

$$\langle\widehat{\Psi}^{(v)}_{p',j',k'}, \widehat{\Psi}^{(-v-N)}_{p'',j'',k''}\rangle_{L^2(\mathbb{R}^N)} = \int_{\mathbb{R}^N} \frac{\widehat{\psi}_{p',j',k'}(\eta)}{|2^{-j'}\eta|^{v+N/2}} |2^{-j''}\eta|^{v+N/2} \overline{\widehat{\psi}_{p'',j'',k''}(\eta)}\, d\eta$$

$$= 2^{(j'-j'')(v+N/2)} \int_{\mathbb{R}^N} \widehat{\psi}_{p',j',k'}(\eta)\overline{\widehat{\psi}_{p'',j'',k''}(\eta)}\, d\eta$$

$$= 2^{(j'-j'')(v+N/2)} \langle\widehat{\psi}_{p',j',k'}, \widehat{\psi}_{p'',j'',k''}\rangle_{L^2(\mathbb{R}^N)}. \quad (5.32)$$

Thus, using the orthonormality of the functions $\widehat{\psi}_{p,j,k}$ one obtains (5.28).

Having shown that (i) is satisfied, let us now prove that (ii) is satisfied as well. Recall that the notion of a Riesz basis of a Hilbert space has been introduced in Definition 4.2. Let us first show, for every $v \in \mathbb{R}$, that

$$L^2(\mathbb{R}^N) = \overline{\text{span}\Big\{\widehat{\Psi}^{(v)}_{p,j,k} : p \in \{1,\ldots,2^N-1\}, \ j \in \mathbb{Z} \text{ and } k \in \mathbb{Z}^N\Big\}}. \quad (5.33)$$

To this end, for each $J \in \mathbb{Z}_+$, one denotes by $\mathcal{O}_J$ the compact subset of $\mathbb{R}^N$ defined as

$$\mathcal{O}_J := \Big[-\frac{2^{J+3}\pi}{3}, \frac{2^{J+3}\pi}{3}\Big]^N \setminus \Big(-\frac{2^{-J+1}\pi}{3}, \frac{2^{-J+1}\pi}{3}\Big)^N. \quad (5.34)$$

Observe that it follows from (5.7) that

$$\mathcal{O}_J = \bigcup_{j=-J}^{J} \Delta_j. \tag{5.35}$$

Also, observe that

$$\mathcal{O}_J \subset \mathcal{O}_{J+1}, \text{ for every } J \in \mathbb{Z}_+, \text{ and } \mathbb{R}^N \setminus \{0\} = \bigcup_{J=0}^{+\infty} \mathcal{O}_J. \tag{5.36}$$

Next, assume that $f \in L^2(\mathbb{R}^N)$ is arbitrary and that ϵ is an arbitrarily small positive real number. In view of (5.36), there exists a positive integer J_0 (depending on ϵ) such that

$$\int_{\mathbb{R}^N} \left| f(\eta) - f(\eta)\mathbb{1}_{\mathcal{O}_{J_0}}(\eta) \right|^2 d\eta \leq \frac{\epsilon^2}{4}. \tag{5.37}$$

Next, one denotes by g the function of $L^2(\mathbb{R}^N)$ defined as:

$$g(\eta) = |\eta|^{v+N/2} f(\eta)\mathbb{1}_{\mathcal{O}_{J_0}}(\eta), \quad \text{for all } \eta \in \mathbb{R}^N. \tag{5.38}$$

Expanding g on the orthonormal basis

$$\left\{ \widehat{\psi}_{p,j,k} : p \in \{1,\ldots,2^N - 1\}, \ j \in \mathbb{Z} \text{ and } k \in \mathbb{Z}^N \right\}$$

and using the fact that $\text{Supp}\, g \subseteq \mathcal{O}_{J_0}$ as well as (5.6), (5.7) and (5.35), one gets that

$$g = \sum_{p=1}^{2^N-1} \sum_{-J_0-1}^{J_0+1} \sum_{k \in \mathbb{Z}^N} \langle g, \widehat{\psi}_{p,j,k} \rangle_{L^2(\mathbb{R}^N)} \widehat{\psi}_{p,j,k},$$

where the series is unconditionally convergent in the sense of the $L^2(\mathbb{R}^N)$-norm. Therefore, there exists a positive integer K_0 (depending on ϵ) such that one has

$$\int_{\mathbb{R}^N} \left| g(\eta) - \sum_{p=1}^{2^N-1} \sum_{-J_0-1}^{J_0+1} \sum_{|k| \leq K_0} \langle g, \widehat{\psi}_{p,j,k} \rangle_{L^2(\mathbb{R}^N)} \widehat{\psi}_{p,j,k}(\eta) \right|^2 d\eta$$

$$\leq \frac{\epsilon^2}{4\left(2^{J_0+4}\, \pi \, \sqrt{N}\right)^{2|v|+N}}. \tag{5.39}$$

Next, observe that, for every $\eta \in \mathcal{O}_{J_0+1}$, (5.34) implies that

$$\frac{2^{-J_0}\pi}{3} \leq |\eta| \leq \frac{2^{J_0+4}\, \pi \, \sqrt{N}}{3},$$

and consequently that

$$
|\eta|^{-2v-N} \leq \left(|\eta| + |\eta|^{-1}\right)^{|2v+N|} \leq \left(|\eta| + |\eta|^{-1}\right)^{2|v|+N}
$$
$$
\leq \left(\frac{2^{J_0+4}\,\pi\,\sqrt{N}}{3} + \frac{3}{\pi}\,2^{J_0}\right)^{2|v|+N} \leq \left(2^{J_0+4}\,\pi\,\sqrt{N}\right)^{2|v|+N}.
$$

Thus, using (5.39) and the fact that the support of the integrand in it is included in $\mathcal{O}_{J_0+1}$, one obtains that

$$
\int_{\mathbb{R}^N} \left| g(\eta) - \sum_{p=1}^{2^N-1} \sum_{-J_0-1}^{J_0+1} \sum_{|k|\leq K_0} \langle g, \widehat{\psi}_{p,j,k}\rangle_{L^2(\mathbb{R}^N)} \widehat{\psi}_{p,j,k}(\eta) \right|^2 |\eta|^{-2v-N}\,d\eta \leq \frac{\epsilon^2}{4}.
$$

Next, notice that, in view of (5.38) and (5.31) the latter inequality can be expressed as

$$
\int_{\mathbb{R}^N} \left| f(\eta) \mathbb{1}_{\mathcal{O}_{J_0}}(\eta) \right. \tag{5.40}
$$
$$
\left. - \sum_{p=1}^{2^N-1} \sum_{-J_0-1}^{J_0+1} \sum_{|k|\leq K_0} 2^{-j(v+N/2)} \langle g, \widehat{\psi}_{p,j,k}\rangle_{L^2(\mathbb{R}^N)} \widehat{\Psi}_{p,j,k}^{(v)}(\eta) \right|^2 d\eta \leq \frac{\epsilon^2}{4}.
$$

Next, it follows from (5.37), (5.40) and the triangle inequality that

$$
\left\| f - \sum_{p=1}^{2^N-1} \sum_{-J_0-1}^{J_0+1} \sum_{|k|\leq K_0} 2^{-j(v+N/2)} \langle g, \widehat{\psi}_{p,j,k}\rangle_{L^2(\mathbb{R}^N)} \widehat{\Psi}_{p,j,k}^{(v)} \right\|_{L^2(\mathbb{R}^N)} \leq \epsilon,
$$

which shows that (5.33) holds.

Let us now prove that, for every fixed $v \in \mathbb{R}$,

$$
\left\{ \widehat{\Psi}_{p,j,k}^{(v)} : p \in \{1,\ldots,2^N-1\},\ j \in \mathbb{Z} \text{ and } k \in \mathbb{Z}^N \right\}
$$

is a Riesz sequence of $L^2(\mathbb{R}^N)$, which means that there are two constants $0 < c_1(v) \leq c_2(v)$, such that, for all complex-valued sequence $(a_{p,j,k})_{p,j,k}$ with a finite number of non vanishing terms, one has

$$
c_1(v) \sum_{p=1}^{2^N-1} \sum_{(j,k)\in\mathbb{Z}\times\mathbb{Z}^N} |a_{p,j,k}|^2 \leq \left\| \sum_{p=1}^{2^N-1} \sum_{(j,k)\in\mathbb{Z}\times\mathbb{Z}^N} a_{p,j,k} \widehat{\Psi}_{p,j,k}^{(v)} \right\|_{L^2(\mathbb{R}^N)}^2
$$
$$
\leq c_2(v) \sum_{p=1}^{2^N-1} \sum_{(j,k)\in\mathbb{Z}\times\mathbb{Z}^N} |a_{p,j,k}|^2. \tag{5.41}
$$

Using the biorthogonality property (part (i) of the proposition), it turns out that it is enough to show that, for all fixed $v \in \mathbb{R}$, the second inequality in (5.41) holds. Indeed, combining this property with Cauchy-Schwarz inequality and the second inequality in (5.41), one gets that

$$
\sum_{p=1}^{2^N-1} \sum_{(j,k)\in\mathbb{Z}\times\mathbb{Z}^N} |a_{p,j,k}|^2
$$

$$
= \left\langle \sum_{p=1}^{2^N-1} \sum_{(j,k)\in\mathbb{Z}\times\mathbb{Z}^N} a_{p,j,k} \widehat{\Psi}_{p,j,k}^{(v)}, \sum_{p=1}^{2^N-1} \sum_{(j,k)\in\mathbb{Z}\times\mathbb{Z}^N} a_{p,j,k} \widehat{\Psi}_{p,j,k}^{(-v-N)} \right\rangle_{L^2(\mathbb{R}^N)}
$$

$$
\le \left\| \sum_{p=1}^{2^N-1} \sum_{(j,k)\in\mathbb{Z}\times\mathbb{Z}^N} a_{p,j,k} \widehat{\Psi}_{p,j,k}^{(v)} \right\|_{L^2(\mathbb{R}^N)}
$$

$$
\times \left\| \sum_{p=1}^{2^N-1} \sum_{(j,k)\in\mathbb{Z}\times\mathbb{Z}^N} a_{p,j,k} \widehat{\Psi}_{p,j,k}^{(-v-N)} \right\|_{L^2(\mathbb{R}^N)}
$$

$$
\le \left\| \sum_{p=1}^{2^N-1} \sum_{(j,k)\in\mathbb{Z}\times\mathbb{Z}^N} a_{p,j,k} \widehat{\Psi}_{p,j,k}^{(v)} \right\|_{L^2(\mathbb{R}^N)}
$$

$$
\times \left(c_2(-v-N) \sum_{p=1}^{2^N-1} \sum_{(j,k)\in\mathbb{Z}\times\mathbb{Z}^N} |a_{p,j,k}|^2 \right)^{1/2}.
$$

Thus, setting $c_1(v) := 1/c_2(-v-N)$, one obtains the first inequality in (5.41). Finally, let us now show that the second inequality in (5.41) holds. For every $m \in \mathbb{Z}$, let Δ_m be the compact subset of $\mathbb{R}^N$ defined by (5.7), with $j = m$. Observe that

$$
\mathbb{R}^N \setminus \{0\} = \bigcup_{m=-\infty}^{+\infty} \Delta_m.
$$

Therefore, using the definition of the norm $\|\cdot\|_{L^2(\mathbb{R}^N)}$, (5.31), (5.6), the change of variable $\xi = 2^{-m}\eta$, (5.7), (5.5), and the orthonormality property of the sequence of functions

$$
\left\{ \widehat{\psi}_{p,q,k} : (p,q,k) \in \{1,\dots,2^N-1\} \times \{-1,0,1\} \times \mathbb{Z}^N \right\},
$$

one gets that

$$\left\| \sum_{p=1}^{2^N-1} \sum_{(j,k)\in\mathbb{Z}\times\mathbb{Z}^N} a_{p,j,k} \widehat{\Psi}_{p,j,k}^{(v)} \right\|_{L^2(\mathbb{R}^N)}^2$$

$$\leq \sum_{m=-\infty}^{+\infty} \int_{\Delta_m} \left| \sum_{p=1}^{2^N-1} \sum_{(j,k)\in\mathbb{Z}\times\mathbb{Z}^N} a_{p,j,k} 2^{-jN/2} e^{-i2^{-j}k\cdot\eta} \frac{\widehat{\psi}_p(2^{-j}\eta)}{|2^{-j}\eta|^{v+N/2}} \right|^2 d\eta$$

$$= \sum_{m=-\infty}^{+\infty} \int_{\Delta_m} \left| \sum_{p=1}^{2^N-1} \sum_{q=-1}^{1} \sum_{k\in\mathbb{Z}^N} a_{p,m+q,k} 2^{-(m+q)N/2} e^{-i2^{-m-q}k\cdot\eta} \frac{\widehat{\psi}_p(2^{-m-q}\eta)}{|2^{-m-q}\eta|^{v+N/2}} \right|^2 d\eta$$

$$= \sum_{m=-\infty}^{+\infty} \int_{\Delta_0} \left| \sum_{p=1}^{2^N-1} \sum_{q=-1}^{1} \sum_{k\in\mathbb{Z}^N} a_{p,m+q,k} 2^{qv} e^{-i2^{-q}k\cdot\xi} \widehat{\psi}_p(2^{-q}\xi) \right|^2 |\xi|^{-2v-N} d\xi$$

$$\leq c_3(v) \sum_{m=-\infty}^{+\infty} \int_{\mathbb{R}^N} \left| \sum_{p=1}^{2^N-1} \sum_{q=-1}^{1} \sum_{k\in\mathbb{Z}^N} a_{p,m+q,k} 2^{qv} e^{-i2^{-q}k\cdot\eta} \widehat{\psi}_p(2^{-q}\xi) \right|^2 d\xi$$

$$= c_3(v) \sum_{m=-\infty}^{+\infty} \int_{\mathbb{R}^N} \left| \sum_{p=1}^{2^N-1} \sum_{q=-1}^{1} \sum_{k\in\mathbb{Z}^N} a_{p,m+q,k} 2^{q(v+N/2)} \widehat{\psi}_{p,q,k}(\xi) \right|^2 d\xi$$

$$= c_3(v) \sum_{m=-\infty}^{+\infty} \sum_{p=1}^{2^N-1} \sum_{q=-1}^{1} \sum_{k\in\mathbb{Z}^N} |a_{p,m+q,k}|^2 2^{q(2v+N)}$$

$$\leq c_2(v) \sum_{p=1}^{2^N-1} \sum_{(j,k)\in\mathbb{Z}\times\mathbb{Z}^N} |a_{p,j,k}|^2,$$

where the constants $c_3(v) = \left(8\pi\sqrt{N}/3\right)^{2|v|}$ and $c_2(v) := 3\cdot 2^{2|v|+N} c_3(v)$. $\square$

5.3 Optimality of the wavelet series representation

One denotes by I an arbitrary compact interval of $\mathbb{R}^M$, the integer $M \geq 1$ being arbitrary. Let $\mathbb{G} := \{\mathbb{G}(z), z \in I\}$ be an arbitrary real-valued centered Gaussian stochastic field possessing almost surely continuous paths on I. Then one knows (see for instance [Ledoux and Talagrand (1991)] or Exercise 3.11 in [Li and Queffélec (2004)]) that it can be represented as

$$\mathbb{G}(z) = \sum_{q=1}^{+\infty} \widetilde{\varepsilon}_q f_q(z), \tag{5.42}$$

where $\{\widetilde{\varepsilon}_q : q \in \mathbb{N}\}$ is a sequence of independent real-valued $\mathcal{N}(0,1)$ Gaussian random variables, and $\{f_q : q \in \mathbb{N}\}$ a sequence of real-valued deterministic continuous functions over I (recall that $\mathbb{N}$ denotes the set of the (strictly) positive integers). Observe that the series in (5.42) is almost surely convergent uniformly in $z \in I$. That is, for almost all ω, one has

$$\left\| \mathbb{G}(\cdot, \omega) - \sum_{q=1}^{Q-1} \widetilde{\varepsilon}_q(\omega) f_q(\cdot) \right\|_{I,\infty} = \left\| \sum_{q=Q}^{+\infty} \widetilde{\varepsilon}_q(\omega) f_q(\cdot) \right\|_{I,\infty} \xrightarrow[Q \to +\infty]{} 0,$$

where $\|g(\cdot)\|_{I,\infty} := \sup_{z \in I} |g(z)|$ denotes the uniform norm on the interval I. We mention that in the case of Brownian Motion, two well-known examples of representations of the type (5.42) are provided by its expansion on the trigonometric system (see for instance [Kahane (1968 1st edition, 1985 2nd edition)]) and the one on the first variant of the Faber-Schauder system, given by Theorem 3.13. Also, we mention in passing that more generally when the set of indices I is an arbitrary compact metric space then $\{\mathbb{G}(z), z \in I\}$ can still be represented of the form (5.42), where the series is almost surely uniformly convergent in $z \in I$.

In fact, generally speaking, a centered Gaussian field $\{\mathbb{G}(z), z \in I\}$ which possesses almost surely continuous paths on I has not only one representation of the form (5.42) but infinitely many of them, and it seems natural to try to look for representations where the series converge at the optimal rate. In the seminal article [Kühn and Linde (2002)], the optimal rate is defined as the one at which the sequence of the l-numbers of $\{\mathbb{G}(z), z \in I\}$, denoted by $\left(l_Q(\mathbb{G})\right)_{Q \in \mathbb{N}}$, converges to zero; this non-increasing [2] sequence of positive real numbers is defined, for each $Q \in \mathbb{N}$, as

$$l_Q(\mathbb{G}) := \inf \left\{ \mathbb{E}\left(\left\| \sum_{q=Q}^{+\infty} \widetilde{\varepsilon}_q f_q(\cdot) \right\|_{I,\infty} \right) : \mathbb{G}(\cdot) = \sum_{q=1}^{+\infty} \widetilde{\varepsilon}_q f_q(\cdot) \right\}, \qquad (5.43)$$

where the infimum is taken over all the representations of $\{\mathbb{G}(z), z \in I\}$ of the form (5.42).

---

[2]Indeed, using classical properties of conditional expectation and the fact that $\{\widetilde{\varepsilon}_q : q \in \mathbb{N}\}$ is a sequence of independent real-valued $\mathcal{N}(0,1)$ Gaussian random variables, for all $Q \in \mathbb{N}$, one has

$$\mathbb{E}\left(\left\| \sum_{q=Q}^{+\infty} \widetilde{\varepsilon}_q f_q(\cdot) \right\|_{\infty} \right) = \mathbb{E}\left[\mathbb{E}\left(\left\| \sum_{q=Q}^{+\infty} \widetilde{\varepsilon}_q f_q(\cdot) \right\|_{\infty} \Big| \widetilde{\varepsilon}_{Q+1}, \widetilde{\varepsilon}_{Q+2}, \cdots \right) \right]$$

$$\geq \mathbb{E}\left[\left\| \mathbb{E}\left(\sum_{q=Q}^{+\infty} \widetilde{\varepsilon}_q f_q(\cdot) \Big| \widetilde{\varepsilon}_{Q+1}, \widetilde{\varepsilon}_{Q+2}, \cdots \right) \right\|_{\infty} \right] = \mathbb{E}\left(\left\| \sum_{q=Q+1}^{+\infty} \widetilde{\varepsilon}_q f_q(\cdot) \right\|_{\infty} \right).$$

Rates of convergences of sequences of l-numbers of centered continuous Gaussian fields belonging to a general class, which includes Fractional Brownian Field (FBF) (see Definition 1.17 and Proposition 1.28), are determined in [Ayache and Linde (2008)]. In particular, it is shown in this article that when $\{B_H(t), t \in \mathcal{J}\}$ is a FBF of an arbitrary Hurst parameter $H \in (0,1)$, indexed by an arbitrary compact interval $\mathcal{J}$ of $\mathbb{R}^N$ (N being an arbitrary positive integer), there exist two constants $0 < c_1 \leq c_2$ (which a priori depend on H), such that, for all $Q \in \mathbb{N}$, one has

$$c_1 Q^{-H/N} \sqrt{\log(1+Q)} \leq l_Q(B_H) \leq c_2 Q^{-H/N} \sqrt{\log(1+Q)}. \qquad (5.44)$$

Also, it is shown in [Ayache and Linde (2008)] that, for all fixed $H \in (0,1)$, the wavelet series representation of $\{B_H(t), t \in \mathcal{J}\}$ derived from Theorem 5.5 (recall that $X(\cdot, H)$ is a FBF of Hurst parameter H) converges, almost surely, in the uniform norm at the same optimal rate of the sequence $(l_Q(B_H))_{Q \in \mathbb{N}}$.

These are extensions of earlier results in the literature. For Brownian Motion ($N = 1$ and $H = 1/2$) the rate of convergence of the sequence of the l-numbers is determined in the pioneering article [Maiorov and Wasilkowski (1996)], and it has turned out that the two classical expansions of Brownian Motion on the trigonometric system and on the first variant of the Faber-Schauder system [3] allow to reach this optimal rate. Another pioneering paper in this topic is [Kühn and Linde (2002)] which, among other things, provides a sharp estimate of the rate of convergence of the sequence of the l-numbers of a Fractional Brownian Motion ($N = 1$ and H arbitrary) and more generally of a Fractional Brownian Sheet [4]. Let us mention that the article [Ayache and Taqqu (2003)] shows that many of the wavelet series representations of Fractional Brownian Motion, studied in [Meyer *et al.* (1999)], allow to reach the optimal rate of its sequence of l-numbers. This result can rather easily be extended to Fractional Brownian Sheet. Also, we mention that other explicit optimal random series representations of Fractional Brownian Sheet are constructed in the paper [Dzhaparidze and van Zanten (2005)].

Let us now describe the goal of the present section. Recall that $\mathcal{J}$ is an arbitrary compact interval of $\mathbb{R}^N$, on the other hand, one denotes by $\underline{H}$ and $\overline{H}$ two real numbers satisfying $0 < \underline{H} \leq \overline{H} < 1$. Let $(l_Q(X))_{Q \in \mathbb{N}}$ be the sequence of the l-numbers of the continuous Gaussian

---

[3]Notice that the optimality for the Faber-Schauder system is a consequence of Theorem 3.20 in Chapter 3.

[4]This is a centered Gaussian field with continuous paths and non stationary increments whose covariance is a tensor product of covariances of Fractional Brownian Motions.

field $\{X(u,v),(u,v) \in \mathcal{J} \times [\underline{H},\overline{H}]\}$, which consists in the restriction to $\mathcal{J} \times [\underline{H},\overline{H}]$ of the Generator of MuBF (see (5.1)). Then, in view of (5.44) and of the fact that, for each $v \in [\underline{H},\overline{H}]$, $X(\cdot,v)$ is a FBF of Hurst parameter v, by arguing, ad absurdum, one can show that there exists a constant $c_3 > 0$, such that, for all $Q \in \mathbb{N}$, one has $c_3 Q^{-\underline{H}/N}\sqrt{\log(1+Q)} \leq l_Q(X)$. It seems natural to wonder whether the reverse inequality also holds. More precisely, does there exist a (finite) constant c_4, such that, for all $Q \in \mathbb{N}$, one has $l_Q(X) \leq c_4 Q^{-\underline{H}/N}\sqrt{\log(1+Q)}$? Roughly speaking, the main result of the present section allows to provide a positive answer to this question, since one can derive from it that the wavelet series representation of X, given in Theorem 5.5, converges at the rate $Q^{-\underline{H}/N}\sqrt{\log(1+Q)}$, that is the same rate as the one of the sequence $(l_Q(X))_{Q \in \mathbb{N}}$.

From now on, for the sake of simplicity, one assumes that

$$\mathcal{J} = [0,1]^N; \tag{5.45}$$

notice that the main result of the present section can rather easily be extended to the general case where $\mathcal{J}$ is an arbitrary compact interval of $\mathbb{R}^N$. In order to state it in a precise way, first one needs to introduce several notations. Let $J \in \mathbb{Z}_+$ be arbitrary and fixed.

- The finite sets A_J and A_J^N are defined as:

$$A_J := \{m \in \mathbb{Z} : |m| \leq 2^{[J/2N]}\} \tag{5.46}$$

and

$$A_J^N := \{k = (k_1,\ldots,k_N) \in \mathbb{Z}^N : \text{for all } n,\, k_n \in A_J\} \tag{5.47}$$
$$= \{k = (k_1,\ldots,k_N) \in \mathbb{Z}^N : \text{for all } n,\, |k_n| \leq 2^{[J/2N]}\},$$

where $[\cdot]$ denotes the integer part function.
- The centered continuous Gaussian field $X_J^{\mathrm{lf}} := \{X_J^{\mathrm{lf}}(u,v),(u,v) \in \mathbb{R}^N \times (0,1)\}$ is defined, for all $(u,v) \in \mathbb{R}^N \times (0,1)$, as:

$$X_J^{\mathrm{lf}}(u,v) := \sum_{p=1}^{2^N-1}\sum_{j=-2^{[J/2]}}^{-1}\sum_{k \in A_J^N} 2^{-jv}\,\varepsilon_{p,j,k}\big(\Psi_p(2^j u - k, v) - \Psi_p(-k,v)\big). \tag{5.48}$$

- For all integer j satisfying $0 \leq j \leq [J/N]$, the finite sets $B_{J,j}$ and $B_{J,j}^N$ are defined as:

$$B_{J,j} := \{m \in \mathbb{Z} : |m| \leq 2^{\mu(J,j)+1}\} \tag{5.49}$$

and

$$B_{J,j}^N := \{k = (k_1,\ldots,k_N) \in \mathbb{Z}^N : \text{for all } n,\, k_n \in B_{J,j}\} \tag{5.50}$$
$$= \{k = (k_1,\ldots,k_N) \in \mathbb{Z}^N : \text{for all } n,\, |k_n| \leq 2^{\mu(J,j)+1}\},$$

where

$$\mu(J,j) := \max\{[J/N] - j, j\}. \tag{5.51}$$

- The centered continuous Gaussian field $X_J^{\mathrm{hf}} := \{X_J^{\mathrm{hf}}(u,v), (u,v) \in \mathbb{R}^N \times (0,1)\}$ is defined, for all $(u,v) \in \mathbb{R}^N \times (0,1)$, as:

$$X_J^{\mathrm{hf}}(u,v) := \sum_{p=1}^{2^N-1} \sum_{j=0}^{[J/N]} \sum_{k \in B_{J,j}^N} 2^{-jv} \varepsilon_{p,j,k} \big(\Psi_p(2^j u - k, v) - \Psi_p(-k, v)\big). \tag{5.52}$$

- The centered continuous Gaussian field $X_J := \{X_J(u,v), (u,v) \in \mathbb{R}^N \times (0,1)\}$ is defined, for all $(u,v) \in \mathbb{R}^N \times (0,1)$, as:

$$X_J(u,v) := X_J^{\mathrm{lf}}(u,v) + X_J^{\mathrm{hf}}(u,v). \tag{5.53}$$

Remark 5.14. One denote by $c_0(N)$, the finite constant, only depending on N, defined as $c_0(N) := 2^{N+3}(2^N - 1)$. Using (5.47), (5.50) and (5.51), it can easily be shown that in each one of the finite sums (5.48) and (5.52) the number of terms is at most $c_0(N)2^J$. Thus, in view of (5.53), the field X_J is defined through a finite sum which consists in at most $c_0(N)2^{J+1}$ terms.

We are now in position to state the main result of the present section.

Theorem 5.15. *Recall that $\mathcal{J} = [0,1]^N$ and that $\underline{H}$ and $\overline{H}$ and two fixed real numbers satisfying $0 < \underline{H} \le \overline{H} < 1$. Let X be the Generator of MuBF and X_J be the field defined in (5.53). Then, the positive random variable*

$$C := \sup \left\{ \frac{|X(u,v) - X_J(u,v)|}{2^{-Jv/N}\sqrt{1+J}} : (J,u,v) \in \mathbb{Z}_+ \times \mathcal{J} \times [\underline{H}, \overline{H}] \right\},$$

has a finite moment of any order, which in particular implies that C is finite almost surely.

Theorem 5.15 is a straightforward consequence of (5.16), (5.53), the triangle inequality, and the following two propositions, in which $\mathcal{J}$, $\underline{H}$ and $\overline{H}$ are as in the theorem.

Proposition 5.16. *Let X^{lf} and X_J^{lf} be the fields introduced in Definition 5.7 and in (5.48). Then, for any fixed arbitrarily large positive real number γ, the positive random variable*

$$C_\gamma^{\mathrm{lf}} := \sup \left\{ \frac{|X^{\mathrm{lf}}(u,v) - X_J^{\mathrm{lf}}(u,v)|}{2^{-J\gamma}} : (J,u,v) \in \mathbb{Z}_+ \times \mathcal{J} \times [\underline{H}, \overline{H}] \right\},$$

has a finite moment of any order.

Proposition 5.17. *Let X^{hf} and $X_{\mathcal{J}}^{\mathrm{hf}}$ be the fields introduced in Definition 5.8 and in (5.52). Then, the positive random variable*

$$C^{\mathrm{hf}} := \sup \left\{ \frac{\left| X^{\mathrm{hf}}(u,v) - X_{\mathcal{J}}^{\mathrm{hf}}(u,v) \right|}{2^{-Jv/N}\sqrt{1+J}} : (J,u,v) \in \mathbb{Z}_+ \times \mathcal{J} \times [\underline{H}, \overline{H}] \right\},$$

has a finite moment of any order.

In order to prove these two propositions, one needs some preliminary results. For simplifying the notations, one sets:

$$\mathcal{H} := [\underline{H}, \overline{H}]. \tag{5.54}$$

Lemma 5.18. *Denote by L an arbitrary positive integer. Let $\mathcal{J}$ be as in (5.45) and $\mathcal{H}$ as in (5.54). There exists a deterministic constant $c > 0$, such that, for all $p \in \{1, \ldots 2^N - 1\}$ and $(j,k) \in \mathbb{N} \times \mathbb{Z}^N$, one has*

$$\sup_{(u,v) \in \mathcal{J} \times \mathcal{H}} \left| \Psi_p(2^{-j}u - k, v) - \Psi_p(-k, v) \right| \le c \, 2^{-j} \prod_{n=1}^{N} \left(2 + |k_n| \right)^{-L}. \tag{5.55}$$

Proof of Lemma 5.18. Assume that $(u,v) \in \mathcal{J} \times \mathcal{H}$, $p \in \{1, \ldots, 2^N - 1\}$ and $(j,k) \in \mathbb{N} \times \mathbb{Z}^N$ are arbitrary and fixed. One denotes by ϕ the function defined, for each λ belonging to the interval $[0,1]$ of the real line, as:

$$\phi(\lambda) := \Psi_p(\lambda 2^{-j}u - k, v).$$

In view of Proposition 5.10, it is clear that ϕ is an infinitely differentiable function on $[0,1]$. Thus, denoting by ϕ' its first derivative, it follows from the Mean Value Theorem that there exists $\nu \in (0,1)$ such that $\phi(1) - \phi(0) = \phi'(\nu)$. Moreover, it can easily be seen that the latter equality can be expressed as:

$$\Psi_p(2^{-j}u - k, v) - \Psi_p(-k, v) = 2^{-j} \sum_{q=1}^{N} u_q (\partial_{y_q} \Psi_p)(\nu 2^{-j}u - k, v), \tag{5.56}$$

where $\partial_{y_q} \Psi_p$ denotes the first partial derivative of Ψ_p with respect to the q-th coordinate of y. On the other hand, using (5.20), the fact $\nu 2^{-j}u \in \mathcal{J} := [0,1]^N$, and the triangle inequality, one gets that

$$\left| (\partial_{y_q} \Psi_p)(\nu 2^{-j}u - k, v) \right| \le c_1 \prod_{n=1}^{N} \left(3 + |k_n| - \nu 2^{-j}u_n \right)^{-L} \le c_1 \prod_{n=1}^{N} \left(2 + |k_n| \right)^{-L}, \tag{5.57}$$

where c_1 is a constant only depending on the interval $\mathcal{H}$. Finally, setting $c := N c_1$, and putting together (5.56), (5.57) and the inequality $|u_q| \le 1$, for all $q \in \{1, \ldots, N\}$, one obtains (5.55). $\qquad \square$

Lemma 5.19. *For each fixed integer $L \geq 2$, there exists a finite constant $c > 0$, such that, for all real number $\rho \geq 0$, one has*

$$\sum_{m=0}^{+\infty} (2 + m + \rho)^{-L} \sqrt{\log(3 + m + \rho)} \leq c(1 + \rho)^{-(L-1)} \sqrt{\log(2 + \rho)}. \quad (5.58)$$

Proof of Lemma 5.19. Standard computations allow to show that, for each fixed $\rho \in [0, +\infty)$,

$$z \mapsto (1 + z + \rho)^{-L} \sqrt{\log(2 + z + \rho)}$$

is a nonincreasing function on $[0, +\infty)$. Thus, one obtains that

$$\sum_{m=0}^{+\infty} (2 + m + \rho)^{-L} \sqrt{\log(3 + m + \rho)} \leq \int_0^{+\infty} (1 + z + \rho)^{-L} \sqrt{\log(2 + z + \rho)} \, dz. \quad (5.59)$$

Moreover, setting $t = (1 + \rho)^{-1} z$ in the last integral, it follows that

$$\int_0^{+\infty} (1 + z + \rho)^{-L} \sqrt{\log(2 + z + \rho)} \, dz$$

$$= (1 + \rho) \int_0^{+\infty} \left(1 + (1 + \rho)t + \rho\right)^{-L} \sqrt{\log\left(2 + (1 + \rho)t + \rho\right)} \, dt$$

$$\leq (1 + \rho)^{-(L-1)} \int_0^{+\infty} (1 + t)^{-L} \left(\sqrt{\log(2 + \rho)} + \sqrt{\log(1 + t)}\right) dt$$

$$\leq c(1 + \rho)^{-(L-1)} \sqrt{\log(2 + \rho)}, \quad (5.60)$$

where c is a constant only depending on L. Finally, combining (5.59) and (5.60), one obtains (5.58). $\quad\square$

Lemma 5.20. *Let a be an arbitrary positive real number, and let $\tilde{c}(a)$ be the finite constant defined as:*

$$\tilde{c}(a) := \frac{2}{\sqrt{\log 2}} \sum_{m=0}^{+\infty} 2^{-ma} \sqrt{\log(3 + m)}. \quad (5.61)$$

Then, for every $r \in \mathbb{Z}_+$, one has

$$\sum_{j=r}^{+\infty} 2^{-ja} \sqrt{\log(2 + j)} \leq \tilde{c}(a) 2^{-ra} \sqrt{\log(2 + r)}. \quad (5.62)$$

Proof of Lemma 5.20. Using the inequalities $2 + m + r \leq (2 + m)(2 + r)$ and

$$\sqrt{\alpha + \beta} \leq \sqrt{\alpha} + \sqrt{\beta}, \quad \text{for all } (\alpha, \beta) \in \mathbb{R}_+^2, \quad (5.63)$$

one has

$$\sum_{j=r}^{+\infty} 2^{-ja} \sqrt{\log(2+j)} = \sum_{m=0}^{+\infty} 2^{-(m+r)a} \sqrt{\log(2+m+r)}$$

$$\leq 2^{-ra} \sum_{m=0}^{+\infty} 2^{-ma} \left(\sqrt{\log(2+m)} + \sqrt{\log(2+r)} \right)$$

$$\leq \widetilde{c}(a) 2^{-ra} \sqrt{\log(2+r)}.$$

$\square$

We are now in position to prove Proposition 5.16.

Proof of Proposition 5.16. For each $J \in \mathbb{Z}_+$ and $q \in \{1, \ldots, N\}$, one denotes by $G_J^{N,q}$ the set defined as

$$G_J^{N,q} := \{ k = (k_1, \ldots, k_N) \in \mathbb{Z}^N : k_q \notin A_J \} \qquad (5.64)$$
$$= \{ k = (k_1, \ldots, k_N) \in \mathbb{Z}^N : |k_q| \geq 2^{[J/2N]} + 1 \}.$$

Therefore, using (5.47) one gets that

$$\mathbb{Z}^N \setminus A_J^N = \bigcup_{q=1}^{N} G_J^{N,q}. \qquad (5.65)$$

It follows from (5.14), (5.48), (5.65) and the triangle inequality that, for every $(J, u, v) \in \mathbb{Z}_+ \times \mathcal{J} \times \mathcal{H}$, one has

$$\left| X^{\mathrm{lf}}(u, v) - X_J^{\mathrm{lf}}(u, v) \right| \leq R_{1,J}(u, v) + R_{2,J}(u, v), \qquad (5.66)$$

where

$$R_{1,J}(u, v) := \sum_{p=1}^{2^N - 1} \sum_{j=2^{[J/2]}+1}^{+\infty} \sum_{k \in \mathbb{Z}^N} 2^{jv} \left| \varepsilon_{p,-j,k} \right| \left| \Psi_p(2^{-j}u - k, v) - \Psi_p(-k, v) \right|$$

and

$$R_{2,J}(u, v) := \sum_{p=1}^{2^N - 1} \sum_{q=1}^{N} \sum_{j=1}^{2^{[J/2]}} \sum_{k \in G_J^{N,q}} 2^{jv} \left| \varepsilon_{p,-j,k} \right| \left| \Psi_p(2^{-j}u - k, v) - \Psi_p(-k, v) \right|.$$

Next, using Lemmas A.23, 5.18 and 5.19, using (5.54), and using Lemma 5.20, one obtains, almost surely, for each fixed integer $L \geq 2$, that

$$R_{1,J}(u, v) \leq C_1 \sum_{j=2^{[J/2]}+1}^{+\infty} 2^{-j(1-\overline{H})} \sqrt{\log(2+j)}$$

$$\leq C_2 2^{-(2^{[J/2]}+1)(1-\overline{H})} \sqrt{\log \left(3 + 2^{[J/2]} \right)} \qquad (5.67)$$

and

$$R_{2,J}(u,v)$$

$$\leq C_3 \left(\sum_{j=1}^{2^{[J/2]}} 2^{-j(1-\overline{H})} \sqrt{\log(2+j)} \right) \left(2 + 2^{[J/2N]} \right)^{-(L-1)} \sqrt{\log\left(3 + 2^{[J/2N]}\right)}$$

$$\leq C_4 \left(2 + 2^{[J/2N]} \right)^{-(L-1)} \sqrt{\log\left(3 + 2^{[J/2N]}\right)}, \tag{5.68}$$

where C_1, C_2, C_3 and C_4 are positive random variables not depending on (J, u, v) and of finite moment of any order. Finally, in view of the fact that L can be an arbitrarily large integer, one can derive from (5.66), (5.67) and (5.68) that the proposition is satisfied. $\qquad\square$

Proof of Proposition 5.17. For each $J \in \mathbb{Z}_+$, $j \in \{0, \ldots, J\}$ and $q \in \{1, \ldots, N\}$, one denotes by $S_{J,j}^{N,q}$ the set defined as

$$S_{J,j}^{N,q} := \left\{ k = (k_1, \ldots, k_N) \in \mathbb{Z}^N : k_q \notin B_{J,j} \right\} \tag{5.69}$$

$$= \left\{ k = (k_1, \ldots, k_N) \in \mathbb{Z}^N : |k_q| \geq 2^{\mu(J,j)+1} + 1 \right\},$$

where $\mu(J,j)$ is the same as in (5.51). Therefore, using (5.50) one gets that

$$\mathbb{Z}^N \setminus B_{J,j}^N = \bigcup_{q=1}^{N} S_{J,j}^{N,q}. \tag{5.70}$$

It follows from (5.15), (5.52), (5.70) and the triangle inequality, that, for every $(J, u, v) \in \mathbb{Z}_+ \times \mathcal{J} \times \mathcal{H}$, one has

$$\left| X^{\mathrm{lf}}(u,v) - X_J^{\mathrm{lf}}(u,v) \right| \leq \widetilde{R}_{1,J}(u,v) + \widetilde{R}_{2,J}(u,v) + \widetilde{R}_{1,J}(0,v) + \widetilde{R}_{2,J}(0,v), \tag{5.71}$$

where

$$\widetilde{R}_{1,J}(u,v) := \sum_{p=1}^{2^N-1} \sum_{j=[J/N]+1}^{+\infty} \sum_{k\in\mathbb{Z}^N} 2^{-jv} |\varepsilon_{p,j,k}| \left| \Psi_p(2^j u - k, v) \right| \tag{5.72}$$

and

$$\widetilde{R}_{2,J}(u,v) := \sum_{p=1}^{2^N-1} \sum_{q=1}^{N} \sum_{j=0}^{[J/N]} \sum_{k\in S_{J,j}^{N,q}} 2^{-jv} |\varepsilon_{p,j,k}| \left| \Psi_p(2^j u - k, v) \right|. \tag{5.73}$$

Let us first provide an appropriate upper bound for $\widetilde{R}_{1,J}(u,v)$. Assume that the integer $j \geq [J/N] + 1$ is arbitrary and fixed. Using Lemma A.23, Proposition 5.11, (5.45), and the inequalities

$$\log(2 + a + b) \leq \log(2 + a) + \log(2 + b) \tag{5.74}$$

$$\leq 4\log(2 + a)\log(2 + b), \quad \text{for all } (a,b) \in \mathbb{R}_+^2,$$

one obtains, almost surely, for each fixed integer $L \geq 2$, that

$$
\sum_{k \in \mathbb{Z}^N} |\varepsilon_{p,j,k}| |\Psi_p(2^j u - k, v)| \leq C_1 \sum_{k \in \mathbb{Z}^N} \frac{\sqrt{\log \left(2 + j + \sum_{n=1}^{N} |k_n|\right)}}{\prod_{n=1}^{N} \left(3 + |2^j u_n - k_n|\right)^L}
$$

$$
= C_1 \sum_{k \in \mathbb{Z}^N} \frac{\sqrt{\log \left(2 + j + \sum_{n=1}^{N} |k_n + [2^j u_n]|\right)}}{\prod_{n=1}^{N} \left(3 + |2^j u_n - [2^j u_n] - k_n|\right)^L}
$$

$$
\leq C_1 \sum_{k \in \mathbb{Z}^N} \frac{\sqrt{\log \left(2 + (N+1)2^j + \sum_{n=1}^{N} |k_n|\right)}}{\prod_{n=1}^{N} \left(3 + |2^j u_n - [2^j u_n] - k_n|\right)^L}
$$

$$
\leq C_2 \sqrt{\log \left(2 + (N+1)2^j\right)}.
$$

$$
\leq C_3 \sqrt{1 + j}, \tag{5.75}
$$

where C_1,

$$
C_2 := 2\, C_1 \sup_{y \in \mathcal{J}} \sum_{k \in \mathbb{Z}^N} \frac{\sqrt{\log \left(2 + \sum_{n=1}^{N} |k_n|\right)}}{\prod_{n=1}^{N} \left(3 + |y_n - k_n|\right)^L}
$$

and $C_3 := C_2 \sqrt{N+2}$ are three positive random variables of finite moment of any order not depending on (J, j, u, v). Next, setting $C_4 := (2^N - 1)C_3$ and

$$
C_5 := C_4 \sum_{m=0}^{+\infty} 2^{-m\underline{H}} \sqrt{2 + m},
$$

then, it results from (5.72), (5.75) and (5.54) that, one has, almost surely,

$$
\widetilde{R}_{1,J}(u,v) \leq C_4 \sum_{j=[J/N]+1}^{+\infty} 2^{-jv} \sqrt{1+j}
$$

$$
= C_4 \sum_{m=0}^{+\infty} 2^{-(m+[J/N]+1)v} \sqrt{2 + m + [J/N]}
$$

$$
\leq C_4 2^{-Jv/N} \sqrt{1 + [J/N]} \sum_{m=0}^{+\infty} 2^{-m\underline{H}} \sqrt{\frac{1+m}{1+[J/N]} + 1}
$$

$$
\leq C_5 2^{-Jv/N} \sqrt{1 + J}. \tag{5.76}
$$

Let us now provide an appropriate upper bound for $\widetilde{R}_{2,J}(u,v)$. Assume that the integer $j \in \{0, \ldots, [J/N]\}$ is arbitrary and fixed. Using (5.69), Lemma A.23, Proposition 5.11, (5.45), the second inequality in (5.74),

(5.51), the triangle inequality, the first inequality in (5.74), (5.63), the fact that $z \mapsto (2+z)^{-L}$ is a decreasing function on $\mathbb{R}_+$, and Lemma 5.19, one obtains, almost surely, for each fixed integer $L \geq 2$, that

$$\sum_{k \in S_{J,j}^{N,1}} |\varepsilon_{p,j,k}| |\Psi_p(2^j u - k, v)|$$

$$\leq C_1 \sum_{|k_1| > 2^{\mu(J,j)+1}} \sum_{(k_2,\ldots,k_N) \in \mathbb{Z}^{N-1}} \frac{\sqrt{\log\left(2 + j + \sum_{n=1}^{N} |k_n|\right)}}{\prod_{n=1}^{N} \left(3 + |2^j u_n - k_n|\right)^L}$$

$$\leq C_6 \sum_{|k_1| > 2^{\mu(J,j)+1}} \frac{\sqrt{\log\left(2 + N2^j + |k_1|\right)}}{\left(3 + |2^j u_1 - k_1|\right)^L}$$

$$\leq C_6 \sum_{|k_1| > 2^{\mu(J,j)+1}} \frac{\sqrt{\log\left(2 + N2^j + |k_1|\right)}}{\left(3 + |k_1| - 2^j\right)^L}$$

$$\leq C_6 \sum_{|k_1| > 2^{\mu(J,j)+1}} \frac{\sqrt{\log\left(2 + N2^j + |k_1|\right)}}{\left(3 + 2^{-1}|k_1|\right)^L}$$

$$\leq 2^{L+1} C_6 \sum_{m=2^{\mu(J,j)+1}}^{+\infty} \frac{\sqrt{\log\left(2 + N2^j + m\right)}}{\left(3 + m\right)^L}$$

$$\leq 2^{L+1} C_6 \left(\sum_{m=2^{\mu(J,j)+1}}^{+\infty} \frac{\sqrt{\log\left(2 + N2^j\right)}}{\left(3 + m\right)^L} + \sum_{m=2^{\mu(J,j)+1}}^{+\infty} \frac{\sqrt{\log\left(2 + m\right)}}{\left(3 + m\right)^L} \right)$$

$$\leq 2^{L+1} C_7 \left(\sqrt{\log\left(2 + N2^j\right)} \int_{2^{\mu(J,j)+1}}^{+\infty} \frac{dz}{\left(2 + z\right)^L} + \frac{\sqrt{\log\left(3 + 2^{\mu(J,j)+1}\right)}}{\left(1 + 2^{\mu(J,j)+1}\right)^{L-1}} \right)$$

$$\leq C_8 2^{-([J/N]-j)(L-1)} \sqrt{1 + J}, \tag{5.77}$$

where

$$C_6 := 2 C_1 \sup_{(y_2,\ldots,y_N) \in [0,1]^{N-1}} \sum_{k \in \mathbb{Z}^{N-1}} \frac{\sqrt{\log\left(2 + \sum_{n=2}^{N} |k_n|\right)}}{\prod_{n=2}^{N} \left(3 + |y_n - k_n|\right)^L},$$

C_7 and C_8 are three positive random variables of finite moment of any order not depending on (J, j, u, v). Similarly to (5.77), it can be shown that, one has, almost surely, for all $q \in \{1, \ldots, N\}$,

$$\sum_{k \in S_{J,j}^{N,q}} |\varepsilon_{p,j,k}| |\Psi_p(2^j u - k, v)| \leq C_8 2^{-([J/N]-j)(L-1)} \sqrt{1 + J}. \tag{5.78}$$

Next, setting $C_9 := N(2^N - 1)C_8$, and using (5.73) and (5.78), one gets, almost surely, that

$$\widetilde{R}_{2,J}(u,v) \leq C_9 \sum_{j=0}^{[J/N]} 2^{-jv} 2^{-([J/N]-j)(L-1)} \sqrt{1+J}$$

$$\leq C_9 2^{-[J/N](L-1)} \sqrt{1+J} \sum_{j=0}^{[J/N]} 2^{j(L-1-v)}$$

$$\leq C_9 2^{-[J/N](L-1)} \sqrt{1+J}\; 2^{([J/N]+1)(L-1-v)}$$

$$\leq 2^{L-1} C_9 2^{-Jv/N} \sqrt{1+J}. \tag{5.79}$$

Finally, putting together (5.71), (5.76) and (5.79), one obtains the proposition. $\qquad\square$

Chapter 6

Behavior of Multifractional Field and wavelet methods

6.1 Introduction

Recall that the Multifractional Brownian Field (MuBF) $Y := \{Y(t), t \in \mathbb{R}^N\}$, of deterministic Hurst functional parameter $H(\cdot)$ with values in a compact interval $[\underline{H}, \overline{H}]$ included in $(0, 1)$, has been introduced in Definition 1.82 (see also (5.3)) starting from the Generator of MuBF; that is the centered Gaussian field $X := \{X(u, v), (u, v) \in \mathbb{R}^N \times (0, 1)\}$ with continuous paths defined through (5.1) and (5.2) (see also Definition 1.78 and Remark 1.81).

Is it really necessary to impose to the Hurst functional parameter to be deterministic? This is a natural question one can address since Definition 1.82 still makes sense when the deterministic function $H(\cdot)$ in it is replaced by an arbitrary random field $S := \{S(t), t \in \mathbb{R}^N\}$ with values in a random compact interval $[\underline{S}, \overline{S}]$ of $(0, 1)$. One denotes by $Z := \{Z(t), t \in \mathbb{R}^N\}$ the stochastic field defined in this way, that is

$$Z(t) := X(t, S(t)), \quad \text{for all } t \in \mathbb{R}^N. \tag{6.1}$$

Such a field Z was first briefly introduced when $N = 1$ in [Papanicolaou and Sølna (2003)], which also assumed that S and X are independent. Two years later, thanks to wavelet methods, it was studied in detail in [Ayache and Taqqu (2005)], where the restrictive independency assumption on S and X was dropped but the assumption $N = 1$ was kept. Therefore, in the latter article, Z was called Multifractional Process with Random Exponent. After that, wavelet methods, rather reminiscent of those in [Ayache and Taqqu (2005)], were used in [Ayache et al. (2007)] in order to make an extensive theoretical study of path behavior of a considerably more general field than Z; this generalisation was called Generalized Multifractional *Process* with Random Exponent despite the fact that in [Ayache et al. (2007)]

the dimension of the Euclidian space of indices (that is N in our setting) was assumed to be arbitrary. Before ending this brief review of the literature on this topic, it is worth mentioning that when $N = 1$, the Multifractional Process with Random Exponent Z has been used with success by Bianchi and his co-authors as a model for financial assests (see for instance [Bianchi and Pianese (2014); Bianchi *et al.* (2013, 2012)]).

In our book, the real-valued stochastic field $Z := \{Z(t), t \in \mathbb{R}^N\}$, defined through (6.1), is always called Multifractional Field with Random Exponent (MuFRE). The main difficulty for studying it is that, in contrast with the MuBF (see the second equality in (5.3)), the equality

$$Z(t) = \frac{1}{(2\pi)^{N/2}} \int_{\mathbb{R}^N} \frac{e^{it\cdot\xi} - 1}{|\xi|^{S(t)+N/2}} \, d\widehat{\mathbb{W}}(\xi) \quad \text{does not make sense,}$$

unless the field S and the stochastic measure $d\widehat{\mathbb{W}}$ are independent (or equivalently the fields S and X are independent); indeed, when S and $d\widehat{\mathbb{W}}$ are dependent it is not clear at all how one can define the latter stochastic integral. Another important difficulty is that the characteristic function of the random variable $Z(t)$ does not have an explicit form which is easy to handle. Therefore, the methods relying on local times, presented in Chapter 2, can hardly be used in the setting of the MuFRE Z. It is even not completely clear that the local time of the field Z exists.

These two difficulties can be overcomed by using wavelet methods relying on the wavelet series representation of the field $\{Z(t), t \in \mathbb{R}^N\}$ provided by Theorem 5.5 in which (u, v) is replaced $(t, S(t))$. More precisely, one knows from this theorem and from (6.1) that

$$Z(t) = \sum_{p=1}^{2^N-1} \sum_{(j,k)\in\mathbb{Z}\times\mathbb{Z}^N} 2^{-jS(t)} \varepsilon_{p,j,k} \big(\Psi_p(2^j t - k, S(t)) - \Psi_p(-k, S(t))\big),$$

$$(6.2)$$

where the series is, with probability 1, uniformly convergent in t, on each compact interval of $\mathbb{R}^N$. In view of (6.2) and of Proposition 5.10, it turns out that a sufficient condition for having almost sure continuity of paths of the field $\{Z(t), t \in \mathbb{R}^N\}$ is that paths of the corresponding random Hurst functional parameter $S(\cdot)$ be almost surely continuous on $\mathbb{R}^N$. From now on, one always assumes that this condition is fulfilled.

The rest of this chapter is organized in the following way. In Section 6.2 fine estimates on path behavior of the Generator of MuBF X, which among many other things go far beyond Proposition 1.80, are obtained. Then, thanks to them and to the equality (6.1), behavior of paths of the MuFRE

Z is finely estimated globally and locally. Section 6.3 is mainly concerned with the discussion of the optimality of the estimates on path behavior of X and Z; among many other things, it turns out that some of them significantly improve the results on global, local and pointwise Hölder regularity of MuBF which have been previously exposed in Section 1.4.

Before ending this introduction, it is useful to point out that in the present chapter Lemma A.23 will be rather frequently applied to the sequence of independent real-valued $\mathcal{N}(0,1)$ Gaussian random variables

$$\big\{\varepsilon_{p,j,k} : p \in \{1,\ldots,2^N-1\},\ j \in \mathbb{Z} \text{ and } k \in \mathbb{Z}^N\big\},$$

which was introduced in Definition 5.4; thus, Ω_0^* denotes the same event of probability 1 as in Lemma A.23.

6.2 Fine estimates on behavior of MuFRE and its Generator

First one states the main results of the section, and then one gives their proofs.

Proposition 6.1 (Differentiability properties of the Generator X).
The following two results hold, for all ω belonging to the event Ω_0^* of probability 1.

(i) Let $X^{\mathrm{lf}} := \{X^{\mathrm{lf}}(u,v),\ (u,v) \in \mathbb{R}^N \times (0,1)\}$ be the low frequency part of X, which was introduced in Definition 5.7. Then, the function $X^{\mathrm{lf}}(\cdot,\cdot,\omega) : (u,v) \mapsto X^{\mathrm{lf}}(u,v,\omega)$ is infinitely differentiable on $\mathbb{R}^N \times (0,1)$; moreover, for every $\gamma = (\gamma_1,\ldots,\gamma_N) \in \mathbb{Z}_+^N \setminus \{0\}$, $q \in \mathbb{Z}_+$, and $(u,v) \in \mathbb{R}^N \times (0,1)$, one has

$$(\partial_v^q X^{\mathrm{lf}})(u,v,\omega) \tag{6.3}$$

$$= \sum_{m=0}^{q} \sum_{p=0}^{2^N-1} \binom{q}{m} (-\log 2)^m \sum_{j=-\infty}^{-1} \sum_{k \in \mathbb{Z}^N} j^m \, 2^{-jv} \varepsilon_{p,j,k}(\omega)$$

$$\times \big((\partial_v^{q-m}\Psi_p)(2^j u - k, v) - (\partial_v^{q-m}\Psi_p)(-k,v)\big)$$

and

$$(\partial_u^\gamma \partial_v^q X^{\mathrm{lf}})(u,v,\omega) \tag{6.4}$$

$$= \sum_{m=0}^{q} \sum_{p=0}^{2^N-1} \binom{q}{m} (-\log 2)^m \sum_{j=-\infty}^{-1} \sum_{k \in \mathbb{Z}^N} j^m \, 2^{-j(v-\gamma_1-\ldots-\gamma_N)} \varepsilon_{p,j,k}(\omega)$$

$$\times (\partial_y^\gamma \partial_v^{q-m}\Psi_p)(2^j u - k, v),$$

where $\binom{q}{m}$ denotes the binomial coefficient $\frac{q!}{m!(q-m)!}$. Notice that the series in (6.3) and (6.4) are uniformly convergent in (u,v) on each compact interval of $\mathbb{R}^N \times (0,1)$.

(ii) Let $X^{\mathrm{hf}} := \{X^{\mathrm{hf}}(u,v), (u,v) \in \mathbb{R}^N \times (0,1)\}$ be the high frequency part of X, which was introduced in Definition 5.8. Then, for each fixed $u \in \mathbb{R}^N$, the function $X^{\mathrm{hf}}(u,\cdot,\omega) : v \mapsto X^{\mathrm{hf}}(u,v,\omega)$ is infinitely differentiable on $(0,1)$; its derivative of any order $q \in \mathbb{Z}_+$ at all $v \in (0,1)$ is given by

$$(\partial_v^q X^{\mathrm{hf}})(u,v,\omega) \tag{6.5}$$

$$= \sum_{m=0}^{q} \sum_{p=0}^{2^N-1} \binom{q}{m} \left(-\log 2\right)^m \sum_{j=0}^{+\infty} \sum_{k\in\mathbb{Z}^N} j^m \, 2^{-jv} \varepsilon_{p,j,k}(\omega)$$

$$\times \left((\partial_v^{q-m}\Psi_p)(2^j u - k, v) - (\partial_v^{q-m}\Psi_p)(-k,v)\right),$$

where $0^0 := 1$, and the series is uniformly convergent in (u,v) on each compact interval of $\mathbb{R}^N \times (0,1)$. Notice that, for all $q \in \mathbb{Z}_+$, $(\partial_v^q X^{\mathrm{hf}})(\cdot,\cdot,\omega) : (u,v) \mapsto (\partial_v^q X^{\mathrm{hf}})(u,v,\omega)$ is a continuous function on $\mathbb{R}^N \times (0,1)$.

Remark 6.2. Observe that the right-hand side of (6.4) (resp. (6.3) and (6.5)) is simply obtained by applying the partial derivative operator $\partial_u^\gamma \partial_v^q$ (resp. ∂_v^q) to each term of the series in (5.14) (resp. (5.14) and (5.15)), and by using the general Leibniz rule.

Remark 6.3. It easily results from (5.16) and Proposition 6.1 that, for each fixed $(u,\omega) \in \mathbb{R}^N \times \Omega_0^*$, the function $X(u,\cdot,\omega) : v \mapsto X(u,v,\omega)$ is infinitely differentiable on $(0,1)$; its derivative of any order $q \in \mathbb{Z}_+$ at all $v \in (0,1)$ is given by

$$(\partial_v^q X)(u,v,\omega) \tag{6.6}$$

$$= \sum_{m=0}^{q} \sum_{p=0}^{2^N-1} \binom{q}{m} \left(-\log 2\right)^m \sum_{(j,k)\in\mathbb{Z}\times\mathbb{Z}^N} j^m \, 2^{-jv} \varepsilon_{p,j,k}(\omega)$$

$$\times \left((\partial_v^{q-m}\Psi_p)(2^j u - k, v) - (\partial_v^{q-m}\Psi_p)(-k,v)\right),$$

where $0^0 := 1$, and the series is uniformly convergent in (u,v) on each compact interval of $\mathbb{R}^N \times (0,1)$. Notice that, for all $q \in \mathbb{Z}_+$, $(\partial_v^q X)(\cdot,\cdot,\omega) : (u,v) \mapsto (\partial_v^q X)(u,v,\omega)$ is a continuous function on $\mathbb{R}^N \times (0,1)$.

Theorem 6.4 (Global modulus of continuity for $\partial_v^q X$). *For each $(\omega, q) \in \Omega_0^* \times \mathbb{Z}_+$ and, for every compact intervals $\mathcal{J} \subset \mathbb{R}^N$ and $\mathcal{H} \subset (0,1)$, one has*

$$\sup_{(u^1,u^2,v_1,v_2)\in\mathcal{J}^2\times\mathcal{H}^2} \left\{ \tag{6.7} \right.$$

$$\left. \frac{\left|(\partial_v^q X)(u^1, v_1, \omega) - (\partial_v^q X)(u^2, v_2, \omega)\right|}{|u^1 - u^2|^{v_1 \vee v_2}\left(1 + \left|\log|u^1 - u^2|\right|\right)^{q+\frac{1}{2}} + |v_1 - v_2|} \right\} < +\infty,$$

where $v_1 \vee v_2 := \max\{v_1, v_2\}$.

Corollary 6.5 (Global modulus of continuity for the MuFRE Z). *Let $I \subset \mathbb{R}^N$ be an arbitrary compact interval. One sets*

$$\underline{S}(I) := \min\{S(t) : t \in I\} \quad and \quad \overline{S}(I) := \max\{S(t) : t \in I\}. \tag{6.8}$$

Assuming that, almost surely,

$$S(\cdot) \in \mathcal{C}^{\overline{S}(I)}(I), \tag{6.9}$$

where $\mathcal{C}^{\overline{S}(I)}(I)$ denotes the global Hölder space on I of order $\overline{S}(I)$ (see Definition 1.33). Then, one has, almost surely, that

$$\sup_{(t^1,t^2)\in I^2} \left\{ \frac{|Z(t^1) - Z(t^2)|}{|t^1 - t^2|^{S(t^1)\vee S(t^2)}\left(1 + \left|\log|t^1 - t^2|\right|\right)^{\frac{1}{2}}} \right\} < +\infty, \tag{6.10}$$

and, consequently, that

$$\sup_{(t^1,t^2)\in I^2} \left\{ \frac{|Z(t^1) - Z(t^2)|}{|t^1 - t^2|^{\underline{S}(I)}\left(1 + \left|\log|t^1 - t^2|\right|\right)^{\frac{1}{2}}} \right\} < +\infty. \tag{6.11}$$

Remark 6.6. A careful inspection of the proof of Corollary 6.5, which will be given in the sequel, shows that (6.11) remains valid under the weaker assumption that one has, almost surely,

$$S(\cdot) \in \mathcal{C}^{\underline{S}(I)}(I). \tag{6.12}$$

Theorem 6.7 (Pointwise modulus of continuity for $\partial_v^q X$). *Let $\tau \in \mathbb{R}^N$ be an arbitrary fixed point. Then, one has, almost surely, for every compact interval $\mathcal{H} \subset (0,1)$,*

$$\sup_{(u,v)\in\mathcal{B}(\tau)\times\mathcal{H}} \left\{ \tag{6.13} \right.$$

$$\left. \frac{\left|(\partial_v^q X)(u, v) - (\partial_v^q X)(\tau, v)\right|}{|u - \tau|^v\left(1 + \left|\log|u - \tau|\right|\right)^q \log^{\frac{1}{2}}\left(2 + \left|\log|u - \tau|\right|\right)} \right\} < +\infty,$$

where $\mathcal{B}(\tau)$ denotes the closed ball of $\mathbb{R}^N$ centered at τ and of radius 1. Observe that the exceptional negligible event on which (6.13) fails to be true depends on τ but does not depend on $\mathcal{H}$.

Corollary 6.8 (Pointwise modulus of continuity for Z). *Let $\tau \in \mathbb{R}^N$ be an arbitrary fixed point. Assume that, almost surely,*

$$S(\cdot) \in \mathcal{C}^{S(\tau)}(\tau), \tag{6.14}$$

where $\mathcal{C}^{S(\tau)}(\tau)$ denotes the pointwise Hölder space at τ of order $S(\tau)$ (see Definition 1.42). Then, one has, almost surely, that

$$\sup_{t \in \mathcal{B}(\tau)} \left\{ \frac{\left| Z(t) - Z(\tau) \right|}{|t - \tau|^{S(\tau)} \log^{\frac{1}{2}} \left(2 + \left| \log |t - \tau| \right| \right)} \right\} < +\infty. \tag{6.15}$$

Observe that the exceptional negligible event on which (6.15) fails to be true depends on τ.

Theorem 6.9 (Behavior of $\partial_v^q X$ at infinity). *For any fixed positive real number δ, one sets*

$$\mathbb{R}_{\geq \delta}^N := \left\{ u \in \mathbb{R}^N : |u| \geq \delta \right\}. \tag{6.16}$$

Then, for each $(\omega, q) \in \Omega_0^* \times \mathbb{Z}_+$, for all $\delta > 0$, and for every compact interval $\mathcal{H} \subset (0,1)$, one has

$$\sup_{(u,v) \in \mathbb{R}_{\geq \delta}^N \times \mathcal{H}} \left\{ \frac{\left| (\partial_v^q X^{\mathrm{lf}})(u, v, \omega) \right|}{|u|^v \left(1 + \left| \log |u| \right| \right)^q \log^{\frac{1}{2}} \left(2 + \left| \log |u| \right| \right)} \right\} < +\infty, \tag{6.17}$$

$$\sup_{(u,v) \in \mathbb{R}_{\geq \delta}^N \times \mathcal{H}} \left\{ \frac{\left| (\partial_v^q X^{\mathrm{hf}})(u, v, \omega) \right|}{\left(1 + \left| \log |u| \right| \right)^{q + \frac{1}{2}}} \right\} < +\infty, \tag{6.18}$$

and

$$\sup_{(u,v) \in \mathbb{R}_{\geq \delta}^N \times \mathcal{H}} \left\{ \frac{\left| (\partial_v^q X)(u, v, \omega) \right|}{|u|^v \left(1 + \left| \log |u| \right| \right)^q \log^{\frac{1}{2}} \left(2 + \left| \log |u| \right| \right)} \right\} < +\infty. \tag{6.19}$$

Corollary 6.10 (Behavior of Z at infinity). *For any fixed positive real number δ, let $\mathbb{R}_{\geq \delta}^N$ be as in (6.16). Then, for each $\omega \in \Omega_0^*$ and for all $\delta > 0$, one has*

$$\sup_{t \in \mathbb{R}_{\geq \delta}^N} \left\{ \frac{\left| Z(t, \omega) \right|}{|t|^{S(t, \omega)} \log^{\frac{1}{2}} \left(2 + \left| \log |t| \right| \right)} \right\} < +\infty. \tag{6.20}$$

Let us now turn to the proofs of the main results of the section. To this end, one needs several preliminary results.

Lemma 6.11. *For all fixed* $(j, \gamma, r, \omega) \in \mathbb{Z} \times \mathbb{Z}_+^N \times \mathbb{Z}_+ \times \Omega_0^*$, *the series*

$$\sum_{p=1}^{2^N-1} \sum_{k \in \mathbb{Z}^N} |\varepsilon_{p,j,k}(\omega)| \left| (\partial_y^\gamma \partial_v^r \Psi_p)(y-k, v) \right| \tag{6.21}$$

is uniformly convergent in $(y, v) \in \mathcal{J} \times [0, 1]$, *where* $\mathcal{J}$ *is an arbitrary compact interval in* $\mathbb{R}^N$. *Moreover, there exists* $C_{\gamma, r}$, *a positive random variable of finite moment of any order only depending* (γ, r) *and* N, *such that, for every* $(j, y) \in \mathbb{Z} \times \mathbb{R}^N$, *one has*

$$\sum_{p=1}^{2^N-1} \sum_{k \in \mathbb{Z}^N} |\varepsilon_{p,j,k}(\omega)| \sup_{v \in [0,1]} \left| (\partial_y^\gamma \partial_v^r \Psi_p)(y-k, v) \right| \leq C_{\gamma, r}(\omega) \sqrt{\log \left(2 + |j| + |y|\right)}. \tag{6.22}$$

Proof of Lemma 6.11. In order to show that the series in (6.21) is uniformly convergent in $(u, v) \in \mathcal{J} \times [0, 1]$, it is enough to prove that

$$\sum_{p=1}^{2^N-1} \sum_{k \in \mathbb{Z}^N} |\varepsilon_{p,j,k}(\omega)| \sup_{(y,v) \in \mathcal{J} \times [0,1]} \left| (\partial_y^\gamma \partial_v^r \Psi_p)(y-k, v) \right| < +\infty. \tag{6.23}$$

Observe that there is no restriction to assume that $\mathcal{J} = [-\nu, \nu]^N$, where ν is an arbitrary positive real number. Also, observe that it easily follows from (5.20) that there exist two finite deterministic constants $c_1 < c_2$, only depending on ν and (γ, r), such that, for all $z = (z_1, \ldots, z_N) \in \mathbb{R}^N$, one has

$$\sup_{v \in [0,1]} \left| (\partial_y^\gamma \partial_v^r \Psi_p)(z, v) \right| \leq c_1 \prod_{n=1}^N \left(3 + |z_n|\right)^{-2} \leq c_2 \prod_{n=1}^N \left(3 + \nu + |z_n|\right)^{-2}. \tag{6.24}$$

Notice that c_1 only depends on (γ, r), and c_2 only depends on c_1 and ν. Next, using the second inequality in (6.24) and the triangle inequality, one gets, for every $k = (k_1, \ldots, k_N) \in \mathbb{Z}^N$, that

$$\sup_{(y,v) \in \mathcal{J} \times [0,1]} \left| (\partial_y^\gamma \partial_v^r \Psi_p)(y-k, v) \right| \leq c_2 \prod_{n=1}^N \left(3 + |k_n|\right)^{-2}. \tag{6.25}$$

Thus, combining Lemma A.23 and (6.25), one obtains (6.23).

On the other hand, one can derive from the same lemma, the first inequality in (6.24), the second inequality in (5.74), and the triangle inequality, that there is $C'_{\gamma, r}$, a positive random variable of finite moment of any order only depending on (γ, r), such that, for all $j \in \mathbb{Z}$ and $y = (y_1, \ldots, y_N) \in \mathbb{R}^N$, one has

$$
\sum_{p=1}^{2^N-1} \sum_{k \in \mathbb{Z}^N} |\varepsilon_{p,j,k}(\omega)| \sup_{v \in [0,1]} \left| (\partial_y^\gamma \partial_v^r \Psi_p)(y - k, v) \right|
$$

$$
\leq C'_{\gamma,r}(\omega) \sum_{k \in \mathbb{Z}^N} \frac{\sqrt{\log\left(2 + |j| + \sum_{n=1}^N |k_n|\right)}}{\prod_{n=1}^N \left(3 + |y_n - k_n|\right)^2}
$$

$$
= C'_{\gamma,r}(\omega) \sum_{k \in \mathbb{Z}^N} \frac{\sqrt{\log\left(2 + |j| + \sum_{n=1}^N |k_n + [y_n]|\right)}}{\prod_{n=1}^N \left(3 + |y_n - [y_n] - k_n|\right)^2}
$$

$$
\leq C'_{\gamma,r}(\omega) \sum_{k \in \mathbb{Z}^N} \frac{\sqrt{\log\left(2 + N + |j| + \sqrt{N}|y| + \sum_{n=1}^N |k_n|\right)}}{\prod_{n=1}^N \left(3 + |y_n - [y_n] - k_n|\right)^2}
$$

$$
\leq c_3 \, C'_{\gamma,r}(\omega) \sqrt{\log\left(2 + |j| + \sqrt{N}|y|\right)},
$$

where

$$
c_3 := 2 \sum_{k \in \mathbb{Z}^N} \frac{\sqrt{\log\left(2 + N + \sum_{n=1}^N |k_n|\right)}}{\prod_{n=1}^N \left(2 + |k_n|\right)^2} < +\infty,
$$

which shows that (6.22) is satisfied. $\qquad\square$

Lemma 6.12. *For all fixed* $(j, y, v, \omega) \in \mathbb{Z} \times \mathbb{R}^N \times (0,1) \times \Omega_0^*$, *one sets*

$$
B_j(y, v, \omega) := \sum_{p=1}^{2^N-1} \sum_{k \in \mathbb{Z}^N} \varepsilon_{p,j,k}(\omega) \Psi_p(y - k, v). \tag{6.26}
$$

Then, $B_j(\cdot, \cdot, \omega) : (y, v) \mapsto B_j(y, v, \omega)$ *is a well-defined real-valued infinitely differentiable function on* $\mathbb{R}^N \times (0,1)$. *Moreover, for each* $(\gamma, r) \in \mathbb{Z}_+^N \times \mathbb{Z}_+$, *one has*

$$
(\partial_y^\gamma \partial_v^r B_j)(y, v, \omega) = \sum_{p=1}^{2^N-1} \sum_{k \in \mathbb{Z}^N} \varepsilon_{p,j,k}(\omega) (\partial_y^\gamma \partial_v^r \Psi_p)(y - k, v), \tag{6.27}
$$

where the series is uniformly convergent in (y, v), *on each compact interval of* $\mathbb{R}^N \times (0,1)$. *One also has, for every* $(j, y) \in \mathbb{Z} \times \mathbb{R}^N$, *that*

$$
\sup_{v \in (0,1)} \left| (\partial_y^\gamma \partial_v^r B_j)(y, v, \omega) \right| \leq C_{\gamma,r}(\omega) \sqrt{\log\left(2 + |j| + |y|\right)}, \tag{6.28}
$$

where $C_{\gamma,r}$ *is the same random variable (not depending on* (j, y)*) as in Lemma 6.11.*

Proof of Lemma 6.12. In order to show that $B_j(\cdot,\cdot,\omega)$ is a well-defined real-valued infinitely differentiable function on $\mathbb{R}^N \times (0,1)$ with partial derivatives satisfying (6.27), it is enough to prove that all the term by term partial derivatives of the series in (6.26) are uniformly convergent in (y,v), on each compact interval of $\mathbb{R}^N \times (0,1)$. This can easily be done by using Lemma 6.11, which also allows to obtain (6.28). □

We are now in position to prove Proposition 6.1.

Proof of Proposition 6.1. First notice that it follows from Lemmas 6.12 and 6.11, and from a careful inspection of the proofs of Propositions 5.16 and 5.17, that the fields $\{X^{\mathrm{lf}}(u,v),\ (u,v) \in \mathbb{R}^N \times (0,1)\}$ and $\{X^{\mathrm{hf}}(u,v),\ (u,v) \in \mathbb{R}^N \times (0,1)\}$ can, for every $\omega \in \Omega_0^*$, be expressed as:

$$X^{\mathrm{lf}}(u,v) = \sum_{j=-\infty}^{-1} A_j(u,v,\omega) \tag{6.29}$$

and

$$X^{\mathrm{hf}}(u,v) = \sum_{j=0}^{+\infty} A_j(u,v,\omega), \tag{6.30}$$

where, for all $(j,u,v) \in \mathbb{Z} \times \mathbb{R}^N \times (0,1)$,

$$A_j(u,v,\omega) := 2^{-jv}\big(B_j(2^j u,v,\omega) - B_j(0,v,\omega)\big). \tag{6.31}$$

Observe that the series in (6.29) and (6.30) are uniformly convergent in (u,v) on each compact interval of $\mathbb{R}^N \times (0,1)$. Also, observe that one knows from (6.31), Lemma 6.12 and the general Leibniz rule, that the functions $A_j(\cdot,\cdot,\omega)$, $j \in \mathbb{Z}$ and $\omega \in \Omega_0^*$, are infinitely differentiable on $\mathbb{R}^N \times (0,1)$, with partial derivatives satisfying, for every $\gamma \in \mathbb{Z}_+^N \setminus \{0\}$, $q \in \mathbb{Z}_+$ and $(u,v) \in \mathbb{R}^N \times (0,1)$,

$$(\partial_v^q A_j)(u,v,\omega) \tag{6.32}$$

$$= \sum_{m=0}^{q} \binom{q}{m} \big(-j\log 2\big)^m 2^{-jv}\big((\partial_v^{q-m} B_j)(2^j u,v,\omega) - (\partial_v^{q-m} B_j)(0,v,\omega)\big)$$

$$= \sum_{m=0}^{q} \sum_{p=1}^{2^N-1} \binom{q}{m} \big(-\log 2\big)^m \sum_{k \in \mathbb{Z}^N} j^m\, 2^{-jv} \varepsilon_{p,j,k}(\omega)$$

$$\times \big((\partial_v^{q-m} \Psi_p)(2^j u - k,v) - (\partial_v^{q-m} \Psi_p)(-k,v)\big)$$

and

$$(\partial_u^\gamma \partial_v^q A_j)(u, v, \omega) \tag{6.33}$$

$$= \sum_{m=0}^{q} \binom{q}{m} \left(- j \log 2 \right)^m 2^{-j(v - \gamma_1 - \ldots - \gamma_N)} (\partial_u^\gamma \partial_v^{q-m} B_j)(2^j u, v, \omega)$$

$$= \sum_{m=0}^{q} \sum_{p=1}^{2^N - 1} \binom{q}{m} \left(- \log 2 \right)^m \sum_{k \in \mathbb{Z}^N} j^m \, 2^{-j(v - \gamma_1 - \ldots - \gamma_N)} \varepsilon_{p,j,k}(\omega)$$

$$\times (\partial_y^\gamma \partial_v^{q-m} \Psi_p)(2^j u - k, v).$$

Next, one assumes that the real numbers $\nu > 0$ and $0 < \underline{H} < \overline{H} < 1$ are arbitrary and fixed. In view of (6.29), in order to show that part (i) of the proposition holds, it is enough to prove that, for every $(\gamma, q) \in \mathbb{Z}_+^N \times \mathbb{Z}_+$, one has

$$\sum_{j=-\infty}^{-1} \sup_{(u,v) \in [-\nu, \nu]^N \times [\underline{H}, \overline{H}]} \left| (\partial_u^\gamma \partial_v^q A_j)(u, v, \omega) \right| < +\infty. \tag{6.34}$$

Moreover, in view of (6.30), in order to show that part (ii) of the proposition holds, it is enough to prove that, for every $q \in \mathbb{Z}_+$, one has

$$\sum_{j=0}^{+\infty} \sup_{(u,v) \in [-\nu, \nu]^N \times [\underline{H}, \overline{H}]} \left| (\partial_v^q A_j)(u, v, \omega) \right| < +\infty. \tag{6.35}$$

Using (6.32), the Mean Value Theorem, and the triangle inequality, for all $q \in \mathbb{Z}_+$ and $j \in \mathbb{Z}_- \setminus \{0\}$, it can be shown that

$$\sup_{(u,v) \in [-\nu, \nu]^N \times [\underline{H}, \overline{H}]} \left| (\partial_v^q A_j)(u, v, \omega) \right| \tag{6.36}$$

$$\leq \nu \sum_{m=0}^{q} \sum_{n=1}^{N} \binom{q}{m} |j|^m \, 2^{-|j|(1 - \overline{H})} \sup_{(y,v) \in [-\nu, \nu]^N \times [\underline{H}, \overline{H}]} \left| (\partial_{y_n} \partial_v^{q-m} B_j)(y, v, \omega) \right|.$$

Moreover, for any $\gamma \in \mathbb{Z}_+^N \setminus \{0\}$, it can easily be derived from (6.33) and the triangle inequality that

$$\sup_{(u,v) \in [-\nu, \nu]^N \times [\underline{H}, \overline{H}]} \left| (\partial_u^\gamma \partial_v^q A_j)(u, v, \omega) \right| \leq \sum_{m=0}^{q} \binom{q}{m} \tag{6.37}$$

$$\times |j|^m \, 2^{-|j|(\gamma_1 + \ldots + \gamma_N - \overline{H})} \sup_{(y,v) \in [-\nu, \nu]^N \times [\underline{H}, \overline{H}]} \left| (\partial_u^\gamma \partial_v^{q-m} B_j)(y, v, \omega) \right|.$$

Next, putting together (6.28), (6.36) and (6.37), it follows that (6.34) is satisfied.

Finally, let us show that (6.35) holds. Using (6.32) and the triangle inequality, for all $q \in \mathbb{Z}_+$ and $j \in \mathbb{Z}_+$, it can be shown that

$$\sup_{(u,v) \in [-\nu,\nu]^N \times [\underline{H},\overline{H}]} \left| (\partial_v^q A_j)(u,v,\omega) \right|$$

$$\leq 2 \sum_{m=0}^{q} \binom{q}{m} j^m \, 2^{-j\underline{H}} \sup_{(y,v) \in [-2^j\nu, 2^j\nu]^N \times [\underline{H},\overline{H}]} \left| (\partial_v^{q-m} B_j)(y,v,\omega) \right|.$$

Thus, (6.35) results from (6.28). $\qquad\square$

The following lemma is a straightforward consequence of of Remark 6.3.

Lemma 6.13. *For each* $(\omega, q) \in \Omega_0^* \times \mathbb{Z}_+$ *and, for every compact intervals* $\mathcal{J} \subset \mathbb{R}^N$ *and* $\mathcal{H} \subset (0,1)$*, one has*

$$\sup_{(u,v_1,v_2) \in \mathcal{J} \times \mathcal{H}^2} \left\{ \frac{\left| (\partial_v^q X)(u,v_1,\omega) - (\partial_v^q X)(u,v_2,\omega) \right|}{|v_1 - v_2|} \right\} < +\infty. \qquad (6.38)$$

Lemma 6.14. *For each* $(\omega, q) \in \Omega_0^* \times \mathbb{Z}_+$ *and, for every compact intervals* $\mathcal{J} \subset \mathbb{R}^N$ *and* $\mathcal{H} \subset (0,1)$*, one has*

$$\sup_{(u^1,u^2,v) \in \mathcal{J}^2 \times \mathcal{H}} \left\{ \frac{\left| (\partial_v^q X)(u^1,v,\omega) - (\partial_v^q X)(u^2,v,\omega) \right|}{|u^1 - u^2|^v \left(1 + \left| \log |u^1 - u^2| \right| \right)^{q+\frac{1}{2}}} \right\} < +\infty. \qquad (6.39)$$

Proof of Lemma 6.14. First notice that part (i) of Proposition 6.1 implies that the lemma holds when X is replaced by its low frequency part X^{lf}. Thus, in view of (5.16), one only has to prove that the lemma is true when X is replaced by its high frequency part X^{hf}. Moreover, one knows from part (ii) of Proposition 6.1, that, for each $q \in \mathbb{Z}_+$, the partial derivative function $(u,v) \mapsto (\partial_v^q X^{\mathrm{hf}})(u,v,\omega)$ is continuous on $\mathbb{R}^N \times (0,1)$. Therefore, using the fact that $\mathcal{K}' := \left\{ (u^1,u^2,v) \in \mathcal{J}^2 \times \mathcal{H} : |u^1 - u^2| \geq 1 \right\}$ is a compact subset of $\mathbb{R}^N \times (0,1)$, it can immediately be seen that

$$\sup_{(u^1,u^2,v) \in \mathcal{K}'} \left\{ \frac{\left| (\partial_v^q X^{\mathrm{hf}})(u^1,v,\omega) - (\partial_v^q X^{\mathrm{hf}})(u^2,v,\omega) \right|}{|u^1 - u^2|^v \left(1 + \left| \log |u^1 - u^2| \right| \right)^{q+\frac{1}{2}}} \right\} < +\infty,$$

with the usual convention that a supremum on an empty set equals to $-\infty$. Thus, in order to derive the lemma, it is enough to prove that

$$\sup_{(u^1,u^2,v) \in \mathcal{K}} \left\{ \frac{\left| (\partial_v^q X^{\mathrm{hf}})(u^1,v,\omega) - (\partial_v^q X^{\mathrm{hf}})(u^2,v,\omega) \right|}{|u^1 - u^2|^v \left(1 + \left| \log |u^1 - u^2| \right| \right)^{q+\frac{1}{2}}} \right\} < +\infty, \qquad (6.40)$$

where $\mathcal{K} := \{(u^1, u^2, v) \in \mathcal{J}^2 \times \mathcal{H} : |u^1 - u^2| \leq 1\}$. Let $(u^1, u^2, v) \in \mathcal{K}$ be arbitrary and fixed. There is no restriction to assume that $u^1 \neq u^2$, thus, one can set

$$j_0 := \left[\frac{\log\left(|u^1 - u^2|^{-1}\right)}{\log 2} \right], \tag{6.41}$$

where $[\cdot]$ denotes the integer part function. Observe that j_0 is a nonnegative integer and that one has

$$2^{-j_0-1} < |u^1 - u^2| \leq 2^{-j_0}. \tag{6.42}$$

Next, for all $m \in \{0, \ldots, q\}$ and $(u, v, \omega) \in \mathbb{R}^N \times (0, 1) \times \Omega_0^*$, one sets

$$F_m(u, v, \omega) := \sum_{j=0}^{+\infty} j^m \, 2^{-jv} (\partial_v^{q-m} B_j)(2^j u, v, \omega). \tag{6.43}$$

Notice that $(\partial_v^{q-m} B_j)(2^j u, v, \omega)$ is given by (6.27), and that one knows from (6.28) that the series in (6.43) is uniformly convergent in (u, v), on each compact interval of $\mathbb{R}^N \times (0, 1)$. Also, notice that, in view of (6.5), for any $(u, v, \omega) \in \mathbb{R}^N \times (0, 1) \times \Omega_0^*$, one has

$$(\partial_v^q X^{\mathrm{hf}})(u, v, \omega) = \sum_{m=0}^{q} \binom{q}{m} (-\log 2)^m \big(F_m(u, v, \omega) - F_m(0, v, \omega)\big). \tag{6.44}$$

In view of (6.43) and (6.44), in order to show that (6.40) is satisfied, it is enough to prove that there exist two positive random variables C' and C'' of finite moment of any order, which do not depend on (u^1, u^2, v) and on j_0, such that

$$\sum_{j=j_0+1}^{+\infty} j^m \, 2^{-jv} \sup_{(u,v) \in [-\nu,\nu]^N \times [\underline{H},\overline{H}]} \left| (\partial_v^{q-m} B_j)(2^j u, v, \omega) \right| \tag{6.45}$$

$$\leq C'(\omega) |u^1 - u^2|^v \big(1 + \big| \log|u^1 - u^2| \big|\big)^{m+\frac{1}{2}}$$

and

$$\sum_{j=0}^{j_0} j^m \, 2^{-jv} \left| (\partial_v^{q-m} B_j)(2^j u^1, v, \omega) - (\partial_v^{q-m} B_j)(2^j u^2, v, \omega) \right| \tag{6.46}$$

$$\leq C''(\omega) |u^1 - u^2|^v \big(1 + \big| \log|u^1 - u^2| \big|\big)^{m+\frac{1}{2}},$$

where ν is a constant such that $\mathcal{J} \subseteq [-\nu, \nu]^N$, and $\mathcal{H} = [\underline{H}, \overline{H}]$. Using (6.28), (6.41) and (6.42), one has that

$$\sum_{j=j_0+1}^{+\infty} j^m \, 2^{-jv} \sup_{(u,v)\in[-\nu,\nu]^N\times[\underline{H},\overline{H}]} \left|(\partial_v^{q-m} B_j)(2^j u, v, \omega)\right|$$

$$\leq C_1'(\omega) \sum_{l=0}^{+\infty} 2^{-(l+j_0+1)v}(l + j_0 + 1)^{m+\frac{1}{2}}$$

$$\leq C_2'(\omega)2^{-(j_0+1)v}(j_0 + 1)^{m+\frac{1}{2}} \leq C'(\omega)|u^1 - u^2|^v\left(1 + \big|\log|u^1 - u^2|\big|\right)^{m+\frac{1}{2}},$$

where C_1' is a positive random variable of finite moment of any order not depending on (u^1, u^2, v),

$$C_2' := C_1' \sum_{l=0}^{+\infty} 2^{-l\underline{H}}(l+1)^{m+\frac{1}{2}} < +\infty,$$

and $C' := (\log 2)^{-m-\frac{1}{2}}C_2'$. This shows that (6.45) holds. On the other hand, using the Mean Value Theorem, the triangle inequality, (6.28), (6.41) and (6.42), one has that

$$\sum_{j=0}^{j_0} j^m \, 2^{-jv}\left|(\partial_v^{q-m} B_j)(2^j u^1, v, \omega) - (\partial_v^{q-m} B_j)(2^j u^2, v, \omega)\right|$$

$$\leq \sum_{j=0}^{j_0} j^m \, 2^{j(1-v)}|u^1 - u^2|$$

$$\times \sum_{n=1}^{n} \sup_{(u,v)\in[-\nu,\nu]^N\times[\underline{H},\overline{H}]} \left|(\partial_{y_n}\partial_v^{q-m} B_j)(2^j u, v, \omega)\right|$$

$$\leq C_1''(\omega)|u^1 - u^2|\sum_{j=0}^{j_0}(j + 1)^{m+\frac{1}{2}} \, 2^{j(1-v)}$$

$$\leq C_2''(\omega)|u^1 - u^2|(j_0 + 1)^{m+\frac{1}{2}} \, 2^{j_0(1-v)}$$

$$\leq C''(\omega)|u^1 - u^2|^v\left(1 + \big|\log|u^1 - u^2|\big|\right)^{m+\frac{1}{2}},$$

where C_1'' is a positive random variable of finite moment of any order not depending on (u^1, u^2, v), $C_2'' := 2^{1-\underline{H}}\left(2^{1-\overline{H}} - 1\right)^{-1}$ and $C'' := (\log 2)^{-m-\frac{1}{2}}C_2''$. This shows that (6.46) holds. $\square$

We are now in position to prove Theorems and Corollaries 6.4 to 6.10.

Proof of Theorem 6.4. For all $(u^1, u^2, v_1, v_2) \in \mathcal{J}^2 \times \mathcal{H}^2$, one sets

$$f(u^1, u^2, v_1, v_2) := \frac{\left| \left(\partial_v^q X \right)(u^1, v_1, \omega) - \left(\partial_v^q X \right)(u^2, v_2, \omega) \right|}{\left| u^1 - u^2 \right|^{v_1 \vee v_2} \left(1 + \left| \log |u^1 - u^2| \right| \right)^{q + \frac{1}{2}} + |v_1 - v_2|},$$

with the convention that $0/0 = 0$. Observe that one has $f(u_1, u_2, v_1, v_2) = f(u_2, u_1, v_2, v_1)$. Thus, one gets that

$$\sup_{(u^1, u^2, v_1, v_2) \in \mathcal{J}^2 \times \mathcal{H}^2} \left\{ \frac{\left| \left(\partial_v^q X \right)(u^1, v_1, \omega) - \left(\partial_v^q X \right)(u^2, v_2, \omega) \right|}{\left| u^1 - u^2 \right|^{v_1 \vee v_2} \left(1 + \left| \log |u^1 - u^2| \right| \right)^{q + \frac{1}{2}} + |v_1 - v_2|} \right\}$$

$$= \sup_{(u^1, u^2, v_1, v_2) \in \mathcal{J}^2 \times \mathcal{H}^2} \left\{ \frac{\left| \left(\partial_v^q X \right)(u^1, v_1 \vee v_2, \omega) - \left(\partial_v^q X \right)(u^2, v_1 \wedge v_2, \omega) \right|}{\left| u^1 - u^2 \right|^{v_1 \vee v_2} \left(1 + \left| \log |u^1 - u^2| \right| \right)^{q + \frac{1}{2}} + |v_1 - v_2|} \right\}. \quad (6.47)$$

Moreover, using the triangle inequality, and the inequality, for every $(u^1, u^2, v_1, v_2) \in \mathcal{J}^2 \times \mathcal{H}^2$,

$$\max \left\{ \left| u^1 - u^2 \right|^{v_1 \vee v_2} \left(1 + \left| \log |u^1 - u^2| \right| \right)^{q + \frac{1}{2}}, |v_1 - v_2| \right\}$$

$$\leq \left| u^1 - u^2 \right|^{v_1 \vee v_2} \left(1 + \left| \log |u^1 - u^2| \right| \right)^{q + \frac{1}{2}} + |v_1 - v_2|,$$

one obtains that

$$\sup_{(u^1, u^2, v_1, v_2) \in \mathcal{J}^2 \times \mathcal{H}^2} \left\{ \frac{\left| \left(\partial_v^q X \right)(u^1, v_1 \vee v_2, \omega) - \left(\partial_v^q X \right)(u^2, v_1 \wedge v_2, \omega) \right|}{\left| u^1 - u^2 \right|^{v_1 \vee v_2} \left(1 + \left| \log |u^1 - u^2| \right| \right)^{q + \frac{1}{2}} + |v_1 - v_2|} \right\}$$

$$\leq \sup_{(u^1, u^2, v_1, v_2) \in \mathcal{J}^2 \times \mathcal{H}^2} \left\{ \frac{\left| \left(\partial_v^q X \right)(u^1, v_1 \vee v_2, \omega) - \left(\partial_v^q X \right)(u^2, v_1 \vee v_2, \omega) \right|}{\left| u^1 - u^2 \right|^{v_1 \vee v_2} \left(1 + \left| \log |u_1 - u_2| \right| \right)^{q + \frac{1}{2}} + |v_1 - v_2|} \right\}$$

$$+ \sup_{(u^1, u^2, v_1, v_2) \in \mathcal{J}^2 \times \mathcal{H}^2} \left\{ \frac{\left| \left(\partial_v^q X \right)(u^2, v_1 \vee v_2, \omega) - \left(\partial_v^q X \right)(u^2, v_1 \wedge v_2, \omega) \right|}{\left| u^1 - u^2 \right|^{v_1 \vee v_2} \left(1 + \left| \log |u^1 - u^2| \right| \right)^{q + \frac{1}{2}} + |v_1 - v_2|} \right\}$$

$$\leq \sup_{(u^1, u^2, v) \in \mathcal{J}^2 \times \mathcal{H}} \left\{ \frac{\left| \left(\partial_v^q X \right)(u^1, v, \omega) - \left(\partial_v^q X \right)(u^2, v, \omega) \right|}{\left| u^1 - u^2 \right|^v \left(1 + \left| \log |u^1 - u^2| \right| \right)^{q + \frac{1}{2}}} \right\} \quad (6.48)$$

$$+ \sup_{(u, v_1, v_2) \in \mathcal{J} \times \mathcal{H}^2} \left\{ \frac{\left| \left(\partial_v^q X \right)(u, v_1, \omega) - \left(\partial_v^q X \right)(u, v_2, \omega) \right|}{|v_1 - v_2|} \right\}.$$

Finally, putting together (6.47), (6.48), Lemma 6.14, and Lemma 6.13, it follows that (6.7) holds. □

Proof of Corollary 6.5. Applying (6.7) in the case where $q = 0$, $\mathcal{J} = I$ and $\mathcal{H} = [\underline{S}(I), \overline{S}(I)]$, and using (6.1), one obtains, almost surely, that

$$\sup_{(t^1, t^2) \in I^2} \left\{ \frac{\left| Z(t^1) - Z(t^2) \right|}{\left| t^1 - t^2 \right|^{S(t^1) \vee S(t^2)} \left(1 + \left| \log \left| t^1 - t^2 \right| \right| \right)^{\frac{1}{2}} + \left| S(t^1) - S(t^2) \right|} \right\} < +\infty.$$

(6.49)

Moreover, one knows from (6.9), the second equality in (6.8), and Definition 1.33 that one has, almost surely,

$$\sup_{(t^1, t^2) \in I^2} \left\{ \frac{\left| t^1 - t^2 \right|^{S(t^1) \vee S(t^2)} \left(1 + \left| \log \left| t^1 - t^2 \right| \right| \right)^{\frac{1}{2}} + \left| S(t^1) - S(t^2) \right|}{\left| t^1 - t^2 \right|^{S(t^1) \vee S(t^2)} \left(1 + \left| \log \left| t^1 - t^2 \right| \right| \right)^{\frac{1}{2}}} \right\} < +\infty.$$

(6.50)

Combining (6.49) and (6.50) one gets (6.10). □

Proof of Theorem 6.7. For the sake of simplicity one only gives the proof in the case where $q = 0$; it can be done in a rather similar way in the general case where q is an arbitrary nonnegative integer.

First notice that part (i) of Proposition 6.1 implies that the theorem holds when X is replaced by its low frequency part X^{lf}. Thus, in view of (5.16), one only has to prove that the theorem is true when X is replaced by its high frequency part X^{hf}. Moreover, one is allowed to assume that the sum over j in (5.15) starts from $j = 1$ instead of $j = 0$, since the term corresponding to $j = 0$ is almost surely an infinitely differentiable function in (u, v) over $\mathbb{R}^N \times (0, 1)$ (see Lemma 6.12).

Let $\underline{H}, \overline{H} \in (0, 1)$ be such that $\mathcal{H} = [\underline{H}, \overline{H}]$, let $(u, v) \in \mathcal{B}(\tau) \times [\underline{H}, \overline{H}]$ be arbitrary and fixed. There is no restriction to assume that $0 < |u - \tau| \leq 2^{-15}$. One sets

$$j_1 := 1 + \left[\frac{\log \left(|u - \tau|^{-1} \right)}{\log(2)} \right] \quad \text{and} \quad j_2 := j_1 + \left[\frac{\log(j_1)}{2\underline{H} \log(2)} \right], \quad (6.51)$$

where $[\cdot]$ denotes the integer part function. Observe that one has $j_1 \geq 16$, $j_2 \geq j_1 + 2$,

$$j^{1/\underline{H}} \geq 2^{j - j_1 + 2}, \quad \text{for all } j \in \{j_1 + 1, j_1 + 2, \ldots, j_2\}, \quad (6.52)$$

and

$$2^{-j_1} < |u - \tau| \leq 2^{1 - j_1}. \quad (6.53)$$

Next, for all $j \in \mathbb{N}$, one assumes that the finite set $\mathcal{D}_j(\tau, 1/\underline{H})$ is defined through (A.33) with $\theta = 1/\underline{H}$, and one denotes by $\mathcal{D}_{\pi,j}(\tau)$ the finite set defined as the Cartesian product

$$\mathcal{D}_{\pi,j}(\tau) := \{1, \ldots, 2^{N-1}\} \times \mathcal{D}_j(\tau, 1/\underline{H}) \quad (6.54)$$

$$= \left\{ (p, k) \in \{1, \ldots, 2^{N-1}\} \times \mathbb{Z}^N : |\tau - 2^{-j}k| \leq N^{1/2} j^{1/\underline{H}} 2^{-j} \right\}.$$

Observe that, one knows from Lemma A.28 that there is C_1 a positive random variable of finite moment of any order such that one has, almost surely,

$$\max_{(p,k)\in\mathcal{D}_{\pi,j}(\tau)} |\varepsilon_{p,j,k}| \leq C_1 \sqrt{\log(2+j)}, \quad \text{for every } j \in \mathbb{N}. \qquad (6.55)$$

Moreover, one denotes by $\mathcal{D}_{\pi,j}^{\mathrm{co}}(\tau)$ the complement of $\mathcal{D}_{\pi,j}(\tau)$, that is

$$\mathcal{D}_{\pi,j}^{\mathrm{co}}(\tau) := \{1,\ldots,2^{N-1}\} \times \mathbb{Z}^N \setminus \mathcal{D}_{\pi,j}(\tau) \qquad (6.56)$$

$$= \{(p,k) \in \{1,\ldots,2^{N-1}\} \times \mathbb{Z}^N \ : \ |\tau - 2^{-j}k| > N^{1/2} j^{1/\underline{H}} 2^{-j}\}.$$

Observe that one has

$$\mathcal{D}_{\pi,j}^{\mathrm{co}}(\tau) \subseteq \bigcup_{q=1}^{N} \mathcal{G}_{\pi,j}^{q}(\tau), \qquad (6.57)$$

where, for each $q \in \{1,\ldots,N\}$,

$$\mathcal{G}_{\pi,j}^{q}(\tau) := \{(p,k) \in \{1,\ldots,2^{N-1}\} \times \mathbb{Z}^N \ : \ |\tau_q - 2^{-j}k_q| > j^{1/\underline{H}} 2^{-j}\}, \qquad (6.58)$$

τ_q and k_q being the q-th coordinates of the vectors τ and k. Using (5.15) as well as our assumption that the sum over j in it starts from $j = 1$, in view of (6.54) and (6.56), one gets that

$$|X^{\mathrm{hf}}(u,v) - X^{\mathrm{hf}}(\tau,v)| \qquad (6.59)$$

$$\leq T_1(u,\tau,v) + T_2(u,\tau,v) + T_3(u,\tau,v) + T_4(u,\tau,v),$$

where

$$T_1(u,\tau,v) := \sum_{j=1}^{j_1} \sum_{(p,k)\in\mathcal{D}_{\pi,j}(\tau)} 2^{-jv} |\varepsilon_{p,j,k}| |\Psi_p(2^j u - k, v) - \Psi_p(2^j \tau - k, v)|, \qquad (6.60)$$

$$T_2(u,\tau,v) := \sum_{j=j_1+1}^{+\infty} \sum_{(p,k)\in\mathcal{D}_{\pi,j}(\tau)} 2^{-jv} |\varepsilon_{p,j,k}| |\Psi_p(2^j u - k, v) - \Psi_p(2^j \tau - k, v)|, \qquad (6.61)$$

$$T_3(u,\tau,v) := \sum_{j=1}^{j_2} \sum_{(p,k)\in\mathcal{D}_{\pi,j}^{\mathrm{co}}(\tau)} 2^{-jv} |\varepsilon_{p,j,k}| |\Psi_p(2^j u - k, v) - \Psi_p(2^j \tau - k, v)|, \qquad (6.62)$$

and

$$T_4(u,\tau,v) := \sum_{j=j_2+1}^{+\infty} \sum_{(p,k)\in\mathcal{D}_{\pi,j}^{\mathrm{co}}(\tau)} 2^{-jv} |\varepsilon_{p,j,k}| |\Psi_p(2^j u - k, v) - \Psi_p(2^j \tau - k, v)|. \qquad (6.63)$$

From now on, our goal is to derive an appropriate upper bound for each term in the right-hand side of (6.59). In all the sequel, one assumes that L is an arbitrarily large fixed positive integer. Therefore, one has that

$$c_2 := \sup_{y \in \mathbb{R}^N} \left\{ \sum_{k \in \mathbb{Z}^N} \prod_{m=1}^{N} \left(1 + |y_m - k_m| \right)^{-L} \right\} < +\infty. \tag{6.64}$$

Putting together (6.61), (6.55), (5.20), (6.64), Lemma 5.20, (6.53), and (6.51), one obtains, almost surely, that

$$T_2(u, \tau, v) \le C_3 \sum_{j=j_1+1}^{+\infty} 2^{-jv} \sqrt{\log(2+j)} \tag{6.65}$$

$$\le C_3 \, \widetilde{c}(\underline{H}) 2^{-(j_1+1)v} \sqrt{\log(3+j_1)} \le C_4 |u - \tau|^v \log^{\frac{1}{2}} \left(2 + \left| \log |u - \tau| \right| \right),$$

where C_3 and C_4 are two positive random variables of finite moment of any order not depending on (u, v), and where the finite deterministic constant $\widetilde{c}(\underline{H})$ is defined through (5.61) with $a = \underline{H}$.

Next, putting to together (6.63), (6.22), the inequality $|u| \le |\tau| + 1$, the inequality $2^j > j$, for all $j \in \mathbb{N}$, (6.51), the inequality $v \ge \underline{H}$, and (6.53), it follows, almost surely, that

$$T_4(u, \tau, v) \le C_5 \sum_{j=j_2+1}^{+\infty} 2^{-jv} \sqrt{\log \left(2 + (|\tau| + 2)2^j \right)}$$

$$\le C_6 2^{-(j_2+1)v} \sqrt{j_2 + 1}$$

$$= C_6 2^{-j_1 v} \exp \left(-v \log(2) \left(1 + \left[\frac{\log(j_1)}{2\underline{H} \log(2)} \right] \right) + \frac{\log(j_2 + 1)}{2} \right)$$

$$\le C_6 |u - \tau|^v \exp \left(2^{-1} \log \left(\frac{j_2 + 1}{j_1} \right) \right)$$

$$\le C_6 \left(2 + \underline{H}^{-1} \right)^{\frac{1}{2}} |u - \tau|^v, \tag{6.66}$$

where C_5 and C_6 are two positive random variables of finite moment of any order not depending on (u, v).

Next, observe that, using the Mean Value Theorem, it can be shown that, for every fixed $(p, j, k) \in \{1, \ldots, 2^{N-1}\} \times \mathbb{N} \times \mathbb{Z}^N$, there exists a real number λ belonging to the interval $(0, 1)$ such that

$$\Psi_p(2^j u - k, v) - \Psi_p(2^j \tau - k, v) = 2^j \sum_{n=1}^{N} (\partial_{y_n} \Psi_p) \left(2^j \tau - k + \lambda 2^j (u - \tau), v \right) (u_n - \tau_n).$$
$$\tag{6.67}$$

Moreover notice that, when $j \in \{1, 2, \ldots, j_1\}$, it follows from (5.20), (6.53) and the triangle inequality that, for all $n \in \{1, \ldots, N\}$,

$$\left| (\partial_{y_n} \Psi_p) \big(2^j \tau - k + \lambda 2^j (u - \tau), v \big) \right|$$

$$\leq c_7 \prod_{m=1}^{N} \left(3 + |2^j \tau_m - k_m + \lambda 2^j (u_m - \tau_m)| \right)^{-L}$$

$$\leq c_7 \prod_{m=1}^{N} \left(3 + |2^j \tau_m - k_m| - 2^j |u_m - \tau_m| \right)^{-L}$$

$$\leq c_7 \prod_{m=1}^{N} \left(1 + |2^j \tau_m - k_m| \right)^{-L}, \tag{6.68}$$

where c_7 is a deterministic finite constant only depending on L. Also, notice that, when $j \in \{j_1 + 1, j_1 + 2, \ldots, j_2\}$ and $(p, k) \in \mathcal{D}_{\tau,j}^{\mathrm{co}}(\tau)$, using (5.20), (6.51), (6.53), (6.52) and (6.56), one has, for all $n \in \{1, \ldots, N\}$,

$$\left| (\partial_{y_n} \Psi_p) \big(2^j \tau - k + \lambda 2^j (u - \tau), v \big) \right|$$

$$\leq c_8 \big(3 + |2^j \tau - k + \lambda 2^j (u - \tau)| \big)^{-NL}$$

$$\leq c_8 \big(3 + 2^j |\tau - k 2^{-j}| - 2^j |u - \tau| \big)^{-NL}$$

$$\leq c_8 \big(3 + 2^{j-1} |\tau - k 2^{-j}| + 2^{-1} N^{1/2} j^{1/\underline{H}} - 2^{j+1-j_1} \big)^{-NL}$$

$$\leq c_9 \prod_{m=1}^{N} \left(1 + |2^j \tau_m - k_m| \right)^{-L}, \tag{6.69}$$

where c_8 is a deterministic finite constant only depending on L and $c_9 := 2^{NL} c_8$.

Next, putting together (6.60), (6.55), (6.67), (6.68), (6.64), (6.53), and (6.51), one obtains, almost surely, that

$$T_1(u, \tau, v) \leq C_{10} |u - \tau| \sum_{j=1}^{j_1} 2^{j(1-v)} \sqrt{\log(2 + j)}$$

$$\leq 2^{2(1-\underline{H})} \big(2^{1-\overline{H}} - 1 \big)^{-1} C_{10} |u - \tau| \, 2^{(j_1 - 1)(1-v)} \sqrt{\log(2 + j_1)}$$

$$\leq C_{11} |u - \tau|^v \log^{\frac{1}{2}} \big(2 + \big| \log |u - \tau| \big| \big), \tag{6.70}$$

where C_{10} and C_{11} are two positive random variables of finite moment of any order not depending on (u, v).

Next, combining (6.58) and Lemma A.23 and using arguments rather similar to those in the proof of (5.77), as well as the fact that, for all fixed $j \in \mathbb{N}$ and $z \in \mathbb{R}$, the set of nonnegative real numbers

$$\left\{ |2^j z - i| \, : \, i \in \mathbb{Z} \text{ and } |2^j z - i| > j^{1/\underline{H}} \right\}$$

can be expressed as

$$\left\{ l + r_1(j,z) + j^{1/\underline{H}} : l \in \mathbb{Z}_+ \right\} \cup \left\{ l + r_2(j,z) + j^{1/\underline{H}} : l \in \mathbb{Z}_+ \right\},$$

where $r_1(j,z)$ and $r_2(j,z)$ are some fixed real numbers only depending on (j,z) and belonging to the interval $(0,1]$, one obtains almost surely, for all $j \in \mathbb{N}$ and $q \in \{1, \ldots, N\}$, that

$$\sum_{(p,k) \in \mathcal{G}_{\pi,j}^q(\tau)} \frac{|\varepsilon_{p,j,k}|}{\prod_{m=1}^{N} \left(1 + |2^j \tau_m - k_m|\right)^L} \tag{6.71}$$

$$\leq C_{12} \sum_{l=0}^{+\infty} \frac{\sqrt{\log\left(2 + 2^j|\tau| + l + j^{1/\underline{H}}\right)}}{\left(1 + l + j^{1/\underline{H}}\right)^L} \leq C_{13} \, j^{3/2 - L/\underline{H}} \, \sqrt{\log(1+j)},$$

where C_{12} and C_{13} are two positive random variables of finite moment of any order not depending on (u,v). Next, putting together (6.62), (6.67), (6.68), (6.69), (6.57), (6.71), (6.51) and (6.53), it follows, almost surely, that

$$T_3(u,\tau,v) \leq C_{14}|u - \tau| \sum_{j=1}^{j_2} 2^{j(1-v)} \, j^{3/2 - L/\underline{H}} \, \sqrt{\log(1+j)}$$

$$\leq C_{15}|u - \tau| \, 2^{(j_2 - 1)(1-v)} \, j_2^{3/2 - L/\underline{H}} \, \sqrt{\log(1+j_2)}$$

$$\leq C_{15}|u - \tau|^v \, 2^{\frac{(1-\underline{H})\log(j_2)}{2\underline{H}\log(2)}} \, j_2^{3/2 - L/\underline{H}} \, \sqrt{\log(1+j_2)}$$

$$= C_{15}|u - \tau|^v \, j_2^{3/2 + (1-\underline{H})/(2\underline{H}) - L/\underline{H}} \, \sqrt{\log(1+j_2)}$$

$$\leq C_{16}|u - \tau|^v, \tag{6.72}$$

where C_{14}, C_{15} and

$$C_{16} := C_{15} \sup_{j \in \mathbb{N}} \left\{ j^{3/2 + (1-\underline{H})/(2\underline{H}) - L/\underline{H}} \, \sqrt{\log(1+j)} \right\}$$

are three positive random variables of finite moment of any order not depending on (u,v).

Finally, putting together (6.59), (6.65), (6.66), (6.70), (6.72), one gets (6.13) in the case where $q = 0$. $\qquad\Box$

Proof of Corollary 6.8. Using (6.1) and the triangle inequality, one has that

$$\sup_{t \in \mathcal{B}(\tau)} \left\{ \frac{|Z(t) - Z(\tau)|}{|t - \tau|^{S(\tau)} \log^{\frac{1}{2}} \left(2 + \left| \log|t - \tau| \right|\right)} \right\} \leq A_1(\tau) + A_2(\tau), \tag{6.73}$$

where

$$A_1(\tau) := \sup_{t \in \mathcal{B}(\tau)} \left\{ \frac{|X(t, S(t)) - X(t, S(\tau))|}{|t - \tau|^{S(\tau)} \log^{\frac{1}{2}} \left(2 + \left| \log|t - \tau| \right|\right)} \right\} \tag{6.74}$$

and

$$A_2(\tau) := \sup_{t \in \mathcal{B}(\tau)} \left\{ \frac{|X(t, S(\tau)) - X(\tau, S(\tau))|}{|t - \tau|^{S(\tau)} \log^{\frac{1}{2}} \left(2 + \left| \log |t - \tau| \right| \right)} \right\}. \qquad (6.75)$$

Next, one let $\mathcal{J} \subset \mathbb{R}^N$ and $\mathcal{H} \in (0,1)$ be the compact intervals defined as $\mathcal{J} = \left[-|\tau| - 1, |\tau| + 1 \right]^N$ and $\mathcal{H} := [\underline{S}, \overline{S}]$; observe that $\mathcal{B}(\tau) \subset \mathcal{J}$, and recall that the random Hurst functional parameter $S(\cdot)$ is always assumed to be with values in this random interval $\mathcal{H}$. Using the assumption (6.14), one has almost surely that

$$\sup_{t \in \mathcal{J}} \left\{ \frac{|S(t) - S(\tau)|}{|t - \tau|^{S(\tau)} \log^{\frac{1}{2}} \left(2 + \left| \log |t - \tau| \right| \right)} \right\} < +\infty. \qquad (6.76)$$

On the other hand, it is clear that

$$\sup_{t \in \mathcal{J}} \left\{ \frac{|X(t, S(t)) - X(t, S(\tau))|}{|S(t) - S(\tau)|} \right\} \qquad (6.77)$$

$$\leq \sup_{(u, v_1, v_2) \in \mathcal{J} \times \mathcal{H}^2} \left\{ \frac{|X(u, v_1) - X(u, v_2)|}{|v_1 - v_2|} \right\}.$$

Thus combining (6.74), (6.76), (6.77) and Lemma 6.13, with $q = 0$, one gets, almost surely, that

$$A_1(\tau) < +\infty. \qquad (6.78)$$

On the other hand, it easily follows from (6.75) that

$$A_2(\tau) \leq \sup_{(u, v) \in \mathcal{B}(\tau) \times \mathcal{H}} \left\{ \frac{|X(u, v) - X(\tau, v)|}{|u - \tau|^v \log^{\frac{1}{2}} \left(2 + \left| \log |u - \tau| \right| \right)} \right\}.$$

Thus, Theorem 6.7, with $q = 0$, implies that one has almost surely

$$A_2(\tau) < +\infty. \qquad (6.79)$$

Finally, combining (6.73), (6.78) and (6.79), one obtains (6.15). □

Proof of Theorem 6.9. For the sake of simplicity one only gives the proof in the case where $q = 0$; it can be done in a rather similar way in the general case where q is an arbitrary nonnegative integer.

As usual, the real numbers $\underline{H}, \overline{H} \in (0,1)$ are such that $\mathcal{H} = [\underline{H}, \overline{H}]$. There is no restriction to assume that $\delta = 2$. Let $(u, v) \in \mathbb{R}^N_{\geq 2} \times \mathcal{H}$ be arbitrary and fixed. The integer j_3 is defined as

$$j_3 := \left[\frac{\log |u|}{\log(2)} \right]. \qquad (6.80)$$

Therefore, one has that

$$2^{j_3} \leq |u| < 2^{j_3+1}. \tag{6.81}$$

Moreover, in view of (6.16), the assumption $u \in \mathbb{R}^N_{\geq 2}$ implies that $j_3 \geq 1$.
Let us first show that (6.17) holds in the case where $q = 0$. One knows
from Lemma 6.11 and from the proof of Proposition 5.16 that, for all $\omega \in$
Ω_0^*, the quantity $X^{\mathrm{lf}}(u, v, \omega)$ can be expressed as the absolutely convergent
series of real numbers

$$X^{\mathrm{lf}}(u, v, \omega) = \sum_{j=1}^{+\infty} 2^{jv} \big(B_{-j}(2^{-j}u, v, \omega) - B_{-j}(0, v, \omega) \big), \tag{6.82}$$

where $B_{-j}(\cdot, \cdot, \omega)$ is the infinitely differentiable function on $\mathbb{R}^N \times (0,1)$,
which was introduced in Lemma 6.12. One denotes by $\mathcal{B}(0, 2^{-j}|u|)$ the
closed ball of $\mathbb{R}^N$ centered at 0 and of radius $2^{-j}|u|$. Using the Mean Value
Theorem, (6.28), (6.80), (6.81) and Lemma 5.20, one gets that

$$\sum_{j=j_3+1}^{+\infty} 2^{jv} \big| B_{-j}(2^{-j}u, v, \omega) - B_{-j}(0, v, \omega) \big|$$
$$\leq |u| \sum_{j=j_3+1}^{+\infty} 2^{-j(1-v)} \Big(\sum_{n=1}^{N} \sup_{(y,v)\in\mathcal{B}(0,2^{-j}|u|)\times\mathcal{H}} \big| (\partial_{y_n} B_{-j})(y, v, \omega) \big| \Big)$$
$$\leq C_1(\omega)|u| \sum_{j=j_3+1}^{+\infty} 2^{-j(1-v)} \sqrt{\log(3+j)}$$
$$\leq 2^{1-\underline{H}} \widetilde{c}(1 - \overline{H}) C_1(\omega)|u| \, 2^{-(j_3+2)(1-v)} \sqrt{\log(4+j_3)}$$
$$\leq C_2(\omega)|u|^v \log^{\frac{1}{2}} \big(2 + |\log|u|| \big), \tag{6.83}$$

where C_1 and C_2 are two positive random variables of finite moment of any
order not depending on (u, v, j_3), and the deterministic constant $\widetilde{c}(1 - \overline{H})$,
which does not depend on (u, v, j_3) as well, is defined through (5.61) with
$a = 1 - \overline{H}$. On the other hand, it follows from the triangle inequality,

(6.28), (6.81), the second inequality in (5.74), and (6.80), that

$$
\sum_{j=1}^{j_3} 2^{jv} \left| B_{-j}(2^{-j}u, v, \omega) - B_{-j}(0, v, \omega) \right|
$$

$$
\leq C_3(\omega) \sum_{j=1}^{j_3} 2^{jv} \sqrt{\log \left(2 + j + 2^{-j}|u| \right)}
$$

$$
= C_3(\omega) 2^{(j_3+1)v} \sum_{j=1}^{j_3} 2^{-(j_3+1-j)v} \sqrt{\log \left(2 + j + 2^{-j}|u| \right)}
$$

$$
\leq 2^{\overline{H}} C_3(\omega)|u|^v \sum_{m=1}^{j_3} 2^{-mv} \sqrt{\log \left(2 + (j_3 + 1 - m) + 2^{-(j_3+1-m)}|u| \right)}
$$

$$
\leq C_4(\omega)|u|^v \sum_{m=1}^{j_3} 2^{-mv} \sqrt{\log \left(2 + j_3 + 2^m \right)}
$$

$$
\leq 2C_4(\omega)|u|^v \sqrt{\log \left(2 + j_3 \right)} \sum_{m=1}^{j_3} 2^{-mv} \sqrt{\log \left(2 + 2^m \right)}
$$

$$
\leq C_5(\omega)|u|^v \log^{\frac{1}{2}} \left(2 + \left| \log |u| \right| \right), \tag{6.84}
$$

where C_3, $C_4 := 2^{\overline{H}} C_3$, and

$$
C_5 := C_4 \left(2\sqrt{2} \sum_{m=1}^{+\infty} 2^{-m\underline{H}} \sqrt{\log \left(2 + 2^m \right)} \right)
$$

are three positive random variables of finite moment of any order not depending on (u, v, j_3). Next, putting together (6.82), the triangle inequality, (6.83) and (6.84), one obtains (6.17), in the case where $q = 0$.

Let us now show that (6.18) holds in the case where $q = 0$. One knows from Lemma 6.11 and from the proof of Proposition 5.17, that, for all $\omega \in \Omega_0^*$, the quantity $X^{\text{hf}}(u, v, \omega)$ can be expressed as the absolutely convergent series of real numbers

$$
X^{\text{hf}}(u, v, \omega) = \sum_{j=0}^{+\infty} 2^{-jv} \left(B_j(2^j u, v, \omega) - B_j(0, v, \omega) \right).
$$

Thus, using the triangle inequality, (6.28) and (5.63), one gets that

$$\left| X^{\mathrm{hf}}(u,v,\omega) \right|$$

$$\leq C_6(\omega) \sum_{j=0}^{+\infty} 2^{-j\underline{H}} \sqrt{\log\left(2 + j + 2^j |u|\right)}$$

$$\leq C_6(\omega) \sum_{j=0}^{+\infty} 2^{-j\underline{H}} \sqrt{\log\left(2 + j + 2^j\right) + \log\left(2 + |u|\right)}$$

$$\leq C_7(\omega) \sqrt{\log\left(2 + |u|\right)}, \qquad (6.85)$$

where C_6 and

$$C_7 := 2 C_6 \sum_{j=0}^{+\infty} 2^{-j\underline{H}} \sqrt{\log\left(2 + j + 2^j\right)}$$

are two positive random variables of finite moment of any order not depending on (u,v,j_3). It is clear that (6.85) implies that (6.18) holds in the case where $q = 0$.

Finally, (6.19) easily results from (5.16), (6.17) and (6.18). $\qquad \square$

Proof of Corollary 6.10. This corollary is a straightforward consequence of (6.1) and (6.19), where $q = 0$ and $\mathcal{H} = [\underline{S}, \overline{S}]$.

$\qquad \square$

6.3 Study of the optimality of some of the estimates

First one states the main results of the section and then one gives their proofs.

The following theorem can be viewed as a counterpart to Corollary 6.5 and more particularly to (6.11) in it.

Theorem 6.15. *Let $I \subset \mathbb{R}^N$ be a compact interval such that one has almost surely*

$$S(\cdot) \in \mathcal{C}^{\gamma_I}(I), \quad \text{for some } \gamma_I \in (\underline{S}(I), 1), \qquad (6.86)$$

where $\underline{S}(I)$ is as in (6.8); observe that (6.86) is a bit stronger condition than (6.12) since $\gamma_I > \underline{S}(I)$. One assumes that, almost surely, there exists a random point, denoted by $\underline{t}^0$, belonging to the topological interior $\overset{\circ}{I}$ of I, such that one has

$$\underline{S}(I) = S(\underline{t}^0). \qquad (6.87)$$

For each $(t^1, t^2) \in I^2$, *one sets*

$$\mathcal{T}_I(t^1, t^2) := \frac{\left| Z(t^1) - Z(t^2) \right|}{\left| t^1 - t^2 \right|^{\underline{S}^{(I)}} \left(1 + \left| \log |t^1 - t^2| \right| \right)^{\frac{1}{2}}}, \tag{6.88}$$

with the convention that $\mathcal{T}_I(t^1, t^2) := 0$ *when* $t^1 = t^2$. *Moreover, for all positive real number* ρ, *one denotes by* $\mathcal{M}_I(\rho)$ *the positive almost surely finite random variable defined as*

$$\mathcal{M}_I(\rho) := \sup \left\{ \mathcal{T}_I(t^1, t^2) \; : \; (t^1, t^2) \in I^2(\rho) \right\}, \tag{6.89}$$

where

$$I^2(\rho) := \left\{ (t^1, t^2) \in I^2 \; : \; |t^1 - t^2| \leq \rho \right\}. \tag{6.90}$$

Then, there exists a deterministic constant $0 < b < +\infty$, *not depending on* I, *such that one has almost surely*

$$\limsup_{\rho \to 0^+} \left\{ \sqrt{\log(\rho^{-1})} \, \mathcal{M}_I(\rho) \right\} \geq b > 0. \tag{6.91}$$

Theorem 6.15 can be improved in the case where the functional Hurst parameter $S(\cdot)$ is deterministic, that is in the case where Z is a Multifractional Brownian Field (MuBF). More precisely, the following result holds.

Theorem 6.16. *Under the same assumptions as in Theorem 6.15 and the additional assumption that the functional Hurst parameter* S *is deterministic, there exists a deterministic constant* $0 < b' < +\infty$, *such that one has almost surely*

$$\lim_{\rho \to 0^+} \mathcal{M}_I(\rho) \geq b' > 0. \tag{6.92}$$

Notice that the limit in (6.92) exists since $\mathcal{M}_I$ *is a nondecreasing random function in the variable* ρ.

The following theorem can be viewed as a counterpart to Corollary 6.8 as well as an extension of Theorem 3.15 to the setting of the Multifractional Field with Random Exponent Z.

Theorem 6.17. *Assume that, one has, almost surely for all point* $\tau \in \mathbb{R}^N$,

$$\lim_{t \to \tau, t \neq \tau} \frac{|S(t) - S(\tau)|}{|t - \tau|^{S(\tau)}} = 0, \tag{6.93}$$

where the equality holds on a event of probability 1 not depending on τ; *observe that this is a stronger condition than (6.14). Let* $\tau \in \mathbb{R}^N$ *be an arbitrary fixed point. For each* $t \in \mathbb{R}^N$, *one sets*

$$\breve{\mathcal{T}}_\tau(t) := \frac{|Z(t) - Z(\tau)|}{|t - \tau|^{S(\tau)}}, \tag{6.94}$$

with the convention that $\check{\mathcal{T}}_\tau(\tau) := 0$. Moreover, for all positive real number ρ, one denotes by $\check{\mathcal{M}}_\tau(\rho)$ the random variable with values in $\overline{\mathbb{R}}_+ = \mathbb{R}_+ \cup \{+\infty\}$ defined as

$$\check{\mathcal{M}}_\tau(\rho) := \sup\left\{\check{\mathcal{T}}_\tau(t) \,:\, t \in \mathcal{B}(\tau,\rho)\right\}, \quad (6.95)$$

where $\mathcal{B}(\tau,\rho)$ is the closed ball of $\mathbb{R}^N$ centered at τ and of radius ρ. Then, there exists a deterministic constant $0 < b'' < +\infty$, not depending on τ, such that one has, almost surely,

$$\lim_{\rho\to 0^+} \check{\mathcal{M}}_\tau(\rho) \geq b'' > 0. \quad (6.96)$$

It is worth noticing that (6.96) is valid on an event of probability 1 not depending on τ. Also notice that the limit $\lim_{\rho\to 0^+} \check{\mathcal{M}}_\tau(\rho)$, which may be frequently equal to $+\infty$, nevertheless, exists since $\check{\mathcal{M}}_\tau$ is a nondecreasing random function in the variable ρ.

Remark 6.18. Assume that, for every dyadic interval [1] I of $\mathbb{R}^N$, one has, almost surely, $S(\cdot) \in \mathcal{C}^{\gamma_I}(I)$, for some random exponent $\gamma_I \in (\overline{S}(I), 1)$ (see (6.8)). Then, the condition (6.93) is satisfied. Thus, combining Definitions 1.34, 1.39 and 1.45 with Remark 1.46, Corollary 6.5 and Theorem 6.17, one obtains, almost surely, that

$$\widetilde{\alpha}_Z(\tau) = \alpha_Z(\tau) = S(\tau), \quad \text{for all } \tau \in \mathbb{R}^N, \quad (6.97)$$

where $\widetilde{\alpha}_Z(\tau)$ and $\alpha_Z(\tau)$ respectively denote the local and pointwise Hölder exponents at τ of the MuFRE Z with a functional random Hurst parameter $S(\cdot)$. Observe that one also knows from Corollary 6.5 and Theorem 6.17 that the equality (6.97) holds on an event of probability 1 not depending on τ.

Remark 6.19. Assume that, for every dyadic interval I of $\mathbb{R}^N$, one has, almost surely, $S(\cdot) \in \mathcal{C}^{\gamma_I}(I)$, for some random exponent $\gamma_I \in (\overline{S}(I), 1)$. Also, assume that the Hurst functional parameter $S(\cdot)$ of the MuFRE Z is not deterministic. Then, it results from Remark 6.18 and from Proposition 1.67 that Z does not satisfy the zero-one law described in Definition 1.64. This in particular means that such a MuFRE Z is not a Gaussian field and even that it does not belong to any finite-order chaos associated with a Gaussian space.

---

[1] Here, by a dyadic interval of $\mathbb{R}^N$, we mean an interval of the form $\prod_{n=1}^{N}[\delta'_n, \delta''_n]$, where, for all $n \in \{1,\ldots,N\}$, δ'_n and δ''_n are two dyadic numbers of $\mathbb{R}$ (see Definition 1.66) such that $\delta'_n < \delta''_n$. Notice that the set of the dyadic intervals of $\mathbb{R}^N$ is countable.

Let us now turn to the proofs of the main results of the section. To this end, one needs several preliminary results.

One skips the proof of the following lemma since it is rather similar to those of Theorems 6.7 and 6.9.

Lemma 6.20. *For each $\omega \in \Omega_0^*$, for every compact interval $\mathcal{H} \in (0,1)$, and for all $(u,v) \in \mathbb{R}^N \times \mathcal{H}$, one has*

$$
\left| X(u,v,\omega) \right|
$$
$$
\leq \sum_{p=1}^{2^N-1} \sum_{(j,k)\in\mathbb{Z}\times\mathbb{Z}^N} 2^{-jv} \left| \varepsilon_{p,j,k}(\omega) \right| \left| \Psi_p(2^j u - k, v) - \Psi_p(-k,v) \right|
$$
$$
\leq C_{\mathcal{H}}(\omega)|u|^v \log^{\frac{1}{2}} \left(2 + \left| \log |u| \right| \right), \tag{6.98}
$$

where $C_{\mathcal{H}}$ is a positive and finite random variable only depending on $\mathcal{H}$.

Remark 6.21. For each compact interval $\mathcal{H} \subset (0,1)$, let $C_{\mathcal{H}}$ be as in (6.98). One denotes by C' the positive and finite random variable, defined, for every $\omega \in \Omega_0^*$, as

$$
C'(\omega) := C_{[\underline{S}(\omega), \overline{S}(\omega)]}(\omega).
$$

Then, combining (6.1) and (6.98), one gets, for all $(t,\omega) \in \mathbb{R}^N \times \Omega_0^*$, that

$$
\left| Z(t,\omega) \right| \leq C'(\omega)|t|^{S(t,\omega)} \log^{\frac{1}{2}} \left(2 + \left| \log |t| \right| \right). \tag{6.99}
$$

Lemma 6.22. *For each $\omega \in \Omega_0^*$, for every $v \in (0,1)$, and for all $(j,k) \in \mathbb{Z} \times \mathbb{Z}^N$, one has*

$$
2^{-jv} \varepsilon_{j,k}(\omega) = 2^{jN} \int_{\mathbb{R}^N} X\left(u,v,\omega\right) \Psi_1\left(2^j u - k, -v - N\right) du, \tag{6.100}
$$

where $\varepsilon_{j,k} := \varepsilon_{1,j,k}$ (see (5.9) and (5.10)).

Proof of Lemma 6.22. First notice that one knows from (6.98) and Proposition 5.11 that the pathwise Lebesgue integral in the right-hand side of (6.100) exists and is finite. For each fixed $J \in \mathbb{Z}_+$, let $X_J := \{X_J(u,v), (u,v) \in \mathbb{R}^N \times (0,1)\}$ be the centered Gaussian field which was defined in (5.53). Using arguments similar to those in the proofs of Theorem 5.15 and Propositions 5.16 and 5.17, it can be shown that, for all $(\omega, u, v) \in \Omega_0^* \times \mathbb{R}^N \times (0,1)$, one has

$$
\left| X(u,v,\omega) - X_J(u,v,\omega) \right| \left| \Psi_1\left(2^j u - k, -v - N\right) \right| \xrightarrow[J \to +\infty]{} 0.
$$

Moreover, one can easily derive from (5.10), (5.53), (5.48), (5.52), (6.98) and (5.20), that

$$\left|X(u,v,\omega) - X_J(u,v,\omega)\right|\left|\Psi_1\big(2^j u - k, -v - N\big)\right| \le F(u,v,j,\omega),$$

where the function $u \mapsto F(u,v,j,\omega)$ does not depend on J and belongs to the Lebesgue space $L^1(\mathbb{R}^N)$. Thus, the dominated convergence theorem entails that

$$\lim_{J\to+\infty} \int_{\mathbb{R}^N} \left|X(u,v,\omega) - X_J(u,v,\omega)\right|\left|\Psi_1\big(2^j u - k, -v - N\big)\right| du = 0$$

and consequently that

$$\lim_{J\to+\infty} \int_{\mathbb{R}^N} X_J(u,v,\omega)\Psi_1\big(2^j u - k, -v - N\big) du \qquad (6.101)$$

$$= \int_{\mathbb{R}^N} X(u,v,\omega)\Psi_1\big(2^j u - k, -v - N\big) du\,.$$

Finally, using (6.101), (5.53), (5.48), (5.52), part (i) of Proposition 5.13, and Remark 5.12 with $(\lambda,\gamma,m,p) = (0,0,0,1)$, one obtains (6.100). □

The following lemma easily results from Lemma 6.20, Proposition 5.11 and classical calculations.

Lemma 6.23. *For each fixed $\omega \in \Omega_0^*$, arbitrarily large positive integer L, and compact intervals $I' \subset \mathbb{R}^N$ and $\mathcal{H} \subset (0,1)$, one has*

$$\sup\left\{\theta^L \int_{\{|y|\ge\theta\}} \left|X\big(\delta + \theta^{-2}\,y, v, \omega\big)\right|\left|\Psi_1\big(y, -v - N\big)\right| dy \right.$$

$$\left. : (\theta,\delta,v) \in [1,+\infty) \times I' \times \mathcal{H}\right\} < +\infty. \qquad (6.102)$$

Lemma 6.24. *For each $\omega \in \Omega_0^*$ and for all $(j,k) \in \mathbb{N} \times \mathbb{Z}^N$, one sets*

$$\mathcal{Z}_{j,k}(\omega) := 2^{jN} \int_{\mathbb{R}^N} Z(u,\omega)\Psi_1\big(2^j u - k, -S(2^{-j}k,\omega) - N\big) du. \qquad (6.103)$$

Let $I \subset \mathbb{R}^N$ be a compact interval for which the condition (6.86) holds, and let I' be an arbitrary compact interval of $\mathbb{R}^N$ which is included in $\overset{\circ}{I}$ the topological interior of I. Then, for almost every $\omega \in \Omega_0^*$, one has

$$\lim_{j\to+\infty}\left\{2^{j\underline{S}(I,\omega)}\sup\left\{\left|\mathcal{Z}_{j,k}(\omega) - 2^{-jS(2^{-j}k,\omega)}\varepsilon_{j,k}(\omega)\right| : k \in \mathcal{K}_j(I')\right\}\right\} = 0,$$

$$(6.104)$$

where, for each fixed $j \in \mathbb{N}$,

$$\mathcal{K}_j(I') := \big\{k \in \mathbb{Z}^N : 2^{-j}k \in I'\big\}. \qquad (6.105)$$

Proof of Lemma 6.24. First, observe that Proposition 5.11 and (6.99) imply that the pathwise Lebesgue integral in (6.103) exists and is finite. Also, observe that the fact that $I' \subset \overset{\circ}{I}$ entails that there exists a constant $a > 0$ such that, for all $x \in I'$ and all $s \in \mathbb{R}^N$ satisfying $|s| \leq a$, one has $x + s \in I$.

Next, one assumes that the positive integer j is arbitrary and large enough so that $2^{-j/2} \leq a$ and $\mathcal{K}_j(I')$ is a nonempty set. Let $k \in \mathcal{K}_j(I')$ be arbitrary. Using (6.100) with $v = S(2^{-j}k, \omega)$, (6.103), (6.1), the change of variable $y = 2^j u - k$, (6.105), Lemma 6.13, (6.86), Lemma 6.23, and Proposition 5.11, one obtains that

$$\left| \mathcal{Z}_{j,k}(\omega) - 2^{-jS(2^{-j}k,\omega)} \varepsilon_{j,k}(\omega) \right|$$

$$\leq \int_{\mathbb{R}^N} \left| X\big(2^{-j}k + 2^{-j}y, S(2^{-j}k + 2^{-j}y, \omega), \omega\big) \right.$$

$$\left. - X\big(2^{-j}k + 2^{-j}y, S(2^{-j}k, \omega), \omega\big) \right| \left| \Psi_1\big(y, -S(2^{-j}k) - N\big) \right| dy$$

$$\leq \int_{\{|y| \leq 2^{j/2}\}} \left| X\big(2^{-j}k + 2^{-j}y, S(2^{-j}k + 2^{-j}y, \omega), \omega\big) \right.$$

$$\left. - X\big(2^{-j}k + 2^{-j}y, S(2^{-j}k, \omega), \omega\big) \right| \left| \Psi_1\big(y, -S(2^{-j}k) - N\big) \right| dy$$

$$+ \int_{\{|y| \geq 2^{j/2}\}} \left| X\big(2^{-j}k + 2^{-j}y, S(2^{-j}k + 2^{-j}y, \omega), \omega\big) \right.$$

$$\left. - X\big(2^{-j}k + 2^{-j}y, S(2^{-j}k, \omega), \omega\big) \right| \left| \Psi_1\big(y, -S(2^{-j}k) - N\big) \right| dy$$

$$\leq C_1(\omega)\, 2^{-j\gamma_I(\omega)} \int_{\{|y| \leq 2^{j/2}\}} |y|^{\gamma_I(\omega)} \sup_{v \in \mathcal{H}(\omega)} \left| \Psi_1(y, -v - N) \right| dy$$

$$+ 2 \sup_{(\delta,v) \in I' \times \mathcal{H}(\omega)} \int_{\{|y| \geq 2^{j/2}\}} \left| X\big(\delta + 2^{-j}y, v, \omega\big) \right| \left| \Psi_1\big(y, -v - N\big) \right| dy$$

$$\leq C_2(\omega)\, 2^{-j\gamma_I(\omega)} + C_3(\omega)\, 2^{-jL/2}, \tag{6.106}$$

where $\mathcal{H}(\omega) := \big[\underline{S}(\omega), \overline{S}(\omega)\big]$, L is a fixed arbitrarily large integer, and $C_1(\omega)$,

$$C_2(\omega) := C_1(\omega) \int_{\mathbb{R}^N} |y|^{\gamma_I(\omega)} \sup_{v \in \mathcal{H}(\omega)} \left| \Psi_1(y, -v - N) \right| dy,$$

and $C_3(\omega)$ are three positive and finite constants not depending on (j, k). Finally, it is clear that (6.106) and the inequality $\gamma_I(\omega) > \underline{S}(I, \omega)$ imply that (6.104) holds. $\qquad\square$

Lemma 6.25. *One assumes that $I \subset \mathbb{R}^N$ is a compact interval for which the condition (6.86) holds, and that $\underline{S}(I)$ is as in (6.8). Let I' be an ar-*

bitrary compact interval of $\mathbb{R}^N$ which is included in $\mathring{I}$, and let L be an arbitrarily large fixed positive integer. Then, there exists $j_0 \in \mathbb{N}$, such that, for all integer $j \geq j_0$ and for almost all $\omega \in \Omega_0^*$, one has

$$\sup\left\{\left|\mathcal{Z}_{j,k}(\omega)\right| : k \in \mathcal{K}_j(I')\right\} \leq c_1 \mathcal{M}_I(2^{-j/2}, \omega)\, 2^{-j\underline{S}(I,\omega)}\,\sqrt{j} + C_2(\omega)2^{-jL}\,,$$
(6.107)

where c_1 is a finite deterministic constant not depending on (I,j,ω), and $C_2(\omega)$ is a finite random constant not depending on j. Notice that the finite set $\mathcal{K}_j(I')$ was introduced in (6.105), and that the positive almost surely finite random variable $\mathcal{M}_I(2^{-j/2})$ is defined through (6.89) with $\rho = 2^{-j/2}$.

Proof of Lemma 6.25. First, one mentions that the positive integer j_0 is chosen so that $\mathcal{K}_{j_0}(I')$ is a nonempty set and, for all $(x,s) \in I' \times \mathbb{R}^N$ satisfying $|s| \leq 2^{-j_0/2}$, one has $x + s \in I$.

Next, one assumes that $(j,k) \in \mathbb{N} \times \mathbb{Z}^N$ is arbitrary and such that $j \geq j_0$ and $k \in \mathcal{K}_j(I')$. Using (6.103), Remark 5.12 with $(\lambda,\gamma,m,p,v) = \left(0,0,0,1,-S(2^{-j}k,\omega)-N\right)$, the change of variable $y = 2^j u - k$, (6.89), (6.88), the triangle inequality, (6.1), Lemma 6.23 with $2L$ in place of L, the fact that

$$\sup_{\delta \in I'}\left\{|Z(\delta,\omega)|\right\} < +\infty\,, \quad \text{for almost all } \omega \in \Omega_0^*,$$
(6.108)

and Proposition 5.11, one obtains that

$$\left|\mathcal{Z}_{j,k}(\omega)\right|$$
$$= \left|\int_{\mathbb{R}^N}\left(Z\left(2^{-j}k + 2^{-j}y,\omega\right) - Z\left(2^{-j}k,\omega\right)\right)\Psi_1\left(y,-S(2^{-j}k,\omega)-N\right)du\right|$$
$$\leq \mathcal{M}_I(2^{-j/2},\omega)\int_{\{|y|\leq 2^{j/2}\}}\left|2^{-j}y\right|^{\underline{S}(I,\omega)}\left(1+\left|\log|2^{-j}y|\right|\right)^{\frac{1}{2}}$$
$$\times \sup_{v'\in[N,N+1]}\left|\Psi_1(y,-v')\right|dy$$
$$+ \sup_{(\delta,v)\in I'\times\mathcal{H}(\omega)}\left\{\int_{\{|y|\geq 2^{j/2}\}}\left|X\left(\delta + 2^{-j}y,v,\omega\right)\right|\left|\Psi_1\left(y,-v-N\right)\right|dy\right\}$$
$$+ \sup_{\delta\in I'}\left\{|Z(\delta,\omega)|\right\}\int_{\{|y|\geq 2^{j/2}\}}\sup_{v'\in[N,N+1]}\left|\Psi_1(y,-v')\right|dy$$
$$\leq c_1\mathcal{M}_I(2^{-j/2},\omega)\,2^{-j\underline{S}(I,\omega)}\,\sqrt{j} + C_2(\omega)2^{-jL}\,,$$

where $\mathcal{H}(\omega) := \left[\underline{S}(\omega),\overline{S}(\omega)\right]$,

$$c_1 := \int_{\mathbb{R}^N}\left(1+|y|\right)\left(2+\left|\log|y|\right|\right)^{\frac{1}{2}}\sup_{v'\in[N,N+1]}\left|\Psi_1(y,-v')\right|dy < +\infty\,,$$

and $C_2(\omega)$ is a finite random constant not depending on j. Notice that (6.108) holds since $t \mapsto Z(t, \omega)$ is a continuous function on $\mathbb{R}^N$ and I' a compact interval of $\mathbb{R}^N$. □

We are now in position to prove Theorems 6.15 and 6.16.

Proof of Theorem 6.15. One denotes by I' a random compact interval of $\mathbb{R}^N$, such that one has, almost surely,

$$\underline{t}^0 \in \overset{\circ}{I'} \subset I' \subset \overset{\circ}{I}, \tag{6.109}$$

where the random point $\underline{t}^0$ is as in (6.87). Recall that, for each $j \in \mathbb{N}$, the finite set $\mathcal{K}_j(I')$ has been defined in (6.105). Using the triangle inequality, one gets that

$$\sup_{k \in \mathcal{K}_j(I')} \left| 2^{-jS(2^{-j}k)} \varepsilon_{j,k} \right| \le \sup_{k \in \mathcal{K}_j(I')} \left| \mathcal{Z}_{j,k} \right| + \sup_{k \in \mathcal{K}_j(I')} \left| \mathcal{Z}_{j,k} - 2^{-jS(2^{-j}k)} \varepsilon_{j,k} \right|.$$

Thus, it follows from Lemma 6.25 that there exists a random positive integer j_0, such that, one has almost surely, for all integer $j \ge j_0$,

$$\sup_{k \in \mathcal{K}_j(I')} \left| 2^{-jS(2^{-j}k)} \varepsilon_{j,k} \right| \tag{6.110}$$

$$\le c_1 \mathcal{M}_I(2^{-j/2}) \, 2^{-j\underline{S}(I)} \sqrt{j} + C_2 2^{-jL} + \sup_{k \in \mathcal{K}_j(I')} \left| \mathcal{Z}_{j,k} - 2^{-jS(2^{-j}k)} \varepsilon_{j,k} \right|,$$

where L is an arbitarily large fixed integer, c_1 is a deterministic strictly positive finite constant not depending on I and C_2 is a random strictly positive finite constant. Notice that none of these two constants depends on j. Next letting c_3 be the deterministic strictly positive finite constant defined as $c_3 := 1/c_1$, one can derive from (6.110), (6.87) and Lemma 6.24, that, one has, almost surely,

$$c_3 \limsup_{j \to +\infty} \left\{ \sup_{k \in \mathcal{K}_j(I')} 2^{-j(S(2^{-j}k) - S(\underline{t}^0))} |\varepsilon_{j,k}| \right\} \le \limsup_{j \to +\infty} \left\{ \sqrt{j} \, \mathcal{M}_I(2^{-j/2}) \right\}. \tag{6.111}$$

Next, one sets $[2^j \underline{t}^0] := \left([2^j \underline{t}_1^0], \dots, [2^j \underline{t}_N^0] \right)$, where $\underline{t}_1^0, \dots, \underline{t}_N^0$ are the coordinates of $\underline{t}^0$, and $[\cdot]$ denotes the integer part function. Therefore, one has, for all $j \in \mathbb{N}$, that

$$\left| 2^{-j} [2^j \underline{t}^0] - \underline{t}^0 \right| < \sqrt{N} \, 2^{-j}. \tag{6.112}$$

It follows from (6.112) and (6.109) that $2^{-j}[2^j \underline{t}^0] \in \overset{\circ}{I'}$, for all j large enough. Thus, it results from (6.105) that

$$[2^j \underline{t}^0] \in \mathcal{K}_j(I'). \tag{6.113}$$

Moreover, one can derive from (6.86) and (6.112) that, one has, almost surely,

$$\lim_{j \to +\infty} 2^{-j(S(2^{-j}[2^j \underline{t}^0]) - S(\underline{t}^0))} = 1. \tag{6.114}$$

Finally, putting together (6.111), (6.113) and (6.114), one gets, almost surely, that

$$c_3 \limsup_{j \to +\infty} \left\{ |\varepsilon_{j,[2^j \underline{t}^0]}| \right\} \leq \limsup_{j \to +\infty} \left\{ \sqrt{j}\, \mathcal{M}_I(2^{-j/2}) \right\}.$$

Therefore, Lemma A.27 implies that (6.91) holds. □

Proof of Theorem 6.16. Using the same arguments as in the proof of Theorem 6.15, it can be shown that there exists a finite deterministic constant [2] $c_1 > 0$, such that one has, almost surely,

$$c_1 \limsup_{j \to +\infty} \left\{ j^{-\frac{1}{2}} \sup_{k \in \mathcal{K}_j(I')} 2^{-j(S(2^{-j}k) - S(\underline{t}^0))} |\varepsilon_{j,k}| \right\} \leq \lim_{j \to +\infty} \mathcal{M}_I(2^{-j/2}). \tag{6.115}$$

Yet, there is a major difference with respect to the proof of Theorem 6.15: the point $\underline{t}^0$, the compact interval I' containing it, and the exponent γ_I in (6.86) are now assumed to be deterministic. Therefore, the set

$$\mathcal{K}_j(\underline{t}^0) := \left\{ k \in \mathcal{K}_j(I') : |2^{-j}k - \underline{t}^0| \leq j^{-1/\gamma_I} \right\} \tag{6.116}$$

is deterministic as well. Observe that, for all j large enough, this is a nonempty finite set, and that its cardinality satisfies

$$c_2\, j^{-N/\gamma_I}\, 2^{jN} \leq \mathrm{card}\big(\mathcal{K}_j(\underline{t}^0)\big) \leq c_3\, j^{-N/\gamma_I}\, 2^{jN}. \tag{6.117}$$

where $0 < c_2 \leq c_3 < +\infty$ are two deterministic constants not depending on j. Also, observe that one can easily derive from (6.116) and (6.86) that

$$c_4 := \inf_{j \in \mathbb{N}} \left\{ \inf_{k \in \mathcal{K}_j(\underline{t}^0)} \left\{ 2^{-j(S(2^{-j}k) - S(\underline{t}^0))} \right\} \right\} > 0, \tag{6.118}$$

with the usual convention that an infimum over an empty set is equal to $+\infty$. Next, using (6.115), the inclusion $\mathcal{K}_j(\underline{t}^0) \subset \mathcal{K}_j(I')$ and (6.118), one obtains that

$$c_1 c_4 \limsup_{j \to +\infty} \left\{ j^{-\frac{1}{2}} \sup_{k \in \mathcal{K}_j(\underline{t}^0)} |\varepsilon_{j,k}| \right\} \leq \lim_{j \to +\infty} \mathcal{M}_I(2^{-j/2}).$$

Thus, in order to complete our proof, it remains to us to show that there exists a deterministic constant $c_5 > 0$, such that one has

$$\limsup_{j \to +\infty} \left\{ j^{-\frac{1}{2}} \sup_{k \in \mathcal{K}_j(\underline{t}^0)} |\varepsilon_{j,k}| \right\} \geq c_5, \quad \text{almost surely.}$$

---

[2]Notice that this constant c_1 is the constant c_3 in (6.111).

To this end, in view of the Borel-Cantelli Lemma, it is enough to show that, for some finite deterministic constant $c_5 \geq 1$, one has

$$\sum_{j=j_0}^{+\infty} \mathbb{P}\Big(\sup_{k \in \mathcal{K}_j(\underline{t}^0)} |\varepsilon_{j,k}| < c_5\, j^{\frac{1}{2}} \Big) < +\infty. \tag{6.119}$$

where the positive deterministic integer j_0 is chosen so that (6.117) holds, for every $j \geq j_0$. Using the independence property of the $\mathcal{N}(0,1)$ Gaussian random variables $\varepsilon_{j,k}$, $k \in \mathcal{K}_j(\underline{t}^0)$, as well as the first inequalities in (6.117) and in (A.14), one obtains that, for all j large enough,

$$\mathbb{P}\Big(\sup_{k \in \mathcal{K}_j(\underline{t}^0)} |\varepsilon_{j,k}| < c_5\, j^{\frac{1}{2}} \Big)$$

$$= \prod_{k \in \mathcal{K}_j(\underline{t}^0)} \Big(1 - \mathbb{P}\Big(|\varepsilon_{j,k}| \geq c_5\, j^{\frac{1}{2}} \Big) \Big)$$

$$\leq \left(1 - c_6 \frac{\exp\big(-2^{-1} c_5^2\, j \big)}{c_5\, j^{\frac{1}{2}}} \right)^{c_2\, j^{-N/\gamma_I}\, 2^{jN}}$$

$$= \exp\left(c_2\, j^{-N/\gamma_I}\, 2^{jN} \log\left(1 - c_6 \frac{\exp\big(-2^{-1} c_5^2\, j \big)}{c_5\, j^{\frac{1}{2}}} \right) \right), \tag{6.120}$$

where $c_6 \in (0,1)$ denotes the constant c' in (A.14). Moreover, one has that

$$j^{-N/\gamma_I}\, 2^{jN} \log\left(1 - c_6 \frac{\exp\big(-2^{-1} c_5^2\, j \big)}{c_5\, j^{\frac{1}{2}}} \right) \tag{6.121}$$

$$\sim -c_6 c_5^{-1}\, j^{-N/\gamma_I - 1/2} \exp\Big(j\big(N \log(2) - 2^{-1} c_5^2 \big) \Big), \quad \text{when } j \to +\infty.$$

Now, let us assume that c_5 is any arbitrary fixed constant belonging to the interval $\big[1, \sqrt{\log(4)} \big)$. Therefore, one has that

$$N \log(2) - 2^{-1} c_5^2 > 0.$$

Thus, (6.120) and (6.121) entail that (6.119) holds. $\qquad\square$

Lemma 6.26. *One assumes that $\tau := (\tau_1, \ldots, \tau_N) \in \mathbb{R}^N$ is an arbitrary fixed point. For every positive integer j, one sets*

$$[2^j \tau] := \big([2^j \tau_1], \ldots, [2^j \tau_N] \big), \tag{6.122}$$

where $[\cdot]$ denotes the integer part function. Let L be an arbitrarily large fixed positive integer. Then, for all positive integer j and $\omega \in \Omega_0^*$, one has

$$\big| \mathcal{Z}_{j,[2^j \tau]}(\omega) \big| \leq c_1 \breve{\mathcal{M}}_\tau\big(c_0\, 2^{-j/2}, \omega \big)\, 2^{-jS(\tau,\omega)} + C_2(\omega) 2^{-jL}, \tag{6.123}$$

where $c_0 := N^{1/2} + 1$, c_1 is a deterministic finite constant not depending on (τ, j, ω), and $C_2(\omega)$ is a deterministic finite random constant not depending on j. Notice that the positive random variable $\breve{\mathcal{M}}_\tau(c_0\, 2^{-j/2})$ is defined through (6.95) with $\rho = c_0\, 2^{-j/2}$.

Proof of Lemma 6.26. First, observe that, in view of (6.122), for all positive integer j, one has

$$\left| 2^{-j}[2^j \tau] - \tau \right| < N^{1/2} \, 2^{-j}. \tag{6.124}$$

Next, using (6.103), Remark 5.12 with

$$(\lambda, \gamma, m, p, v) = \left(0, 0, 0, 1, -S(2^{-j}[2^j \tau], \omega) - N\right),$$

the change of variable $y = 2^j u - [2^j \tau]$, (6.124), (6.95), (6.94), (6.1), the triangle inequality, Lemma 6.23 with $2L$ in place of L, and Proposition 5.11, one obtains that

$$\left| \mathcal{Z}_{j,[2^j \tau]}(\omega) \right|$$

$$= \left| \int_{\mathbb{R}^N} \left(Z\left(2^{-j}[2^j \tau] + 2^{-j} y, \omega\right) - Z\left(\tau, \omega\right)\right) \Psi_1\left(y, -S(2^{-j}[2^j \tau], \omega) - N\right) du \right|$$

$$\leq \breve{\mathcal{M}}_\tau(c_0 \, 2^{-j/2}, \omega) \int_{\{|y| \leq 2^{j/2}\}} \left| \left(2^{-j}[2^j \tau] - \tau\right) + 2^{-j} y \right|^{S(\tau, \omega)}$$

$$\times \sup_{v' \in [N, N+1]} \left| \Psi_1(y, -v') \right| dy$$

$$+ \sup_{(\delta, v) \in I' \times \mathcal{H}(\omega)} \left\{ \int_{\{|y| \geq 2^{j/2}\}} \left| X\left(\delta + 2^{-j} y, v, \omega\right) \right| \left| \Psi_1\left(y, -v - N\right) \right| dy \right\}$$

$$+ |Z(\tau, \omega)| \int_{\{|y| \geq 2^{j/2}\}} \sup_{v' \in [N, N+1]} \left| \Psi_1(y, -v') \right| dy$$

$$\leq c_1 \breve{\mathcal{M}}_\tau(c_0 \, 2^{-j/2}, \omega) \, 2^{-jS(\tau, \omega)} + C_2(\omega) 2^{-jL},$$

where $I' := \prod_{n=1}^N [\tau_n - 1, \tau_n]$, $\mathcal{H}(\omega) := \left[\underline{S}(\omega), \overline{S}(\omega)\right]$, and

$$c_1 := \int_{\mathbb{R}^N} \left(N^{1/2} + |y|\right) \sup_{v' \in [N, N+1]} \left| \Psi_1(y, -v') \right| dy < +\infty,$$

and $C_2(\omega)$ is a finite constant not depending on j. □

Lemma 6.27. *Assume that the condition (6.93) is satisfied. Then, one has, almost surely for all $\tau \in \mathbb{R}^N$,*

$$\lim_{j \to +\infty} \left\{ 2^{jS(\tau)} \left| \mathcal{Z}_{j,[2^j \tau]} - 2^{-jS(2^{-j}[2^j \tau])} \varepsilon_{j,[2^j \tau]} \right| \right\} = 0, \tag{6.125}$$

where the equality holds on an event of probability 1 not depending on τ.

Proof of Lemma 6.27. For each nonnegative real number ρ, one sets

$$\breve{\mathcal{M}}_\tau^S(\rho) := \left\{ \frac{|S(t) - S(\tau)|}{|t - \tau|^{S(\tau)}} \; : \; t \in \mathcal{B}(\tau, \rho) \right\}, \tag{6.126}$$

with the convention that $0/0 = 0$. Observe that the condition (6.93) implies that, one has, on an event of probability 1 not depending on τ,

$$\lim_{\rho \to 0^+} \breve{\mathcal{M}}_\tau^S(\rho) = 0. \tag{6.127}$$

One sets $c_0 := N^{1/2} + 1$. Using (6.100) with $v = S(2^{-j}[2^j\tau], \omega)$, (6.103), (6.1), the change of variable $y = 2^j u - [2^j\tau]$, Lemma 6.13, the inequality

$$\left| S(2^{-j}[2^j\tau] + 2^{-j}y) - S(2^{-j}[2^j\tau]) \right|$$

$$\leq \left| S(2^{-j}[2^j\tau] + 2^{-j}y) - S(\tau) \right| + \left| S(2^{-j}[2^j\tau]) - S(\tau) \right|,$$

(6.124), (6.126), the triangle inequality, Lemma 6.23, and Proposition 5.11, one obtains, on an event of probability 1 not depending on τ, that

$$\left| \mathcal{Z}_{j,[2^j\tau]} - 2^{-jS(2^{-j}[2^j\tau])} \varepsilon_{j,[2^j\tau]} \right|$$

$$\leq \int_{\mathbb{R}^N} \left| X\big(2^{-j}[2^j\tau] + 2^{-j}y, S(2^{-j}[2^j\tau] + 2^{-j}y)\big) \right.$$

$$\left. - X\big(2^{-j}[2^j\tau] + 2^{-j}y, S(2^{-j}[2^j\tau])\big) \right| \left| \Psi_1\big(y, -S(2^{-j}[2^j\tau]) - N\big) \right| dy$$

$$\leq \int_{\{|y| \leq 2^{j/2}\}} \left| X\big(2^{-j}[2^j\tau] + 2^{-j}y, S(2^{-j}[2^j\tau] + 2^{-j}y)\big) \right.$$

$$\left. - X\big(2^{-j}[2^j\tau] + 2^{-j}y, S(2^{-j}[2^j\tau])\big) \right| \left| \Psi_1\big(y, -S(2^{-j}[2^j\tau]) - N\big) \right| dy$$

$$+ \int_{\{|y| \geq 2^{j/2}\}} \left| X\big(2^{-j}[2^j\tau] + 2^{-j}y, S(2^{-j}[2^j\tau] + 2^{-j}y)\big) \right.$$

$$\left. - X\big(2^{-j}[2^j\tau] + 2^{-j}y, S(2^{-j}[2^j\tau])\big) \right| \left| \Psi_1\big(y, -S(2^{-j}[2^j\tau]) - N\big) \right| dy$$

$$\leq C_1 \int_{\{|y| \leq 2^{j/2}\}} \left| S(2^{-j}[2^j\tau] + 2^{-j}y) - S(2^{-j}[2^j\tau]) \right| \sup_{v \in \mathcal{H}} \left| \Psi_1(y, -v - N) \right| dy$$

$$+ 2 \sup_{(\delta, v) \in I' \times \mathcal{H}} \left\{ \int_{\{|y| \geq 2^{j/2}\}} \left| X\big(\delta + 2^{-j}y, v, \omega\big) \right| \left| \Psi_1\big(y, -v - N\big) \right| dy \right\}$$

$$\leq C_2 \, \breve{\mathcal{M}}_\tau^S(c_0 2^{-j/2}) \, 2^{-jS(\tau)} \int_{\{|y| \leq 2^{j/2}\}} \left(2N^{S(\tau)/2} + |y|^{S(\tau)} \right)$$

$$\times \sup_{v \in \mathcal{H}} \left| \Psi_1(y, -v - N) \right| dy + C_3 \, 2^{-jL/2}$$

$$\leq C_4 \, \breve{\mathcal{M}}_\tau^S(c_0 2^{-j/2}) \, 2^{-jS(\tau)} + C_3 \, 2^{-jL/2}, \tag{6.128}$$

where $I' := \prod_{n=1}^{N}[\tau_n - 1, \tau_n]$, $\mathcal{H} := [\underline{S}, \overline{S}]$, L is a fixed arbitrarily large integer, and C_1, C_2, C_3 and

$$C_4 := C_2 \int_{\mathbb{R}^N} \left(2N^{S(\tau)/2} + |y|^{S(\tau)} \right) \sup_{v \in \mathcal{H}} \left| \Psi_1(y, -v - N) \right| dy$$

are four positive and finite random variables not depending j. Finally, it is clear that (6.128) and (6.127) imply that (6.125) holds. $\square$

We are now in position to proof Theorem 6.17.

Proof of Theorem 6.17. It follows from the triangle inequality and Lemmas 6.26 and 6.27 that, one has, almost surely for all $\tau \in \mathbb{R}^N$,

$$\limsup_{j \to +\infty} \left\{ 2^{j(S(\tau) - S(2^{-j}[2^j \tau]))} \left| \varepsilon_{j,[2^j \tau]} \right| \right\} \le c_1 \lim_{j \to +\infty} \check{\mathcal{M}}_\tau(c_0 \, 2^{-j/2}), \quad (6.129)$$

where $0 < c_0 < +\infty$ and $0 < c_1 < +\infty$ are the same deterministic constants as in Lemma 6.26. Recall that these two constants do not depend on τ. On another hand, (6.93) and (6.124) imply, almost surely for all $\tau \in \mathbb{R}^N$, that

$$\lim_{j \to +\infty} 2^{j(S(\tau) - S(2^{-j}[2^j \tau]))} = 1. \quad (6.130)$$

Therefore, (6.129) entails that

$$\limsup_{j \to +\infty} \left| \varepsilon_{j,[2^j \tau]} \right| \le c_1 \lim_{j \to +\infty} \check{\mathcal{M}}_\tau(c_0 \, 2^{-j/2}). \quad (6.131)$$

Observe that (6.129), (6.130) and (6.131) hold on an event of probabilty 1 not depending on τ. Finally, using Lemma A.27 and (6.131), one obtains (6.96). $\square$

Appendix A

Appendix

A.1 Continuity of Gaussian fields

Definition A.1 (continuity in quadratic mean). Let $\{Y(t), t \in \mathbb{R}^N\}$ be a real-valued Gaussian field. It is said to be *continuous in quadratic mean at* $\tau \in \mathbb{R}^N$ if

$$\lim_{t \to \tau} \mathbb{E}\big(|Y(t) - Y(\tau)|^2\big) = 0.$$

Moreover, when the latter property holds at any arbitrary $\tau \in \mathbb{R}^N$, then $\{Y(t), t \in \mathbb{R}^N\}$ is said to be *continuous in quadratic mean*.

Definition A.2 (path continuity). One says that a real-valued Gaussian field $\{Y(t), t \in \mathbb{R}^N\}$ is *almost surely with continuous paths* if with probability 1 a path of $\{Y(t), t \in \mathbb{R}^N\}$ is a continuous function on $\mathbb{R}^N$.

Let us now draw some connections between these two notions of continuity.

Proposition A.3. *When a real-valued centered Gaussian field* $\{Y(t), t \in \mathbb{R}^N\}$ *is almost surely with continuous paths, then it is also continuous in quadratic mean.*

Proposition A.3 easily results from Part (ii) of the following lemma.

Lemma A.4.

(i) *Let* $\{G_n, n \in \mathbb{N}\}$ *be a sequence of real-valued centered Gaussian random variables which converges in distribution to a random variable* G*; then* G *is necessarily a real-valued centered Gaussian random variable with a variance* $\mathbb{E}(G^2)$ *given by*

$$\mathbb{E}(G^2) = \lim_{n \to +\infty} \mathbb{E}(G_n^2).$$

(ii) *Moreover, when* $\{G_n, n \in \mathbb{N}\}$ *is a Gaussian process* [1] *and the convergence to* G *holds almost surely; then this convergence also holds in quadratic mean, that is*

$$\lim_{n \to +\infty} \mathbb{E}(|G - G_n|^2) = 0.$$

Proof of Lemma A.4. Let us first show that part (i) holds. We denote by χ_{G_n} and χ_G the characteristic functions of G_n and G, where $n \in \mathbb{N}$ is arbitrary. In view of the hypotheses, we know that, for all $\xi \in \mathbb{R}$,

$$\chi_{G_n}(\xi) = \exp\left(- 2^{-1} \mathbb{E}(G_n^2)\, \xi^2 \right) \qquad (\text{A}.1)$$

and

$$\chi_G(\xi) = \lim_{n \to +\infty} \chi_{G_n}(\xi). \qquad (\text{A}.2)$$

The equalities (A.1) and (A.2) imply that, when n goes to $+\infty$, the sequence of nonnegative real-numbers $\left(\mathbb{E}(G_n^2)\right)_{n \in \mathbb{N}}$ has a nonnegative limit, denoted by σ^2, which is given by

$$\sigma^2 = -2 \lim_{n \to +\infty} \log\left(\chi_{G_n}(1)\right). \qquad (\text{A}.3)$$

Notice that σ^2 is necessarily finite. Indeed, assuming that $\sigma^2 = +\infty$, then it would result from (A.1) and (A.2) that $\chi_G(0) = 1$ and $\chi_G(\xi) = 0$ for any $\xi \in \mathbb{R} \setminus \{0\}$; which cannot happen since χ_G (and any other characteristic function of a real-valued random variable) is a continuous function on $\mathbb{R}$. Next, putting together (A.1), (A.2) and (A.3), it follows that, for all $\xi \in \mathbb{R}$, one has

$$\chi_G(\xi) = \exp\left(- 2^{-1} \sigma^2\, \xi^2 \right);$$

which means that G is a real-valued centered Gaussian random variable with variance $\sigma^2 := \lim_{n \to +\infty} \mathbb{E}(G_n^2)$.

Let us now prove that part (ii) holds. The fact that $\{G_n, n \in \mathbb{N}\}$ is a real-valued centered Gaussian process implies that, for every $(p, q) \in \mathbb{N}^2$, $G_p - G_q$ is a real-valued centered Gaussian random variable. On the other hand, we know from the hypotheses that that, for all fixed $q \in \mathbb{N}$, the sequence of random variables $(G_p - G_q)_{p \in \mathbb{N}}$ converges almost surely (and as a consequence in distribution) to $G - G_q$, when p goes to $+\infty$. Thus, part (i) entails that, for any $q \in \mathbb{N}$, $G - G_q$ is a centered Gaussian random variable. Using again the hypotheses, it follows that the sequence $(G - G_q)_{q \in \mathbb{N}}$ converges almost surely (and as a consequence in distribution) to 0. Therefore, one necessarily has that

$$\lim_{q \to +\infty} \mathbb{E}|G - G_q|^2 = 0.$$

$\square$

---

[1] That is any linear combination of G_n's is a Gaussian random variable.

Let us mention that the sole continuity in quadratic mean for a real-valued centered Gaussien field $\{Y(t), t \in \mathbb{R}^N\}$ does not necessarily imply the existence of a modification for $\{Y(t), t \in \mathbb{R}^N\}$ with almost surely continuous paths. Yet, as shown in the following theorem, this becomes true under *a Hölder continuity in quadradic mean assumption* on $\{Y(t), t \in \mathbb{R}^N\}$.

Theorem A.5 (Hölder continuity for Gaussian fields). *Let* $\{Y(t), t \in I\}$ *be a real-valued centered Gaussian field, where I denotes a compact interval* ^{2} *of* $\mathbb{R}^N$. *Assume that there are two constants $c > 0$ and $\nu \in (0,1]$ such that, for all $(t^1, t^2) \in I^2$, the inequality*

$$\mathbb{E}\big(|Y(t^1) - Y(t^2)|^2\big) \leq c|t^1 - t^2|^{2\nu}, \tag{A.4}$$

holds. Then there exists a continuous modification of $\{Y(t), t \in I\}$ whose paths almost surely belong to the global Hölder space $\mathcal{C}^\beta(I)$ (see Definition 1.33), for any $\beta \in [0, \nu)$.

Theorem A.5 can be easily obtained by using the "equivalence of Gaussian moments (see Lemma A.7 below) and the following theorem, due to Kolmogorov and Centsov, which provides a strong version of the classical Kolmogorov's continuity Theorem.

Theorem A.6 (a strong Kolmogorov's continuity Theorem). *Let* $\{Y(t), t \in I\}$ *be a real-valued stochastic field, where I denotes a compact interval of $\mathbb{R}^N$. Assume that there are three constants $c > 0$, $p > 0$ and $\gamma > N$ such that, for all $(t^1, t^2) \in I^2$, the inequality*

$$\mathbb{E}\big(|Y(t^1) - Y(t^2)|^p\big) \leq c|t^1 - t^2|^\gamma, \tag{A.5}$$

holds. Then there exists a continuous modification of $\{Y(t), t \in I\}$ whose paths almost surely belong to the global Hölder space $\mathcal{C}^\beta(I)$ (see Definition 1.33), for any $\beta \in [0,1)$ such that $\beta < p^{-1}(\gamma - N)$.

The proof of Theorem A.6 can be found on page 166 in [Khoshnevisan (2002)] for instance. The proof of the following lemma has been omitted since it is very classical and only requires standard computations.

Lemma A.7 (equivalence of Gaussian moments). *Let Z be a centered real-valued Gaussian random variable. Then, for all positive real number $r > 0$, one has*

$$\mathbb{E}\big(|Z|^r\big) = \frac{2^{r/2}\,\Gamma\left(\frac{r+1}{2}\right)}{\Gamma\left(\frac{1}{2}\right)} \times \big(\mathbb{E}(|Z|^2)\big)^{r/2}, \tag{A.6}$$

where Γ denotes the usual Gamma function.

^{2}That is a compact rectangle.

We recall in passing that:

Proposition A.8 (Gamma function). *The Gamma function is denoted by Γ and defined as:*

$$\Gamma(x) := \int_0^{+\infty} y^{x-1} e^{-y}\, dy, \quad \text{for all } x \in (0, +\infty). \tag{A.7}$$

Among many other nice properties, it satisfies the following two fundamental properties:

(i) One has $\Gamma(x+1) = x\Gamma(x)$ for all $x \in (0, +\infty)$ and as consequence $\Gamma(n) = (n-1)!$ for each $n \in \mathbb{N}$.

(ii) There exists a finite constant $c > 0$ such that, for every $x \in [1, +\infty)$, the inequality

$$\left| \frac{\Gamma(x)}{(2\pi)^{1/2}\, x^{x-1/2}\, e^{-x}} - 1 - \frac{1}{12x} \right| \leq \frac{c}{x^2} \tag{A.8}$$

holds.

Remark A.9 (Stirling's formula). Notice that the Stirling's formula: $n! \sim (2\pi n)^{1/2} (n/e)^n$ when the integer n goes to $+\infty$, that is

$$\lim_{n \to +\infty} \frac{n!}{(2\pi n)^{1/2}\, (n/e)^n} = 1,$$

easily results from the previous properties (i) and (ii) of the Gamma function.

The following theorem is somehow a converse of Theorem A.5.

Theorem A.10. *Let $\{\widetilde{Y}(t), t \in I\}$ be a real-valued centered Gaussian field whose paths almost surely belong to some global Hölder space $\mathcal{C}^\beta(I)$, where $\beta \in [0,1)$. Then, for each $\beta' \in [0, \beta)$ (with the convention that $\beta' = 0$ when $\beta = 0$), there is a constant $c' > 0$ (depending on β') such that, for all $(t^1, t^2) \in I^2$, the inequality*

$$\mathbb{E}\big(|\widetilde{Y}(t^1) - \widetilde{Y}(t^2)|^2\big) \leq c' |t^1 - t^2|^{2\beta'}, \tag{A.9}$$

holds.

Proof of Theorem A.10. We denote by $\widetilde{\Omega}$ an event of probability 1 on which the paths of $\{\widetilde{Y}(t), t \in I\}$ belong to $\mathcal{C}^\beta(I)$. Let $\{X(t^1, t^2), (t^1, t^2) \in I \times I\}$ be the centered real-valued Gaussian field with continuous paths which vanishes outside of $\widetilde{\Omega}$ and which is defined on $\widetilde{\Omega}$ as $X(t^1, t^2) := 0$ when $t^1 = t^2$ and $X(t^1, t^2) := |t^1 - t^2|^{-\beta'} (\widetilde{Y}(t^1) - \widetilde{Y}(t^2))$ else. Applying the following lemma to $\{X(t^1, t^2), (t^1, t^2) \in I \times I\}$ one gets the theorem. $\square$

Lemma A.11 (moments of suprema of Gaussian fields). *Let* $\{X(s), s \in T\}$ *be a real-valued centered Gaussian field with almost surely continuous paths* [3] *on a given compact interval* $T \subset \mathbb{R}^M$, *where* M *is an arbitrary positive integer. Then, for all positive real number* p, *one has*

$$\mathbb{E}\left(\sup_{s \in T} |X(s)|^p\right) < +\infty; \tag{A.10}$$

moreover

$$c'(p)\,\mathbb{E}\left(\sup_{s \in T} |X(s)|\right) \leq \left(\mathbb{E}\left(\sup_{s \in T} |X(s)|^p\right)\right)^{1/p} \leq c''(p)\,\mathbb{E}\left(\sup_{s \in T} |X(s)|\right), \tag{A.11}$$

where $0 < c'(p) < c''(p) < +\infty$ *are two constants only depending on* p.

Proof of Lemma A.11. It can be found in Chapter 3 of [Ledoux and Talagrand (1991)] for instance. □

A.2 Differentiability of Gaussian fields

Let us first mention that while path continuity of Gaussian processes and fields has been studied in many works, only few works in the literature (see [Cramér and Leadbetter (1967); Loève (1978); Potthoff (2010)] and the note by Michel Loève in [Lévy (1948 1st edition, 1965 2nd edition)] pages 367 to 420) deal with their path differentiability.

Throughout this section we denote by $\{Y(t), t \in \mathbb{R}^N\}$ a real-valued centered Gaussian field on a probability space Ω, and we denote by $\mathcal{R}_Y$ its covariance function defined, for all $(s,t) \in \mathbb{R}^N \times \mathbb{R}^N$, as

$$\mathcal{R}_Y(s,t) := \mathbb{E}\big(Y(s)Y(t)\big).$$

Definition A.12 (differentiability in quadratic mean). $\{Y(t), t \in \mathbb{R}^N\}$ is said to be *differentiable in quadratic mean at* $\tau \in \mathbb{R}^N$, *if* it satisfies the following property: there exists a linear map [4] from $\mathbb{R}^N$ into $L^2(\Omega)$, denoted by $(D_{\mathrm{qm}}Y)(\tau)$ and called *the derivative in quadratic mean of* $\{Y(t), t \in \mathbb{R}^N\}$ *at* τ, such that

$$\lim_{t \to \tau} \mathbb{E}\left(|t - \tau|^{-2}\left|Y(t) - Y(\tau) - (D_{\mathrm{qm}}Y)(\tau)(t - \tau)\right|^2\right) = 0.$$

Moreover, when this property holds at any arbitrary $\tau \in \mathbb{R}^N$, then $\{Y(t), t \in \mathbb{R}^N\}$ is said to be *differentiable in quadratic mean*.

---

[3] Notice that this condition can be weakened.

[4] Notice that when it exists such a linear map is necessarily unique.

Definition A.13 (partial derivative in quadratic mean).
Let $\{e_1, \ldots, e_N\}$ be the standard [5] cartesian basis of $\mathbb{R}^N$ and assume that $n \in \{1, \ldots, N\}$ is fixed. $\{Y(t), t \in \mathbb{R}^N\}$ is said to have *a partial derivative in quadratic mean at τ in direction n*, if there exists a random variable, belonging to $L^2(\Omega)$ and denoted by $(\partial_{\mathrm{qm},n}Y)(\tau)$, such that

$$\lim_{\lambda \in \mathbb{R},\, \lambda \to 0} \mathbb{E}\left(\left|\lambda^{-1}\big(Y(\tau + \lambda e_n) - Y(\tau)\big) - (\partial_{\mathrm{qm},n}Y)(\tau)\right|^2\right) = 0.$$

Notice that $(\partial_{\mathrm{qm},n}Y)(\tau)$ is centered Gaussian and unique up to almost sure equality; it is called *the partial derivative in quadratic mean of $\{Y(t), t \in \mathbb{R}^N\}$ at τ in direction n.*

The following remark easily follows from Definitions A.12 and A.13.

Remark A.14. Assume that $\{Y(t), t \in \mathbb{R}^N\}$ is differentiable in quadratic mean at $\tau \in \mathbb{R}^N$ with a derivative denoted by $(D_{\mathrm{qm}}Y)(\tau)$. Then, for all $n \in \{1, \ldots, N\}$, the partial derivative $(\partial_{\mathrm{qm},n}Y)(\tau)$ exists and satisfies almost surely

$$(\partial_{\mathrm{qm},n}Y)(\tau) = (D_{\mathrm{qm}}Y)(\tau)(e_n).$$

As a consequence, one has almost surely for every $h := (h_1, \ldots, h_N) \in \mathbb{R}^N$,

$$(D_{\mathrm{qm}}Y)(\tau)(h) = \sum_{n=1}^{N} h_n (\partial_{\mathrm{qm},n}Y)(\tau).$$

The proofs of the following results can be found in [Potthoff (2010)] in the more general setting of second order random fields.

Proposition A.15 (sufficient condition for differentiability in qm).
Assume that, for all $n \in \{1, \ldots, N\}$ and $\tau \in \mathbb{R}^N$, the partial derivative $(\partial_{\mathrm{qm},n}Y)(\tau)$ exists. Also assume that, for any fixed $n \in \{1, \ldots, N\}$, the real-valued centered Gaussian field $\{(\partial_{\mathrm{qm},n}Y)(\tau), \tau \in \mathbb{R}^N\}$ is continuous in quadratic mean (see Definition A.1). Then $\{Y(t), t \in \mathbb{R}^N\}$ is differentiable in quadratic mean.

Lemma A.16. *Assume that the covariance function $\mathcal{R}_Y$ of $\{Y(t), t \in \mathbb{R}^N\}$ belongs to the function space $\mathcal{C}^{1,1}(\mathbb{R}^N \times \mathbb{R}^N)$ defined below. Then, the hypotheses of Proposition A.15 hold true. Namely, the partial derivative $(\partial_{\mathrm{qm},n}Y)(\tau)$ exists for all $n \in \{1, \ldots, N\}$ and $\tau \in \mathbb{R}^N$; also, for any fixed*

---

[5]That is, for all $n = 1, \ldots, N$, the n-th coordinate of e_n equals 1 and its other coordinates equal 0.

$n \in \{1, \ldots, N\}$ *the real-valued centered Gaussian field* $\{(\partial_{\mathrm{qm},n}Y)(\tau), \tau \in \mathbb{R}^N\}$ *is continuous in quadratic mean. Moreover, for any* $(\theta, \tau) \in \mathbb{R}^N \times \mathbb{R}^N$, *the covariance* $\mathcal{R}_{\partial_{\mathrm{qm},n}Y}(\theta, \tau) := \mathbb{E}\{(\partial_{\mathrm{qm},n}Y)(\theta)(\partial_{\mathrm{qm},n}Y)(\tau)\}$ *satisfies* $\mathcal{R}_{\partial_{\mathrm{qm},n}Y}(\theta, \tau) = (\partial_{s_n} \partial_{t_n} \mathcal{R}_Y)(\theta, \tau)$.

Definition A.17 (the space $\mathcal{C}^{1,1}(\mathbb{R}^N \times \mathbb{R}^N)$). $\mathcal{C}^{1,1}(\mathbb{R}^N \times \mathbb{R}^N)$ is the space of the real-valued functions f defined on $\mathbb{R}^N \times \mathbb{R}^N$ satisfying, for all $(m, n) \in \{1, \ldots, N\}^2$, the following two properties:

(i) For any $(s, t) = (s_1, \ldots, s_N, t_1, \ldots, t_N) \in \mathbb{R}^N \times \mathbb{R}^N$, the partial derivatives $(\partial_{s_m} f)(s, t)$, $(\partial_{t_n} f)(s, t)$ and $(\partial_{s_m} \partial_{t_n} f)(s, t)$ exist;

(ii) $(\partial_{s_m} f)$, $(\partial_{t_n} f)$ and $(\partial_{s_m} \partial_{t_n} f)$ are continuous functions on $\mathbb{R}^N \times \mathbb{R}^N$.

Theorem A.18 (sufficient condition for path differentiability). *Assume that the following two conditions (i) and (ii) hold.*

(i) The covariance function $\mathcal{R}_Y$ of the real-valued centered Gaussian field $\{Y(t), t \in \mathbb{R}^N\}$ belongs to the space $\mathcal{C}^{1,1}(\mathbb{R}^N \times \mathbb{R}^N)$;

(ii) For every $n \in \{1, \ldots, N\}$, there exists a modification with almost surely continuous paths for the partial derivative in quadratic mean real-valued centered Gaussian field $\{(\partial_{\mathrm{qm},n}Y)(\tau), \tau \in \mathbb{R}^N\}$.

Then, $\{Y(t), t \in \mathbb{R}^N\}$ has a modification with almost surely differentiable paths, denoted by $\{\widetilde{Y}(t), t \in \mathbb{R}^N\}$. Moreover, for all $n \in \{1, \ldots, N\}$, the pathwise partial derivative real-valued centered Gaussian field $\{(\partial_n Y)(\tau), \tau \in \mathbb{R}^N\}$ is in fact the continuous modification of $\{(\partial_{\mathrm{qm},n}Y)(\tau), \tau \in \mathbb{R}^N\}$; thus, for any $(\theta, \tau) \in \mathbb{R}^N \times \mathbb{R}^N$, the covariance $\mathcal{R}_{\partial_n \widetilde{Y}}(\theta, \tau) := \mathbb{E}\{(\partial_n \widetilde{Y})(\theta)(\partial_n \widetilde{Y})(\tau)\}$ satisfies

$$\mathcal{R}_{\partial_n \widetilde{Y}}(\theta, \tau) = \mathcal{R}_{\partial_{\mathrm{qm},n}Y}(\theta, \tau) = (\partial_{s_n} \partial_{t_n} \mathcal{R}_Y)(\theta, \tau). \tag{A.12}$$

A.3 Other useful results

The following proposition can easily be obtained by using successively the classical Hölder inequality.

Proposition A.19 (the generalized Hölder inequality). *Assume that the fixed integer $k \geq 2$ is arbitrary. Then, for all $p_1, \ldots, p_k \in [1, +\infty]$ satisfying $\sum_{n=1}^{k} 1/p_n = 1$, and for every complex-valued Borel functions $f_1, \ldots, f_k$ defined on $\mathbb{R}^N$, one has*

$$\int_{\mathbb{R}^N} \Big| \prod_{n=1}^{k} f_n(s) \Big| \, ds \leq \prod_{n=1}^{k} \Big(\int_{\mathbb{R}^N} |f_n(s)|^{p_n} \, ds \Big)^{1/p_n}.$$

Proposition A.20 (differentiation under the integral sign). *Let* $\mathcal{L}$: $(t, \xi) \mapsto \mathcal{L}(t, \xi)$ *be a real-valued function defined on* $\mathbb{R}^M \times \mathbb{R}^N$ *which satisfies the following three properties.*

(i) *For every fixed* $t \in \mathbb{R}^M$, *the function* $\xi \mapsto \mathcal{L}(t, \xi)$ *belongs to the Lebesgue space* $L^1(\mathbb{R}^N)$.

(ii) *For every* $m \in \{1, \ldots, M\}$, *almost all fixed* $\xi \in \mathbb{R}^N$ *and each* $t \in \mathbb{R}^M$, $\frac{\partial}{\partial t_m} \mathcal{L}(t, \xi)$, *the partial derivative with respect to the m-th coordinate of* t, *exists; moreover* $t \mapsto \frac{\partial}{\partial t_m} \mathcal{L}(t, \xi)$ *is a continuous function on* $\mathbb{R}^N$.

(iii) *For any fixed positive real number* B *and* $m \in \{1, \ldots, M\}$, *there exist* $g \in L^1(\mathbb{R}^N)$ *such that the inequality*

$$\left| \frac{\partial}{\partial t_m} \mathcal{L}(t, \xi) \right| \leq g(\xi),$$

holds for almost all $\xi \in \mathbb{R}^N$ *and all* $t \in \mathbb{R}^M$ *with* $|t| \leq B$.

Then the real-valued function

$$\mathcal{V} : t \mapsto \mathcal{V}(t) := \int_{\mathbb{R}^N} \mathcal{L}(t, \xi) \, d\xi,$$

is continuously differentiable on $\mathbb{R}^M$, *and one has, for any* $m \in \{1, \ldots, M\}$ *and* $t \in \mathbb{R}^M$,

$$\frac{\partial}{\partial t_m} \mathcal{V}(t) := \int_{\mathbb{R}^N} \frac{\partial}{\partial t_m} \mathcal{L}(t, \xi) \, d\xi.$$

Lemma A.21. *Let* $\big(\{ Z_n(s), s \in \mathbb{R}^N \} \big)_{n \in \mathbb{N}}$ *be a sequence of centered real-valued Gaussian fields having almost surely continuous paths. Assume that*

(i) *for all* $n \in \mathbb{N}$, *one has almost surely* $Z_n(0) \overset{a.s.}{=} 0$;

(ii) *for each fixed* $\delta > 0$, *there exist two constants* $c > 0$ *and* $\nu \in (0, 1]$, *only depending on* δ, *such that the inequality*

$$\mathbb{E}\big(|Z_n(s^1) - Z_n(s^2)|^2 \big) \leq c|s^1 - s^2|^{2\nu}, \tag{A.13}$$

holds for every $n \in \mathbb{N}$ *and for all* $s^1, s^2 \in \mathbb{R}^N$ *satisfying* $|s^1| \leq \delta, |s^2| \leq \delta$.

Then, as soon as $\big(\{ Z_n(s), s \in \mathbb{R}^N \} \big)_{n \in \mathbb{N}}$ *converges in the sense of finite-dimensional distributions to a Gaussian field* $\{ Z(s), s \in \mathbb{R}^N \}$ *having almost surely continuous paths, the convergence in distribution also holds in the space of the continuous functions over an arbitrary compact subset in* $\mathbb{R}^N$.

Proof of Lemma A.21. Let $\mathcal{C}(\mathcal{B}(0, \delta), \mathbb{R})$ be the Banach space of the real-valued continuous functions on the closed ball $\mathcal{B}(0, \delta)$ of $\mathbb{R}^N$; it is equipped with the uniform norm. For each $n \in \mathbb{N}$, we denote by P_n^δ, the probability measure on the Borel σ-field of $\mathcal{C}(\mathcal{B}(0,\delta), \mathbb{R})$ induced by the Gaussian field $\{Z_n(s), s \in \mathbb{R}^N\}$. In order to derive the lemma, it is enough to show that the sequence $(P_n^\delta)_{n \in \mathbb{N}}$ is relatively compact or equivalently that it is tight (see pages 35 to 37 in [Billingsley (1968)] and more particularly Prohorov's Theorem). The proof of the tightness of $(P_n^\delta)_{n \in \mathbb{N}}$ mainly relies on (A.13) and Lemma A.7; it can be done rather similarly to the proof of Theorem 12.3 in Chapter 2 of [Billingsley (1968)]. □

Theorem A.22 (a Gaussian version of Itô-Nisio Theorem). *Let M and N respectively be a positive real number and a positive integer. One denotes by $\mathcal{C}([-M,M]^N, \mathbb{R})$ the Banach space of the real-valued continuous functions on the cube $[-M,M]^N$ equipped with the supremum norm. Let $(e_q)_{q \in \mathbb{N}}$ and $(\zeta_q)_{q \in \mathbb{N}}$ respectively be a sequence of elements of $\mathcal{C}([-M,M]^N, \mathbb{R})$ and a sequence of real-valued independent $\mathcal{N}(0,1)$ Gaussian random variables defined on a probability space Ω. For all $Q \in \mathbb{N}$, for each $t \in [-M,M]^N$ and for every $\omega \in \Omega$, one sets*

$$X_Q(t,\omega) = \sum_{q=1}^{Q} \zeta_q(\omega)\, e_q(t).$$

Assume that the following three conditions hold:

(i) one has $e_q(0) = 0$, for all $q \in \mathbb{N}$;

(ii) for each fixed $t \in \mathbb{R}$, the sequence of random variables $\big(X_Q(t)\big)_{Q \in \mathbb{N}}$ is convergent in $L^2(\Omega)$;

(iii) there are two constants $c > 0$ and $e > 0$ such that the inequality

$$\mathbb{E}\big(|X_Q(t_1) - X_Q(t_2)|^2\big) \le c|t_1 - t_2|^e$$

is satisfied, for all $Q \in \mathbb{N}$ and for every $t_1, t_2 \in [-M,M]^N$.

Then, there exists $\widetilde{\Omega}$ an event of probability 1, such that, for each fixed $\omega \in \widetilde{\Omega}$, the sequence of continuous functions $\big(X_Q(\cdot, \omega)\big)_{Q \in \mathbb{N}}$ is convergent in the Banach space $\mathcal{C}([-M,M]^N, \mathbb{R})$.

A.4 Asymptotic behavior of $\mathcal{N}(0,1)$ Gaussian sequences

Lemma A.23. *Let $\{\varepsilon_\lambda : \lambda \in \mathbb{Z}^d\}$ be an arbitrary sequence of real-valued $\mathcal{N}(0,1)$ Gaussian random variables* [6] *defined on a probability space Ω.*

---

[6]Notice that one does not necessarily need to impose to these random variables to be independent.

Then, there exist two positive random variables C and C' of finite moments of any order, and there is $\Omega_0^* \subseteq \Omega$, an event of probability 1, such that, for each $\omega \in \Omega_0^*$ and for all $\lambda = (\lambda_1, \ldots, \lambda_d) \in \mathbb{Z}^d$, one has

$$\left|\varepsilon_\lambda(\omega)\right| \leq C(\omega)\sqrt{\log\left(2 + |\lambda_1| + \ldots + |\lambda_d|\right)}$$

$$\leq C'(\omega) \prod_{n=1}^{d} \sqrt{\log\left(2 + |\lambda_n|\right)}.$$

Lemma A.23 has been proved in [Ayache and Taqqu (2003)] in the case where $d = 2$. The proof can be done similarly when the positive integer d is arbitrary; notice that its main ingredient is following classical lemma which provides sharp asymptotic estimates on the tail behavior of a standard Gaussian distribution.

Lemma A.24. *Let ε be an arbitrary real-valued $\mathcal{N}(0,1)$ Gaussian random variable, then one has*

$$\lim_{x \to +\infty} \frac{\mathbb{P}\left(|\varepsilon| > x\right)}{(2\pi^{-1})^{1/2}\, x^{-1}e^{-x^2/2}} = 1\,.$$

Thus, there exist two constants $0 < c' \leq c'' \leq 1$ for which the following two inequalities hold:

$$c'x^{-1}e^{-x^2/2} \leq \mathbb{P}\left(|\varepsilon| > x\right) \leq c''x^{-1}e^{-x^2/2}, \quad \text{for all } x \in [1, +\infty). \quad \text{(A.14)}$$

The following lemma can be viewed as a consequence of Lemma A.23.

Lemma A.25. *There exists a finite constant c, such that, for any arbitrary finite sequence $\{G_\zeta : \zeta \in \mathbb{I}\}$ of centered real-valued Gaussian random variables defined on a probability space Ω, one has*

$$E\left(\sup_{\zeta \in \mathbb{I}} |G_\zeta|\right) \leq c\left(\sup_{\zeta \in \mathbb{I}} \sigma(G_\zeta)\right)\sqrt{\log\left(1 + \operatorname{card}(\mathbb{I})\right)},$$

where $\sigma(G_\zeta)$ is the standard deviation of G_ζ and $\operatorname{card}(\mathbb{I})$ denotes the cardinality of the finite set of indices $\mathbb{I}$.

Lemma A.26. *The integer $N \geq 1$ is arbitrary and fixed. One denotes by $\{\varepsilon_{j,k} : (j,k) \in \mathbb{N} \times \mathbb{Z}^N\}$ an arbitrary sequence of real-valued $\mathcal{N}(0,1)$ Gaussian random variables defined on a probability space Ω. One assumes that, for any fixed $j \in \mathbb{N}$, the $\varepsilon_{j,k}$'s, $k \in \mathbb{Z}^N$, are independent. Then, there exists $\Omega_1^* \subseteq \Omega$, an event of probability 1, such that, for all $\omega \in \Omega_1^*$ and for each $\tau \in \mathbb{R}^N$, one has*

$$\liminf_{j \to +\infty}\left\{\varepsilon_j^{max}(\tau, \omega)\right\} := \lim_{J \to +\infty}\left\{\inf_{j \geq J}\left\{\varepsilon_j^{max}(\tau, \omega)\right\}\right\} \geq 2^{-1}, \quad \text{(A.15)}$$

where

$$\varepsilon_j^{max}(\tau,\omega) := \max\left\{ |\varepsilon_{j,k}(\omega)| : k \in \mathbb{Z}^N \ and \ |\tau - 2^{-j}k| \le N^{1/2}j2^{1-j} \right\}. \tag{A.16}$$

Proof of Lemma A.26. For each fixed integers $m \ge 1$ and $j \ge 1$, one denotes by $[\frac{2^j m}{j}]$ (resp. $[-\frac{2^j m}{j}]$) the integer part of $\frac{2^j m}{j}$ (resp. $-\frac{2^j m}{j}$), and one denotes by $I_{m,j}$ the finite subset of $\mathbb{Z}^N$ defined as

$$I_{m,j} := \left\{ \left[-\frac{2^j m}{j}\right], \left[-\frac{2^j m}{j}\right]+1, \ldots, \left[\frac{2^j m}{j}\right]-1, \left[\frac{2^j m}{j}\right] \right\}^N. \tag{A.17}$$

Moreover, for every fixed $q \in I_{m,j}$, one assumes that $D_{m,j}^q$ is the finite subset of $\mathbb{Z}^N$ defined as

$$D_{m,j}^q := \left\{ jq + r : r \in \{0, \ldots, j-1\}^N \right\}. \tag{A.18}$$

Then, let $\mathcal{E}_{m,j}^q$ be the event defined as

$$\mathcal{E}_{m,j}^q := \left\{ \omega \in \Omega : \max_{k \in D_{m,j}^q} |\varepsilon_{j,k}(\omega)| \ge 2^{-1} \right\}. \tag{A.19}$$

Observe that, in view of (A.16), for all $\tau := (\tau_1, \ldots, \tau_N) \in \mathbb{R}^N$ and integers $j \ge 1$ and $m \ge |\tau| := \left(\tau_1^2 + \ldots + \tau_N^2\right)^{1/2}$, one has

$$\mathcal{E}_{m,j}^{[2^j \tau/j]} \subseteq \left\{ \omega \in \Omega : \varepsilon_j^{max}(\tau,\omega) \ge 2^{-1} \right\}, \tag{A.20}$$

where $[2^j \tau/j] \in I_{m,j}$ is defined as $[2^j \tau/j] := \left([2^j \tau_1/j], \ldots, [2^j \tau_N/j]\right)$. Next, one denotes by Ω_1^*, the event, not depending on τ, defined as

$$\Omega_1^* := \bigcap_{m=1}^{+\infty} \bigcup_{J=1}^{+\infty} \bigcap_{j=J}^{+\infty} \bigcap_{q \in I_{m,j}} \mathcal{E}_{m,j}^q. \tag{A.21}$$

In view of (A.20) and (A.21), in order to show that (A.15) holds, it is enough to prove that $\mathbb{P}(\Omega_1^*) = 1$. That is, one has, for any fixed integer $m \ge 1$,

$$\mathbb{P}\left(\bigcup_{J=1}^{+\infty} \bigcap_{j=J}^{+\infty} \bigcap_{q \in I_{m,j}} \mathcal{E}_{m,j}^q \right) = 1. \tag{A.22}$$

One knows from Borel-Cantelli Lemma that, for deriving (A.22), it is enough to show that

$$\sum_{j=1}^{+\infty} \mathbb{P}\left(\bigcup_{q \in I_{m,j}} \overline{\mathcal{E}}_{m,j}^q \right) < +\infty, \tag{A.23}$$

where $\overline{\mathcal{E}}^q_{m,j}$ is the complement of the event $\mathcal{E}^q_{m,j}$. It results from (A.19) that

$$\overline{\mathcal{E}}^q_{m,j} := \Omega \setminus \mathcal{E}^q_{m,j} = \bigcap_{k \in D^q_{m,j}} \left\{ \omega \in \Omega : |\varepsilon_{j,k}(\omega)| < 2^{-1} \right\}.$$

Observe that, in view of (A.18), the cardinality of $D^q_{m,j}$ is equal to j^N. Thus, using the fact that the $\varepsilon_{j,k}$'s, $k \in D^q_{m,j}$, are independent real-valued $\mathcal{N}(0,1)$ Gaussian random variables, one gets that

$$\mathbb{P}(\overline{\mathcal{E}}^q_{m,j}) \leq (2\pi)^{-j^N/2}. \qquad (\text{A}.24)$$

Next, notice that one knows from (A.17) that the cardinality of $I_{m,j}$ is less than or equal to $\left(4m2^j/j\right)^N$. Thus, one can derive from (A.24) that

$$\mathbb{P}\Big(\bigcup_{q \in I_{m,j}} \overline{\mathcal{E}}^q_{m,j} \Big) \leq \sum_{q \in I_{m,j}} \mathbb{P}(\overline{\mathcal{E}}^q_{m,j}) \leq (4m)^N j^{-N} 2^{Nj} (2\pi)^{-j^N/2},$$

which clearly implies that (A.23) holds. $\qquad\square$

Lemma A.27. *The integer $N \geq 1$ is arbitrary and fixed. One denotes by $\left\{\varepsilon_{j,k} : (j,k) \in \mathbb{N} \times \mathbb{Z}^N\right\}$ an arbitrary sequence of independent real-valued $\mathcal{N}(0,1)$ Gaussian random variables defined on a probability space Ω. Then, there exists $\Omega^*_2 \subseteq \Omega$, an event of probability 1, such that, for all $\omega \in \Omega^*_2$ and for each $\tau := (\tau_1, \ldots, \tau_N) \in \mathbb{R}^N$, one has*

$$\limsup_{j \to +\infty} \left\{ |\varepsilon_{j,[2^j\tau]}(\omega)| \right\} := \lim_{J \to +\infty} \left\{ \sup_{j \geq J} \left\{ |\varepsilon_{j,[2^j\tau]}(\omega)| \right\} \right\} \geq 2^{-N-1/2}\sqrt{\pi},$$
$$(\text{A}.25)$$

where $[2^j\tau] := \left([2^j\tau_1], \ldots, [2^j\tau_N]\right)$. Recall that $[\,\cdot\,]$ denotes the integer part function.

Proof of Lemma A.27. First one mentions that this proof is inspired by some remarks on pages 239 and 240 in [Kahane (1968 1st edition, 1985 2nd edition)]. One uses the same notations as in Definition 1.66. Also, one needs to fix some additional notations. Let an arbitrary couple $(J, K) \in \mathbb{N} \times \mathbb{Z}^N$ and let $\Lambda_{J,K} := [2^{-J}K, 2^{-J}(K + \langle 1 \rangle)]$ be the corresponding dyadic compact interval of $\mathbb{R}^N$ of order J. For any fixed $m \in \mathbb{N}$, one denotes by $\mathcal{S}_{J,K,m}$ the finite set of cardinality 2^{mN} whose elements are the dyadic compact intervals of $\mathbb{R}^N$ of order $J+m$ which are included in $\Lambda_{J,K}$; roughly speaking "$\mathcal{S}_{J,K,m}$ is the set of the descendants of $\Lambda_{J,K}$ at the m-th generation". Notice that, for any $S \in \mathcal{S}_{J,K,m}$, there exists a unique finite sequence $(T_n)_{1 \leq n \leq m}$ of dyadic compact intervals of $\mathbb{R}^N$ which is decreasing in the sense of the inclusion and satisfies $T_1 \subset \Lambda_{J,K}$, $T_m = S$ and $T_n \in \mathcal{S}_{J,K,n}$, for all $n \in \{1, \ldots, m\}$. In the sequel, the sequence $(T_n)_{1 \leq n \leq m}$ is denoted by $\mathcal{T}_{J,K,m}(S)$ and any dyadic interval T_n is sometimes identified with the unique couple $(j_*, k_*) \in \mathbb{N} \times \mathbb{Z}^N$ such that $T_n := [2^{-j_*}k_*, 2^{-j_*}(k_* + \langle 1 \rangle)]$. Let $c_0 := 2^{-N-1/2}\sqrt{\pi}$ be the constant in the right-hand side of (A.25), and let p_0 be the probability that an arbitrary real-valued $\mathcal{N}(0,1)$ Gaussian random variable belongs to the interval $(-c_0, c_0)$. Elementary calculations allow to obtain that

$$0 < p_0 < 2^{-N}. \tag{A.26}$$

For all $S \in \mathcal{S}_{J,K,m}$, one denotes $B_{J,K,m}(S)$ the Bernoulli random variable defined as

$$B_{J,K,m}(S) := \prod_{T \in \mathcal{T}_{J,K,m}(S)} \mathbb{1}_{\{|\varepsilon_T| < c_0\}}. \tag{A.27}$$

Notice that, using the definition of p_0 and the independence property of the random variables $\varepsilon_{j,k}$, one has

$$\mathbb{E}\big(B_{J,K,m}(S)\big) = p_0^m. \tag{A.28}$$

Next, let $G_{J,K,m}$ be the random variable defined as

$$G_{J,K,m} = \sum_{S \in \mathcal{S}_{J,K,m}} B_{J,K,m}(S). \tag{A.29}$$

Notice that $G_{J,K,m}$ is with values in $\{0, 1, \ldots, 2^{mN}\}$ since the cardinality of $\mathcal{S}_{J,K,m}$ equals 2^{mN}. Using the latter fact, (A.28) and (A.29) one gets that $\mathbb{E}\big(G_{J,K,m}\big) = \big(2^N p_0\big)^m$. Thus, it follows from Fatou Lemma and (A.26) that

$$0 \leq \mathbb{E}\Big(\liminf_{m \to +\infty} G_{J,,K,m} \Big) \leq \lim_{m \to +\infty} \mathbb{E}\big(G_{J,K,m}\big) = 0.$$

Hence, for any arbitrary $(J, K) \in \mathbb{N} \times \mathbb{Z}$, the event

$$\Omega^*_{2,J,K} := \Big\{ \omega \in \Omega : \liminf_{m \to +\infty} G_{J,K,m}(\omega) = 0 \Big\} \tag{A.30}$$

has a probability equals to 1. Since $\mathbb{N} \times \mathbb{Z}^N$ is a countable set, the event

$$\Omega^*_2 := \bigcap_{(J,K) \in \mathbb{N} \times \mathbb{Z}^N} \Omega^*_{2,J,K} \tag{A.31}$$

also has a probability equals to 1.

Let us now prove that (A.25) is satisfied, for every $\omega \in \Omega^*_2$ and for all $\tau \in \mathbb{R}^N$. Suppose, ad absurdum, that there are $\omega_0 \in \Omega^*_2$, $\tau^0 \in \mathbb{R}^N$ and $J_0 \in \mathbb{N}$ such that

$$\big|\varepsilon_{J_0+n, [2^{J_0+n}\tau^0]}(\omega_0)\big| < c_0 := 2^{-N-1/2}\sqrt{\pi}, \quad \text{for all } n \in \mathbb{N}. \tag{A.32}$$

Notice that, for all fixed $m \in \mathbb{N}$, one has

$$\mathcal{T}_{J_0,[2^{J_0}\tau^0],m}\big(\Lambda_{J_0+m,[2^{J_0+m}\tau^0]}\big) = \big(\Lambda_{J_0+n,[2^{J_0+n}\tau^0]}\big)_{1 \leq n \leq m}.$$

Thus, it follows from (A.27) and (A.32) that

$$B_{J_0,[2^{J_0}\tau^0],m}\big(\Lambda_{J_0+m,[2^{J_0+m}\tau^0]}\big)(\omega_0) = 1.$$

Then, using (A.29), one gets that

$$G_{J_0,[2^{J_0}\tau^0],m}(\omega_0) \geq 1, \quad \text{for all } m \in \mathbb{N}.$$

This implies that

$$\liminf_{m \to +\infty} G_{J_0,[2^{J_0}\tau^0],m}(\omega_0) \geq 1,$$

which contradicts the fact that $\omega_0 \in \Omega_2^*$ (see (A.31) and (A.30)). $\qquad \square$

Lemma A.28. *The integer $N \geq 1$ is arbitrary and fixed. Let $\big\{\varepsilon_{j,k} : (j,k) \in \mathbb{N}\times\mathbb{Z}^N\big\}$ be an arbitrary sequence of independent real-valued $\mathcal{N}(0,1)$ Gaussian random variables defined on a probability space Ω. Assume that the point $\tau \in \mathbb{R}^N$ and the real number $\theta \geq 1$ are arbitrary and fixed. For every $j \in \mathbb{N}$, one denotes by $\mathcal{D}_j(\tau,\theta)$ the finite and nonempty set of indices k defined as*

$$\mathcal{D}_j(\tau,\theta) := \big\{k \in \mathbb{Z}^N \ : \ |\tau - 2^{-j}k| \leq N^{1/2}j^\theta 2^{-j}\big\}. \qquad (A.33)$$

Then, one has almost surely

$$\limsup_{j \to +\infty} \left\{ \frac{\max_{k \in \mathcal{D}_j(\tau,\theta)} |\varepsilon_{j,k}|}{\sqrt{\log(2+j)}} \right\} = \sqrt{2(N\theta + 1)}. \qquad (A.34)$$

Observe that the exceptional negligible event on which (A.34) fails to be satisfied depends on (τ,θ).

Proof of Lemma A.28. One knows from Borel-Cantelli Lemma, that for deriving (A.34), it is enough to prove that, for all fixed real numbers

$$a > \sqrt{2(N\theta + 1)} \quad \text{and} \quad \sqrt{2N\theta} \leq b < \sqrt{2(N\theta + 1)}, \qquad (A.35)$$

one has

$$\sum_{j=1}^{+\infty} \mathbb{P}\Big(\max_{k \in \mathcal{D}_j(\tau,\theta)} |\varepsilon_{j,k}| > a\sqrt{\log(2+j)}\Big) < +\infty \qquad (A.36)$$

and

$$\sum_{j=1}^{+\infty} \mathbb{P}\Big(\max_{k \in \mathcal{D}_j(\tau,\theta)} |\varepsilon_{j,k}| > b\sqrt{\log(2+j)}\Big) = +\infty. \qquad (A.37)$$

Notice that it results from (A.33) that there are two constants $0 < c_1 \leq c_2 < +\infty$, such that, for all $j \in \mathbb{N}$, one has

$$c_1\, j^{N\theta} \leq \operatorname{card}\big(\mathcal{D}_j(\tau)\big) \leq c_2\, j^{N\theta}. \tag{A.38}$$

Next, using the second inequalities in (A.38) and in (A.14), it follows that, for all $j \in \mathbb{N}$,

$$\mathbb{P}\Big(\max_{k \in \mathcal{D}_j(\tau,\theta)} |\varepsilon_{j,k}| > a\sqrt{\log(2+j)} \Big) \leq \sum_{k \in \mathcal{D}_j(\tau,\theta)} \mathbb{P}\Big(|\varepsilon_{j,k}| > a\sqrt{\log(2+j)} \Big)$$
$$\leq c_3\, j^{N\theta} \exp\big(- 2^{-1} a^2 \log(2+j) \big) \leq c_3\, j^{-(2^{-1} a^2 - N\theta)}, \tag{A.39}$$

where $c_3 > 0$ is a finite constant not depending on j. Thus, (A.39) and (A.35) imply that (A.36) is satisfied.

On the other hand, using the independence property of the random variables $\varepsilon_{j,k}$, the first inequalities in (A.38) and in (A.14), and the inequality

$$1 - \exp(-z) \geq 2^{-1}\, z, \quad \text{for every } z \in \big[0, \log(2)\big],$$

one obtains that, for all j large enough,

$$\mathbb{P}\Big(\max_{k \in \mathcal{D}_j(\tau,\theta)} |\varepsilon_{j,k}| > b\sqrt{\log(2+j)} \Big) = 1 - \mathbb{P}\Big(\max_{k \in \mathcal{D}_j(\tau,\theta)} |\varepsilon_{j,k}| \leq b\sqrt{\log(2+j)} \Big)$$
$$= 1 - \prod_{k \in \mathcal{D}_j(\tau,\theta)} \mathbb{P}\Big(|\varepsilon_{j,k}| \leq b\sqrt{\log(2+j)} \Big)$$
$$\geq 1 - \left(1 - c_4 \frac{\exp\big(- 2^{-1} b^2 \log(2+j) \big)}{\sqrt{\log(2+j)}} \right)^{c_1\, j^{N\theta}}$$
$$= 1 - \exp\left(c_1\, j^{N\theta} \log\left(1 - c_4 \frac{(2+j)^{-2^{-1} b^2}}{\sqrt{\log(2+j)}} \right) \right)$$
$$\geq -2^{-1} c_1\, j^{N\theta} \log\left(1 - c_4 \frac{(2+j)^{-2^{-1} b^2}}{\sqrt{\log(2+j)}} \right), \tag{A.40}$$

where $c_4 \in (0,1)$ is a constant not depending on j. Moreover, one has that

$$-j^{N\theta} \log\left(1 - c_4 \frac{(2+j)^{-2^{-1} b^2}}{\sqrt{\log(2+j)}} \right) \sim c_4 \frac{j^{-(2^{-1} b^2 - N\theta)}}{\sqrt{\log(j)}}, \quad \text{when } j \to +\infty.$$

Thus, (A.40) and (A.35) entail that (A.37) holds. $\qquad\square$

A.5 L^p spaces on N-dimensional torus

Throughout this section the integer $N \geq 1$ is arbitrary and fixed. In Fourier analysis, the N-dimensional torus (or more shortly the N-torus), denoted by $\mathbb{T}^N$, is frequently considered to be the Cartesian product of the unit circle N-times with itself. Thus a complex-valued function f over $\mathbb{T}^N$ can be viewed as a function f from $\mathbb{R}^N$ to $\mathbb{C}$ which is $(2\pi\mathbb{Z})^N$-periodic: one has $f(x + 2\pi m) = f(x)$, for every $m \in \mathbb{Z}^N$ and for Lebesgue almost all $x \in \mathbb{R}^N$.

Definition A.29. Let $p \in [1, +\infty)$, the space $L^p(\mathbb{T}^N)$ is defined as the Banach space of the Borel functions f from $\mathbb{R}^N$ to $\mathbb{C}$ which are $(2\pi\mathbb{Z})^N$-periodic and satisfy

$$\int_{\mathbb{T}^N} |f(x)|^p \, dx < +\infty,$$

with the usual convention that $\int_{\mathbb{T}^N}$ is the Lebesgue integral over $[0, 2\pi]^N$ or any other cube of $\mathbb{R}^N$ with edge length 2π. The $L^p(\mathbb{T}^N)$ space is equipped with the norm

$$\|f\|_{L^p(\mathbb{T}^N)} := \left(\int_{\mathbb{T}^N} |f(x)|^p \, dx \right)^{1/p}.$$

Definition A.30. The space $L^\infty(\mathbb{T}^N)$ is defined as the Banach space of the Borel functions f from $\mathbb{R}^N$ to $\mathbb{C}$ which are $(2\pi\mathbb{Z})^N$-periodic and satisfy $|f(x)| \leq c$, for almost all $x \in [0, 2\pi]^N$ (notice that this cube of $\mathbb{R}^N$ can be replaced by any other one with edge length 2π), where c is a finite constant depending on f. The $L^\infty(\mathbb{T}^N)$ space is equipped with the norm

$$\|f\|_{L^\infty(\mathbb{T}^N)} := \inf \left\{ c \in \mathbb{R}_+ \ : \ |f(x)| \leq c \text{ for almost all } x \in [0, 2\pi]^N \right\}.$$

Remark A.31.

(i) It easily follows from Hölder inequality that one has $L^{p_2}(\mathbb{T}^N) \subset L^{p_1}(\mathbb{T}^N)$, for all $p_1, p_2 \in [1, \infty]$ such that $p_1 \leq p_2$.

(ii) $L^2(\mathbb{T}^N)$ is a Hilbert space; it is equipped with the inner product $\langle \cdot, \cdot \rangle_{L^2(\mathbb{T}^N)}$ defined, for all $f, g \in L^2(\mathbb{T}^N)$, as

$$\langle f, g \rangle_{L^2(\mathbb{T}^N)} := \int_{\mathbb{T}^N} f(x)\overline{g(x)} \, dx. \tag{A.41}$$

A very classical orthonormal basis (see Definition 3.1) of $L^2(\mathbb{T}^N)$ is the trigonometric system $\left\{ (2\pi)^{-N/2} e^{il \cdot x} \ : \ l \in \mathbb{Z}^N \right\}$, where $l \cdot x := \sum_{p=1}^N l_p x_p$ denotes the inner product of the vectors $l = (l_1, \dots, l_N)$

and $x = (x_1, \ldots, x_N)$. Thus, in view of Theorem 3.3, any function $f \in L^2(\mathbb{T}^N)$, can be expressed as:

$$f(x) = \sum_{l \in \mathbb{Z}^N} c_l(f) e^{il \cdot x}, \tag{A.42}$$

where the series is unconditionally convergent in the sense of the $L^2(\mathbb{T}^N)$ norm, and, where, for each $l \in \mathbb{Z}^N$, the Fourier coefficient of order l of f, denoted by $c_l(f)$, is defined as:

$$c_l(f) := (2\pi)^{-N} \int_{\mathbb{T}^N} f(x) e^{-il \cdot x} \, dx. \tag{A.43}$$

We mention that the trigonometric system is the first orthonormal basis of $L^2(\mathbb{T}^N)$ to have been discovered. It was initially introduced in 1807 by the celebrated mathematician Joseph Fourier in the univariate case $N = 1$.

Definition A.32 (convolution product in $L^1(\mathbb{T}^N)$). Let f and g in $L^1(\mathbb{T}^N)$. Their convolution product is the function denoted by $f * g$ (or by $g * f$) and defined, for each $x \in \mathbb{T}^N$, as

$$(f * g)(x) := \int_{\mathbb{T}^N} f(y) g(x - y) \, dy = \int_{\mathbb{T}^N} f(x - z) g(z) \, dz. \tag{A.44}$$

Remark A.33.

(i) It easily follows from Tonelli Theorem that the integrals in (A.44) are well-defined and that $f * g$ belongs to $L^1(\mathbb{T}^N)$ and satisfies

$$\|f * g\|_{L^1(\mathbb{T}^N)} \leq \|f\|_{L^1(\mathbb{T}^N)} \|g\|_{L^1(\mathbb{T}^N)}. \tag{A.45}$$

(ii) The inequality (A.45) can be sharpened under the stronger assumption that $f \in L^p(\mathbb{T}^N)$ and $g \in L^q(\mathbb{T}^N)$ with exponents $p, q \in [1, \infty]$ such that $p^{-1} + q^{-1} \geq 1$. Then the function $f * g$ belongs to $L^r(\mathbb{T}^N)$, where r satisfies $1 + r^{-1} = p^{-1} + q^{-1}$, and the so called Young inequality for convolution holds:

$$\|f * g\|_{L^r(\mathbb{T}^N)} \leq \|f\|_{L^p(\mathbb{T}^N)} \|g\|_{L^q(\mathbb{T}^N)}. \tag{A.46}$$

(iii) A sufficient condition for that $f * g$ be a continuous function on $\mathbb{R}^N$ is that f or g be a continuous function on $\mathbb{R}^N$.

The following theorem is the main result of this section.

Theorem A.34. *Let $(\mathbb{K}_m)_{m \geq 1}$ be a sequence of functions with values in $\mathbb{R}_+$ belonging to $L^1(\mathbb{T}^N)$ which satisfy the following two conditions.*

(a) For every m, one has

$$\int_{\mathbb{T}^N} \mathbb{K}_m(y)\, dy = 1. \tag{A.47}$$

(b) There exists a constant $c > 0$ not depending on m, such that, for almost all $y \in [-\pi, \pi]^N$, the following inequality holds:

$$\mathbb{K}_m(y) \le c\, m^N B(my), \tag{A.48}$$

where, for every $s = (s_1, \dots, s_N) \in \mathbb{R}^N$,

$$B(s) := \prod_{l=1}^{N} \left(1 + s_l^2\right)^{-1}. \tag{A.49}$$

Then, for each $p \in [1, \infty)$ and for all $g \in L^p(\mathbb{T}^N)$, the sequence $(\mathbb{K}_m * g)_{m \ge 1}$ converges to g in $L^p(\mathbb{T}^N)$.

Definition A.35 (trigonometric polynomial). Let T be a function from $\mathbb{R}^N$ to $\mathbb{C}$ which is $(2\pi\mathbb{Z})^N$-periodic. One says that T is a trigonometric polynomial over $\mathbb{R}^N$, when it is of the form:

$$T(x) := \sum_{l \in \mathbb{Z}^N} b_l e^{il \cdot x}, \quad \text{for all } x \in \mathbb{R}^N,$$

where $(b_l)_{l \in \mathbb{Z}^N}$ (which does not depend on x) is some complex-valued sequence with only a finite number of non vanishing terms.

Remark A.36. It is clear that trigonometric polynomials are infinitely differentiable over $\mathbb{R}^N$ and that they belong to $L^\infty(\mathbb{T}^N)$, and consequently to $L^p(\mathbb{T}^N)$, for any $p \in [1, \infty]$.

The following proposition provides a classical example of a sequence of trigonometric polynomials $(\mathbb{K}_m)_{m \ge 1}$ with values in $\mathbb{R}_+$ which satisfies the conditions (a) and (b) in Theorem A.34.

Proposition A.37. *For each integer $m \ge 1$, one denotes by κ_m the Fejér kernel of order m, that is the nonnegative trigonometric polynomial over the real line, defined, for all $\theta \in \mathbb{R}$, as*

$$\kappa_m(\theta) := \frac{1}{2\pi m} \left| \sum_{k=0}^{m-1} e^{ik\theta} \right|^2. \tag{A.50}$$

Observe that one has

$$\kappa_m(\theta) = \frac{m}{2\pi}, \quad \text{for every } \theta \in (2\pi\mathbb{Z}), \tag{A.51}$$

and

$$\kappa_m(\theta) = \frac{1}{2\pi m} \frac{\sin^2(m\theta/2)}{\sin^2(\theta/2)}, \quad \textit{for every } \theta \in \mathbb{R} \setminus (2\pi\mathbb{Z}). \tag{A.52}$$

Then, the sequence of functions $(\mathbb{K}_m)_{m\geq 1}$ *defined, for each integer* $m \geq 1$ *and* $x = (x_1, \ldots, x_N) \in \mathbb{R}^N$, *as*

$$\mathbb{K}_m(x) := \prod_{l=1}^{N} \kappa_m(x_l), \tag{A.53}$$

satisfies the conditions (a) *and* (b) *in Theorem A.34.*

The following two corollaries easily result from Theorem A.34 and Proposition A.37.

Corollary A.38. *For each* $p \in [1, \infty)$, *the trigonometric polynomials form a dense subset of* $L^p(\mathbb{T}^N)$.

Corollary A.39. *The sequence* $\left\{ c_l(f) : l \in \mathbb{Z}^N \right\}$ *of the Fourier coefficients of a function* f *from* $\mathbb{R}^N$ *into* $\mathbb{C}$ *can be defined, through (A.43) not only when* $f \in L^2(\mathbb{T}^N)$, *but also, more generally when* $f \in L^1(\mathbb{T}^N)$. *An arbitrary function* $f \in L^1(\mathbb{T}^N)$ *vanishes almost everywhere as soon as all its Fourier coefficients equal zero. On the other hand, we mention that there exists some functions* $f \in L^1(\mathbb{T}^N)$ *for which (A.42) fails to be true.*

Proof of Proposition A.37. In view of (A.53), (A.47), (A.48), and (A.49), it is enough to show that, for every integer $m \geq 1$,

$$\int_{\mathbb{T}} \kappa_m(\theta)\, d\theta = 1 \tag{A.54}$$

and, for all $\theta \in [-\pi, \pi]$,

$$\kappa_m(\theta) \leq c_1 m \left(1 + m^2\theta^2\right)^{-2}, \tag{A.55}$$

where c_1 is a constant not depending on m and θ.

It easily follows from the equality

$$\int_{\mathbb{T}} \kappa_m(\theta)\, d\theta = \frac{1}{m} \Big\langle \sum_{k=0}^{m-1} (2\pi)^{-1/2} e^{ik\cdot}, \sum_{k=0}^{m-1} (2\pi)^{-1/2} e^{ik\cdot} \Big\rangle_{L^2(\mathbb{T})}$$

and from Remark A.31 (ii) that (A.54) is satisfied.

Now we turn to the proof of (A.55). Let $\theta \in [-\pi, \pi]$ be arbitrary and fixed. First, we study the case where $\theta \neq 0$. We assume that c_2 is the finite constant larger than 1 defined as $c_2 := \sup_{\tau \in (0, \pi/2]} \{ \tau^2 \sin^{-2}(\tau) \}$. Then, it results from (A.52) that, for all $m \geq 1$, one has

$$\kappa_m(\theta) \leq \frac{c_2}{m} \frac{\sin^2(m\theta/2)}{(\theta/2)^2}. \tag{A.56}$$

Moreover, the classical inequality, $|\sin(\tau)| \leq \min\{1, |\tau|\}$, for all $\tau \in \mathbb{R}$, implies that

$$\frac{\sin^2(m\theta/2)}{(\theta/2)^2} \leq \frac{\min\{1, 4^{-1}m^2\theta^2\}}{4^{-1}\theta^2} = \tag{A.57}$$

$$\min\{4\theta^{-2}, m^2\} = m^2 \min\{(2^{-1}m\theta)^{-2}, 1\}.$$

Let us now prove that, for any $\tau \in \mathbb{R} \setminus \{0\}$, one has

$$\min\{|\tau|^{-1}, 1\} \leq 2(1 + |\tau|)^{-1}. \tag{A.58}$$

To this end, we study two cases: $0 < |\tau| \leq 1$ and $|\tau| > 1$. When $0 < |\tau| \leq 1$, one has $1 + |\tau| \leq 2$ and thus

$$2(1 + |\tau|)^{-1} \geq 2 \times 2^{-1} = 1 \geq \min\{|\tau|^{-1}, 1\}.$$

When $|\tau| > 1$, one has

$$\min\{|\tau|^{-1}, 1\} = |\tau|^{-1} < \left(\frac{1}{2} + \frac{|\tau|}{2}\right)^{-1} = 2(1 + |\tau|)^{-1}.$$

Next, combining (A.57) and (A.58), we get that

$$\frac{\sin^2(m\theta/2)}{(\theta/2)^2} \leq 2m^2(1 + 4^{-1}m^2\theta^2)^{-1} \leq 8m^2(1 + m^2\theta^2)^{-1}. \tag{A.59}$$

Finally, setting $c_1 := 8c_2$, then it follows from (A.56) and (A.59) that (A.55) holds for all integer $m \geq 1$ and non-vanishing $\theta \in [-\pi, \pi]$. On the other hand, (A.51) and the fact that c_1 has been chosen such $c_1 > 8 > 1/2\pi$ imply that (A.55) remains valid when $\theta = 0$. □

Theorem A.34 is mainly a consequence of Remark A.33 and of the following proposition.

Proposition A.40. *Let $(\mathbb{K}_m)_{m \geq 1}$ be a sequence of functions with values in $\mathbb{R}_+$ belonging to $L^1(\mathbb{T}^N)$ which satisfy the same conditions (a) and (b) as in Theorem A.34. Then, for each function f from $\mathbb{R}^N$ to $\mathbb{C}$ which is continuous and $(2\pi\mathbb{Z})^N$-periodic the sequence $(\mathbb{K}_m * f)_{m \geq 1}$ converges to f in $L^\infty(\mathbb{T}^N)$. In other words, one has that*

$$\lim_{m \to +\infty} \left\{ \sup_{x \in \mathbb{R}^N} |(\mathbb{K}_m * f)(x) - f(x)| \right\} = 0. \tag{A.60}$$

Proof of Proposition A.40. It follows from (A.44), (A.47), (A.48), the change of variable $s = -my$, and (A.49) that, for each fixed integer $m \geq 1$ and $x \in \mathbb{R}^N$, one has

$$
\begin{aligned}
\left|(\mathbb{K}_m * f)(x) - f(x)\right| &= \left| \int_{[-\pi,\pi]^N} \Big(\mathbb{K}_m(y) f(x-y) - \mathbb{K}_m(y) f(x) \Big) \, dy \right| \\
&\leq \int_{[-\pi,\pi]^N} \mathbb{K}_m(y) \big| f(x-y) - f(x) \big| \, dy \\
&\leq c\, m^N \int_{[-\pi,\pi]^N} B(my) \big| f(x-y) - f(x) \big| \, dy \\
&= c \int_{[-m\pi,m\pi]^N} B(-s) \big| f(x+m^{-1}s) - f(x) \big| \, ds \\
&\leq c \int_{\mathbb{R}^N} B(s) \big| f(x+m^{-1}s) - f(x) \big| \, ds. \qquad (\text{A.61})
\end{aligned}
$$

In the sequel, one assumes that ϵ is an arbitrarily small fixed positive real number. Then, in view of (A.49), there exists a positive real number η such that

$$
c \int_{|s| \geq \eta} B(s) \big| f(x+m^{-1}s) - f(x) \big| \, ds \leq 2c\, \|f\|_{L^\infty(\mathbb{T}^N)} \int_{|s| \geq \eta} B(s) \, ds \leq 2^{-1} \epsilon.
$$
$$(\text{A.62})$$

On the other hand, observe that the fact that f is continuous on $\mathbb{R}^N$ and $(2\pi\mathbb{Z})^N$-periodic implies that it is a uniformly continuous function on $\mathbb{R}^N$. Therefore, there is a positive real number γ such that the inequality

$$
\big| f(x+t) - f(x) \big| \leq 2^{-1} c^{-1} \epsilon \left(\int_{\mathbb{R}^N} B(s) \, ds \right)^{-1},
$$

holds, for all $x \in \mathbb{R}^N$ and for all $t \in \mathbb{R}^N$ satisfying $|t| \leq \gamma$. As a consequence, setting $m_0 := [\eta/\gamma] + 1$, where $[\eta/\gamma]$ is the integer part of η/γ, one has, for any integer $m \geq m_0$,

$$
c \int_{|s| < \eta} B(s) \big| f(x+m^{-1}s) - f(x) \big| \, ds \leq 2^{-1} \epsilon. \qquad (\text{A.63})
$$

Finally, combining (A.61) with (A.62) and (A.63), one obtains, for any integer $m \geq m_0$, that

$$
\sup_{x \in \mathbb{R}^N} \big| (\mathbb{K}_m * f)(x) - f(x) \big| \leq \epsilon,
$$

which shows that (A.60) holds. $\qquad \square$

We are now in position to prove Theorem A.34.

Proof of Theorem A.34. Let $p \in [1, \infty)$ and $g \in L^p(\mathbb{T}^N)$ be arbitrary and fixed. It is known that the functions from $\mathbb{R}^N$ to $\mathbb{C}$ which are continuous and $(2\pi\mathbb{Z})^N$-periodic form a dense subset in $L^p(\mathbb{T}^N)$. Thus, for any arbitrarily small fixed $\epsilon > 0$, there exists such a continuous function f_ϵ which further satisfies

$$\|g - f_\epsilon\|_{L^p(\mathbb{T}^N)} \le \epsilon/3. \tag{A.64}$$

Moreover, one can derive from Proposition A.40, that there exists an integer $m_1 \ge 1$ (depending on ϵ) such that, for all integer $m \ge m_1$, one has

$$\left\|\mathbb{K}_m * f_\epsilon - f_\epsilon\right\|_{L^p(\mathbb{T}^N)} \le (2\pi)^{N/p} \left\|\mathbb{K}_m * f_\epsilon - f_\epsilon\right\|_{L^\infty(\mathbb{T}^N)} \le \epsilon/3. \tag{A.65}$$

On the other hand, using (A.46) with $q = 1$ and $r = p$, and using (A.47) and (A.64), one gets, for all $m \ge m_1$,

$$\left\|\mathbb{K}_m * (g - f_\epsilon)\right\|_{L^p(\mathbb{T}^N)} \le \|\mathbb{K}_m\|_{L^1(\mathbb{T}^N)} \|g - f_\epsilon\|_{L^p(\mathbb{T}^N)} \le \epsilon/3. \tag{A.66}$$

Finally, putting together (A.64), (A.65) and (A.66) with the straightforward inequality

$$\left\|\mathbb{K}_m * g - g\right\|_{L^p(\mathbb{T}^N)} \le \left\|\mathbb{K}_m * (g - f_\epsilon)\right\|_{L^p(\mathbb{T}^N)} + \left\|\mathbb{K}_m * f_\epsilon - f_\epsilon\right\|_{L^p(\mathbb{T}^N)} + \|f_\epsilon - g\|_{L^p(\mathbb{T}^N)},$$

one obtains, for every $m \ge m_1$,

$$\left\|\mathbb{K}_m * g - g\right\|_{L^p(\mathbb{T}^N)} \le \epsilon.$$

which shows that the theorem is valid. $\qquad\qquad\square$

Bibliography

Adler, R. (1981). *The geometry of random fields* (Wiley, New York).

Andersson, P. (1997). Characterization of pointwise Hölder regularity, *Applied and Computational Harmonic Analysis* **4**, 4, pp. 429–443.

Ayache, A., Henrich, P., Marsalle, L., and Suquet, C. (2005). Holderian random functions, in J. Lévy Véhel and E. Lutton (eds.), *Fractals in Engineering. New Trends in Theory and Applications.* (Springer-Verlag), pp. 33–56.

Ayache, A. and Jaffard, S. (2010). Hölder exponents of arbitrary functions, *Revista Matemática Iberoamericana* **26**, 1, pp. 77–99.

Ayache, A., Jaffard, S., and Taqqu, M.S. (2007). Wavelet construction of Generalized Multifractional Processes, *Revista Matemática Iberoamericana* **23**, 1, pp. 327–370.

Ayache, A. and Linde, W. (2008). Approximation of Gaussian random fields: general results and optimal wavelet representation of the Lévy fractional motion, *Journal of Theoretical Probability* **21**, 1, pp. 69–96.

Ayache, A., Shieh, N.-R., and Xiao, Y. (2011). Multiparameter Multifractional Brownian Motion: local nondeterminism and joint continuity of the local times, *Annales de l'Institut Henri Poincaré (B) Probabilités et Statistiques* **47**, 4, pp. 1029–1054.

Ayache, A. and Taqqu, M.S. (2003). Rate optimality of wavelet series approximations of Fractional Brownian Motion, *Journal of Fourier Analysis and Applications* **9**, 5, pp. 451–471.

Ayache, A. and Taqqu, M.S. (2005). Multifractional Processes with Random Exponent, *Publicaciones Matemàtiques* **49**, pp. 459–486.

Benassi, A., Jaffard, S., and Roux, D. (1997). Elliptic gaussian random processes, *Revista Matemática Iberoamericana* **13**, 1, pp. 19–90.

Berman, S. (1970). Gaussian processes with stationary increments: local times and sample function properties, *The Annals of Mathematical Statistics* **41**, 4, pp. 1260–1272.

Berman, S. (1972). Gaussian sample functions: uniform dimension and Hölder conditions nowhere, *Nagoya Mathematical Journal* **46**, pp. 63–86.

Berman, S. (1973). Local nondeterminism and local times of Gaussian processes, *Indiana University Mathematics Journal* **23**, 1, pp. 69–94.

Bianchi, S., Pantanella, A., and Pianese, A. (2012). Modeling and simulation of currency exchange rates using MPRE, *International Journal of Modeling and Optimization* **2**, 3, pp. 309–314.

Bianchi, S., Pantanella, A., and Pianese, A. (2013). Modeling stock prices by multifractional Brownian motion: an improved estimation of the pointwise regularity, *Quantitative Finance* **13**, 8, pp. 1317–1330.

Bianchi, S. and Pianese, A. (2014). Multifractional processes in finance, *Risk and Decision Analysis* **5**, 1, pp. 1–22.

Billingsley, P. (1968). *Convergence of probability measures* (John Wiley).

Cramér, H. and Leadbetter, M.R. (1967). *Stationary and related stochastic processes* (John Wiley & Sons).

Daoudi, K., Meyer, Y., and Lévy Véhel, J. (1998). Construction of continuous functions with prescribed local regularity, *Constructive Approximation* **14**, 3, pp. 349–386.

Daubechies, I. (1992). *Ten lectures on wavelets*, Vol. 61 (Society for Industrial Mathematics).

Doukhan, P., Oppenheim, G., and Taqqu, M.S. (eds.) (2003). *Theory and applications of long-range dependence* (Birkhäuser).

Dozzi, M. (2003). Occupation density and sample path properties of N-parameter processes, in V. Capasso, E. Merzbach, B. Ivanoff, M. Dozzi, R. Dalang, and T. Mountford (eds.), *Topics in spatial stochastic processes, Lecture Notes in Mathematics*, Vol. 1802 (Springer), pp. 127–166.

Du Bois-Reymond, P. (1875). Versuch einer classification der wilkürlichen functionen reeller argumente nach ihren aenderungen in den kleinsten intervallen, *Journal für die reine und angewandte Mathematik* **21-37**, 79.

Dvoretzki, A. (1963). On the oscillation of the Brownian motion process, *Israel Journal of Mathematics* **1**, 4, pp. 212–214.

Dzhaparidze, K. and van Zanten, H. (2005). Optimality of an explicit series expansion of the fractional Brownian sheet, *Statistics and Probability Letters* **71**, 4, pp. 295–301.

Ehm, W. (1981). Sample function properties of multi-parameter stable processes, *Zeitschrift für Wahrscheinlichkeitstheorie und Verwandte Gebiete* **56**, 2, pp. 195–228.

Embrechts, P. and Maejima, M. (2002). *Selfsimilar processes* (Princeton University Press).

Falconer, K. (2002). Tangent fields and the local structure of random fields, *Journal of Theoretical Probability* **15**, 3, pp. 731–750.

Falconer, K. (2003). The local structure of random processes, *Journal of the London Mathematical Society* **67**, 3, pp. 657–672.

Geman, D. and Horowitz, J. (1980). Occupation densities, *The Annals of Probability* **8**, 1, pp. 1–67.

Haar, A. (1910). Zur theorie der orthogonalen fuctionnensysteme, *Mathematische Annalen* **69**, 3, pp. 331–371.

Herbin, E. (2006). From N parameter Fractional Brownian Motion to N parameter Multifractional Brownian Motion, *Rocky Mountain Journal of Mathematics* **36**, 4, pp. 1249–1284.

Hernández, E. and Weiss, G. (1996). *A first course on wavelets* (CRC Press).

Jaffard, S. (1995). Functions with prescribed Hölder exponent, *Applied and Computational Harmonic Analysis* **2**, 4, pp. 400–401.

Jaffard, S. and Meyer, Y. (1996). *Wavelet methods for pointwise regularity and local oscillations of functions*, Vol. 587 (American Mathematical Society).

Jaffard, S., Meyer, Y., and Ryan., R. (2001). *Wavelets: tools for science and technology* (Society for Industrial and Applied Mathematics, Philadelphia).

Kahane, J.-P. (1968 1st edition, 1985 2nd edition). *Some Random Series of Functions* (D. C. Heath 1st edition, Cambridge University Press 2nd edition).

Karatzas, I. and Shreve, A. (1987). *Brownian motion and stochastic calculus* (Springer).

Khoshnevisan, D. (2002). *Multiparameter processes: an introduction to random fields* (Springer).

Koralov, L. and Sinai, Y. (2007). *Theory of probability and random processes* (Springer).

Kühn, T. and Linde, W. (2002). Optimal series representation of fractional Brownian sheets, *Bernoulli* **8**, 5, pp. 669–696.

Ledoux, M. and Talagrand, M. (1991). *Probability in Banach spaces* (Springer-Verlag).

Lemarié, P.-G. and Meyer, Y. (1986). Ondelettes et bases hilbertiennes, *Revista Matemática Iberoamericana* **2**, 1-2, pp. 1–18.

Lévy, P. (1937 1st edition, 1954 2nd edition). *Théorie de l'addition des variables aléatoires* (Paris: Gauthier-Villars).

Lévy, P. (1948 1st edition, 1965 2nd edition). *Processus stochastiques et mouvement brownien* (Gauthier-Villars).

Li, D. and Queffélec, H. (2004). *Introduction à l'étude des espaces de Banach*, Vol. 12 (Société Mathématique de France).

Loève, M. (1978). *Probability theory*, Vol. II, 4th edn. (Springer).

Maiorov, V. and Wasilkowski, G. (1996). Probabilistic and average linear widths in L_∞-norm with respect to r-fold Wiener measure, *Journal of Approximation Theory* **84**, 1, pp. 31–40.

Mallat, S. (1998). *A wavelet tour of signal processing* (Academic Press).

Meyer, Y. (1985-1986). Principe d'incertitude, bases hilbertiennes et algèbres d'opérateurs, *Séminaire Bourbaki* **662**.

Meyer, Y. (1990). *Ondelettes et Opérateurs, volume 1* (Hermann, Paris).

Meyer, Y. (1992). *Wavelets and operators*, Vol. 37 (Cambridge University Press).

Meyer, Y. (1997). *Wavelets, vibrations and scalings*, Vol. 9 (American Mathematical Society, Crm Monographs Series).

Meyer, Y., Sellan, F., and Taqqu, M.S. (1999). Wavelets, generalized white noise and fractional integration: the synthesis of Fractional Brownian Motion, *The Journal of Fourier Analysis and Applications* **5**, 5, pp. 465–494.

Nolan, J. (1989). Local nondeterminism and local times for stable processes, *Probability Theory and Related Fields* **82**, pp. 387–410.

Paley, R., Wiener, N., and Zygmund, A. (1933). Notes on random functions, *Mathematische Zeitschrift* **37**, pp. 647–668.

Papanicolaou, G. and Sølna, K. (2003). *Theory and applications of long-range dependence*, chap. Wavelet based estimation of local Kolmogorov turbulence (Birkhäuser).

Peltier, R. and Lévy Véhel, J. (1995). Multifractional Brownian motion: definition and preliminary results, *Rapport de recherche INRIA* , 2645.

Pitt, L. (1978). Local times for Gaussian vector fields, *Indiana University Mathematics Journal* **27**, 2, pp. 309–330.

Potthoff, J. (2010). Sample properties of random fields III: differentiability, *Communications on Stochastic Analysis* **4**, 3, pp. 335–353.

Revuz, D. and Yor, M. (1991). *Continuous martingales and Brownian motion* (Springer).

Rosinsky, J. and Samorodnitsky, G. (1996). Symmetrization and concentration inequalities for multilinear forms with applications to zero-one laws for Lévy chaos, *The Annals of Probability* **24**, 4, pp. 1260–1272.

Rudin, W. (1986). *Real and complex analysis* (McGraw-Hill).

Samorodnitsky, G. and Taqqu, M.S. (1994). *Stable non-Gaussian random processes: stochastic models with infinitive variance* (Chapman and Hall).

Schwartz, L. (1966). *Théorie des distributions* (Hermann).

Seuret, S. and Lévy Véhel, J. (2002). The local Hölder function of a continuous function, *Applied and Computational Harmonic Analysis* **13**, 3, pp. 263–276.

Stein, E. and Weiss, G. (1971). *Introduction to Fourier analysis on Euclidean spaces* (Princeton University Press).

Stromberg, J. (1982). A modified Franklin system and higher order spline systems on $\mathbb{R}^n$ as unconditional bases for Hardy spaces, in *Proceedings of the conference in Chicago 1981 in honor of A. Zygmund, Wadsworth*, pp. 475–493.

Wiener, N. (1923). Differential space, *Journal of Mathematics and Physics* **2**, 1-4, pp. 131–174.

Wojtaszczyk, P. (1997). *A mathematical introduction to wavelets* (Cambridge University Press).

Xiao, Y. (1997). Hölder conditions for the local times and the Hausdorff measure of the level sets of Gaussian random fields, *Probability Theory and Related Fields* **109**, pp. 129–157.

Xiao, Y. (2006). Properties of local-nondeterminism of Gaussian and stable random fields and their applications, *Les Annales de la Faculté des Sciences de Toulouse* **15**, 1, pp. 157–193.

Xiao, Y. (2009). Sample path properties of anisotropic Gaussian random fields, in D. Khoshnevisan and F. Rassoul-Agha (eds.), *A minicourse on stochastic partial differential equations* (Springer, New York), pp. 145–212.

Xiao, Y. (2013). Recent developments on fractal properties of Gaussian random fields, in J. Barral and S. Seuret (eds.), *Further Developments in Fractals and Related Fields* (Springer, New York), pp. 255–288.

Printed in the United States
By Bookmasters